Data Science
Theory, Analysis, and Applications

Data Science

Theory, Analysis, and Applications

Edited by
Qurban A. Memon and
Shakeel Ahmed Khoja

CRC Press is an imprint of the
Taylor & Francis Group, an **informa** business

CRC Press
Taylor & Francis Group
6000 Broken Sound Parkway NW, Suite 300
Boca Raton, FL 33487-2742

ISBN 13: 978-0-367-20861-5 (hbk)

Library of Congress Cataloging-in-Publication Data

Names: Memon, Qurban A. (Qurban Ali), editor. | Khoja, Shakeel Ahmed, editor.
Title: Data science : theory, analysis, and applications / edited by Qurban A Memon, Shakeel Ahmed Khoja.
Description: Boca Raton : CRC Press, [2020] | Includes bibliographical references and index. |
Identifiers: LCCN 2019029260 (print) | LCCN 2019029261 (ebook) | ISBN 9780367208615 (hardback) | ISBN 9780429263798 (ebook)
Subjects: LCSH: Data mining—Statisical methods | Big data—Statisical methods | Quantitative research.
Classification: LCC QA76.9.D343 D3944 2020 (print) | LCC QA76.9.D343 (ebook) | DDC 006.3/12—dc23
LC record available at https://lccn.loc.gov/2019029260
LC ebook record available at https://lccn.loc.gov/2019029261

Visit the Taylor & Francis Web site at
http://www.taylorandfrancis.com

and the CRC Press Web site at
http://www.crcpress.com

Dedicated to our mothers, to whom we owe everything.

Qurban A. Memon

Shakeel Ahmed Khoja

Contents

PART I Data Science: Theory, Concepts, and Algorithms

PART II Data Design and Analysis

Preface

Data Science is an interdisciplinary scientific and technical field that combines techniques and approaches for efficient and effective data management, integration, analysis, visualization, and interaction with vast amounts of data, all as a critical prerequisite for a successful digitized economy. Currently, this science stands at the forefront of new scientific discoveries and is playing a pivotal role largely in our everyday lives.

Availability of multidisciplinary data is due to hyperconnectivity, and is heterogeneous, online, cheap, and ubiquitous. Old data is being digitized and collecting new data from web logs is being added to generate business intelligence. Essentially, the increasing availability of vast libraries of digitized information influences the way in which we comprehend and analyze our environment and realize businesses for our societal benefit. In this situation, data science creates a transformational force with an impact on current innovation potential in industry and academia. Nowadays, people are aware that this data can make a huge difference in business and engineering fields. It promises to revolutionize industries—from business to academics.

Currently, every sector of the economy has access to huge data and accumulates this data at a rate that exceeds their capacity to extract meaningful intelligence from it. The data exists in the form of text, audio, video, images, sensor, blog data, etc., but is unstructured, incomplete in form, and messy. New technologies, for example "big data" have emerged to organize and make sense of this data to create commercial and social value. It seems that every sector of the economy has access to this huge avalanche of data to extract relevant information. The question that still remains is how to use it effectively.

THE OBJECTIVE OF THIS BOOK

The aim of this book is to provide an internationally respected collection of scientific research methods, technologies, and applications in the area of data science. This book enhances the understanding of concepts and technologies in data science and discusses intelligent methods, solutions, and applications in data science. This book can prove useful to researchers, professors, research students, and practitioners as it reports novel research work on challenging topics in the area surrounding data science. In this book, some of the chapters are written in tutorial style concerning machine learning algorithms, data analysis, etc.

THE STRUCTURE

The book is a collection of fourteen chapters written by scholars and experts in this field and organized into three parts. Each part consists of a set of chapters addressing the respective subject outline to provide readers an in-depth and focused understanding of concept and technology related to that part of the book. Some of the chapters in each part are written in tutorial style in chapters concerning the development process of data science and its emerging applications.

The book is structured as follows:

- Part I: Data Science: Theory, Concepts, and Algorithms
- Part II: Data Design and Analysis
- Part III: Applications and New Trends in Data Science

Part I comprises five chapters on data science theory, concepts, techniques, and algorithms.

The first chapter extends the earlier work on Cassandra integrated with Hadoop to a system called GeoMongoSpark and investigates on storage and retrieval of geospatial data using various sharding techniques. Hashed indexing is used to improve the processing performance with less memory.

The purpose of Chapter 2 is to study the different evolutionary algorithms for optimizing neural networks in different ways for image segmentation purposes.

Chapter 3 introduces a new adaptive algorithm called Feature Selection Penguin Search optimization algorithm, which is a metaheuristic feature subset selection method. It is adapted from the natural hunting strategy of penguins, in which a group of penguins take jumps at random depths and come back and share the status of food availability with other penguins, and in this way, the global optimum solution is found, namely Penguin Search Optimization Algorithm. It is combined with different classifiers to find an optimal feature subset.

Currently, graph technology is becoming increasingly important, and graphs are used to model dynamic and complex relationships of data order to generate knowledge. Particularly, Neo4j is a database management system that currently leads the NoSQL system on graph databases. In Chapter 4, the main objective is to propose physical design guidelines that improve query execution time on graph databases in terms of a specific workload in Neo4j. In this work, indexes, path materialization, and query rewriting are considered as guidelines for the physical design on Neo4j databases.

Chapter 5 provides information about the latest techniques of large-scale data collection schemes to readers. A continuous sensor data with different intervals (cycles) as sensor data stream is defined, and the collection methods for distributed sensor data streams as a topic-based pub/sub (TBPS) system are proposed.

Part II comprises five chapters on data design and analysis.

The objective of Chapter 6 in this part is to explain and impart solution for the effective analysis and management of big data in healthcare with four main parts, namely (i) collection of healthcare data; (ii) analysis of healthcare data; (iii) management of Big Data; and (iv) Big Data in healthcare. Then, the ability to query data is primary for reporting and analytics to generate a report that is clear, crisp, and accessible to the target audience when it is offered. This chapter discusses the effective utilization of big data analysis and management in healthcare.

Chapter 7 demonstrates how the analysis of health data, such as blood cholesterol, blood pressure, smoking, and obesity can identify high-risk heart attack patients and how the proactive changes in these high-risk patient lifestyles and use of medication can prevent a heart attack from taking place.

The Brugada syndrome (BrS) is a disease with a great predisposition to sudden cardiac death due to ventricular arrhythmias. Dysfunctional ion channels are considered responsible for this entity. In an effort to characterize proteins, a nonsupervised computational system called the Polarity Index Method® is presented in Chapter 8 to extensively measure the "polar profile" or "electromagnetic balance" of proteins. In the presented work, the method is calibrated with the BrS mutated proteins, searching for an association of these entities with (i) the group of 36 BrS proteins from known 4,388 BrS mutated proteins, (ii) two groups of intrinsically disordered proteins, (iii) six lipoprotein groups, (iv) three main groups of antimicrobial proteins from UniProt and APD2 databases, (v) a set of selective cationic amphipathic antibacterial peptides (SCAAP), and (vi) the group of 557,713 "reviewed" proteins from the UniProt database.

In Chapter 9, the use of machine learning has recently shown that more useful information can be extracted from clinical images and processed accurately, and above all, more reliably than humans. Towards this end, this chapter uses a machine learning algorithm to automatically discriminate healthy skin, superficial burn, and full-thickness burn. After which, the features are used to train a multiclass support vector machine (SVM).

The authors in Chapter 10 classify four classes of emotions (positive, negative, depressed, and harmony) based on electroencephalography signal input collected from brain activities. The proposed state transition system is associated with an increased or decreased value of related channels/electrodes of the specific cortex (frontal, temporal, occipital, etc.).

Part III comprises four chapters on applications and new trends in data science.

The authors of Chapter 11 investigate the performance of different kinds of feature generation schemes, such as local binary pattern (LBP), gradient local binary pattern (GLBP), histogram of oriented gradient (HOG), run length feature (RLF), and pixel density. Various similarity and dissimilarity measures are used to achieve matching in handwritten documents.

The authors of Chapter 12 propose a unique study that combines text mining technique along with machine learning approach to determine the satisfaction status of hotel guests and its determinants with the help of overall ratings, which can be considered as a proxy variable for guest satisfaction. The chapter evaluates the guest satisfaction from 239,859 hotel reviews extracted from Tripadvisor.

Chapter 13 evaluates and compares the performance of naive Bayes and SVM algorithms in classification sentiments embedded in movie reviews. In particular, this study uses the MovieLens review platform to extend the application of sentimental analysis techniques.

In Chapter 14, human emotional states are recognized from full body movements using feedforward deep convolution neural network architecture and Visual Geometric Group (VGG)16 model. Both models are evaluated by emotion action dataset (University of York) with 15 types of emotions.

WHAT TO EXPECT FROM THIS BOOK

This book can prove useful to researchers, professors, research students, and practitioners as it will report novel research work on challenging topics in the area of data science as well as data analysis techniques. This book is useful to professors, research students, and practitioners, as it reports issues in data science theory and novel research work on challenging topics in the area of information design. It enhances the understanding of concepts and technologies in data science and discusses solutions and recent applications such as robotics, business intelligence, bioinformatics, marketing and operation analytics, text and sentiment analysis, healthcare analytics, etc.

WHAT NOT TO EXPECT FROM THIS BOOK

This book is written with the objective to collect latest and focused research works in data science field in mind. The book does not provide detailed introductory information on the basic topics in this field, as it is not written in textbook style. However, it can be used as a good reference to help graduate students and researchers in providing latest concepts, tools, and advancements in data science field and related applications. The readers are required to have at least some basic knowledge about the field.

SPECIAL THANKS

First of all, we would like thank our employer who gave us the time and encouragement in completing this book. We would also like to thank all the authors who contributed to this book by authoring related and focused chapters in the data science field. Their prompt responses, adhering to guidelines, and timely submission of chapters helped us in meeting the book preparation deadlines. A total of 47 authors from 13 countries contributed to the development of this book. We hope that this book will prove beneficial to professionals and researchers, and at the same time, encourage in generating new algorithms, applications, and case studies in this multidisciplinary field.

Editors

Qurban A. Memon has contributed at levels of teaching, research, and community service in the area of electrical and computer engineering. He has authored/coauthored approximately 100 publications over 18 years of his academic career. Additionally, he has written five book chapters: 1. "Intelligent Network System for Process Control: Applications, Challenges, Approaches" in *Robotics, Automation and Control*, 2009; 2. "RFID for Smarter Healthcare Collaborative Network" in *Building Next-Generation Converged Networks: Theory and Practice*, CRC Press, 2012; 3. "Secured and Networked Emergency Notification without GPS Enabled Devices," in *Bio-inspiring Cyber Security and Cloud Services: Trends and Innovations*, Springer, 2014; 4. "Authentication and Error Resilience in Images Transmitted through Open Environment" in *Handbook of Research on Security Considerations in Cloud Computing*, 2015; 5. "JPEG2000 Compatible Layered Block Cipher" in *Multimedia Forensics and Security: Foundations, Innovations, and Applications*, 2017. He has also edited a book titled *Distributed Network Intelligence, Security and Applications*, published by CRC press, 2013. His research project undertaken by UAE-U students won CURA-2009 and CURA-2018 awards.

Shakeel Ahmed Khoja is a professor at Faculty of Computer Science, IBA Karachi, and is a Commonwealth Academic Fellow. He did his PhD and post-doc at the School of Electronics and Computer Science, University of Southampton, UK. His research work includes development of e-learning frameworks, digital libraries, content and concept-based browsing and information retrieval over the Internet.

He has worked as a lead researcher and project manager for two European projects based at Southampton, for using learning technologies to develop web-based tools and to make educational videos easier to access, search, manage, and exploit for students, teachers, and other users through developing and deploying technologies that support the creation of synchronized notes, bookmarks, tags, images, videos, and text captions. He has also carried out a number of data analysis projects at the State Bank of Pakistan and has been involved in developing consumer confidence and business confidence surveys for the bank.

Dr. Khoja has taught computer science courses at the University of Southampton, Karachi Institute of Information Technology, and Bahria University. He has more than 18 years of professional experience and more than 50 research publications to his credit. He is also a reviewer of number of journals and conferences in the field of computer science.

Contributors

Aakash
Department of Operational Research
University of Delhi
New Delhi, India

Aliyu Abubakar
Centre for Visual Computing
University of Bradford
Bradford, United Kingdom

Anu G. Aggarwal
Department of Operational Research
University of Delhi
New Delhi, India

Fahad AlGarni
Department of Computing and Information Technology
University of Bisha
Bisha, Kingdom of Saudi Arabia

Mohammed Al-Khafajiy
Liverpool John Moores University
Liverpool, United Kingdom

Mohamed Alloghani
Liverpool John Moores University
Liverpool, United Kingdom

Thar Baker
Liverpool John Moores University
Liverpool, United Kingdom

Ardhendu Banerjee
Information Technology Department
TechnoIndia
Salt Lake, Kolkata, India

Antara Barman
Information Technology Department
TechnoIndia
Salt Lake, Kolkata, India

Mohamed Lamine Bouibed
Laboratoire d'Ingénierie des Systèmes Intelligents et Communicants, Faculty of Electronics and Computer Science
University of Sciences and Technology
Algiers, Algeria

Thomas Buhse
Centro de Investigaciones Químicas
Universidad Autónoma del Estado de Morelos
Cuernavaca, México

Ali Maina Bukar
Centre for Visual Computing
University of Bradford
Bradford, United Kingdom

Sanjay Chakraborty
Information Technology Department
TechnoIndia
Salt Lake, Kolkata, India

Youcef Chibani
Laboratoire d'Ingénierie des Systèmes Intelligents et Communicants, Faculty of Electronics and Computer Science
University of Sciences and Technology
Algiers, Algeria

Agnip Dasgupta
Information Technology Department
TechnoIndia
Salt Lake, Kolkata, India

Aniket Ghosh Dastidar
Information Technology Department
TechnoIndia
Salt Lake, Kolkata, India

M. Devi
Department of Computer Science
King Khalid University
Abha, Kingdom of Saudi Arabia

R. Dhaya
Department of Computer Science
King Khalid University
Abha, Kingdom of Saudi Arabia

Miguel Arias Estrada
Department of Computer Science
Instituto Nacional de Astrofísica, Óptica y Electrónica
Puebla, México

Vania V. Estrela
Departamento de Engenharia de Telecomunicações
Universidade Federal Fluminence
Rio de Janeiro, Brazil

Anu A. Gokhale
Illinois State University
Normal, Illinois

Marlene Goncalves
Departamento de Computación y Tecnología de Información
Universidad Simón Bolívar
Caracas, Venezuela

Carol Hargreaves
National University of Singapore
Singapore

Abir Hussain
Liverpool John Moores University
Liverpool, United Kingdom

Marcos Jota
Coordinación de Ingeniería en Computación
Universidad Simón Bolívar
Caracas, Venezuela

M. Kalaiselvi Geetha
Department of Computer Science and Engineering
Annamalai University
Chidambaram, Tamil Nadu, India

R. Kanthavel
Department of Computing and Information Technology
University of Bisha
Bisha, Kingdom of Saudi Arabia

Tomoya Kawakami
Nara Institute of Science and Technology
Ikoma, Nara, Japan

Mohammed Khalaf
Almaaref University College
Ramadi, Iraq

Yashwant Kumar
Information Technology Department
TechnoIndia
Salt Lake, Kolkata, India

Hermes J. Loschi
Department of Communications
State University of Campinas – UNICAMP
Campinas, Brazil

Manlio F. Márquez
Department of Electrocardiology
Instituto Nacional de Cardiología Ignacio Chávez
México City, México

Jamila Mustafina
Kazan Federal University
Kazan, Russia

Hassiba Nemmour
Laboratoire d'Ingénierie des Systèmes Intelligents et Communicants, Faculty of Electronics and Computer Science
University of Sciences and Technology
Algiers, Algeria

Subhadip Pal
Information Technology Department
TechnoIndia
Salt Lake, Kolkata, India

Ritces Parra
Coordinación de Ingeniería en Computación
Universidad Simón Bolívar
Caracas, Venezuela

Carlos Polanco
Department of Mathematics, Faculty of Sciences
Universidad Nacional Autónoma de México
México City, México

Navid Razmjooy
Independent Researcher
IEEE Senior member
Belgium

R. Santhosh Kumar
Department of Computer Science and Engineering
Annamalai University
Chidambaram, Tamil Nadu, India

Tarun Saurabh
Information Technology Department
TechnoIndia
Salt Lake, Kolkata, India

Himanshu Sharma
Department of Operational Research
University of Delhi
New Delhi, India

Shailesh Shaw
Information Technology Department
TechnoIndia
Salt Lake, Kolkata, India

Kirsty M. Smith
Plastic Surgery and Burns Research Unit, Centre for Skin Sciences
Bradford Teaching Hospitals
Bradford, United Kingdom

Yuuichi Teranishi
National Institute of Information and Communications Technology
Koganei, Tokyo, Japan

Hassan Ugail
Centre for Visual Computing
University of Bradford
Bradford, United Kingdom

Vladimir N. Uversky
Department of Molecular Medicine and USF Health Byrd Alzheimer's Research Institute
University of South Florida
Tampa, Florida

P. Vamsi Krishna
VR Siddhartha Engineering College
Kanuru, Andhra Pradesh, India

S. Vasavi
VR Siddhartha Engineering College
Kanuru, Andhra Pradesh, India

Tomoki Yoshihisa
Osaka University
Ibaraki, Osaka, Japan

Part I

Data Science

Theory, Concepts, and Algorithms

1 Framework for Visualization of GeoSpatial Query Processing by Integrating MongoDB with Spark

S. Vasavi and P. Vamsi Krishna
VR Siddhartha Engineering College

Anu A. Gokhale
Illinois State University

CONTENTS

1.1 INTRODUCTION

Companies that use big data for business challenges can gain advantages by integrating MongoDB with Spark. The Spark framework provides support for analytics, where process execution is fast because of in-memory optimization. Out of various NoSQL databases, MongoDB provides a document data model that suits to variety and voluminous amounts of data. As such, when integrated, MongoDB and Spark together can index data efficiently and help in analytics of data-driven applications.

Geospatial data can be analyzed to serve the needs of various applications such as tourism, healthcare, geomarketing, and intelligent transportation systems. Spatial data occurs in two forms: vector and raster that store latitude and longitude of objects. Keyhole markup language (KML) is a tag-based structure and can be used to display geographic data.

Tableau uses various file formats such as KML, ERSI shape files, GeoJSON (JavaScript Object Notation) files, and MapInfo Interchange formats for geographic data analysis, and display. Traditional databases process on structured data that guarantees ACID (Atomicity, Consistency, Isolation, and Durability) properties. NoSQL databases are developed to store and process unstructured data that guarantees CAP (Consistency, Availability and Partition Tolerance) properties with less response time. MongoDB has no query language support, but data can be indexed as in relational databases, structured as JSON fragments and can process social networking applications, where latency has to be optimized. Cassandra monitors nodes, handles redundancy, and avoids lazy nodes, whereas MongoDB monitors these activities at a higher granular level. Even though some works are reported for labeling and retrieving data from MongoDB, they are inefficient either at indexing or at retrieval. This chapter aims at adding the functionality of spatial querying for MongoDB database by integrating it with Spark.

Every location on this globe is a representation of intersection of latitude and longitude coordinates. Spatial operations are classified by the authors of [1]: proximity analysis queries such as "Which parcels are within 100 meters of the subway?," contiguity analysis queries such as "Which states share a border with Coorg?," and neighborhood analysis queries such as "Calculate an output value at a location from the values at nearby locations?" Geohashing techniques are used to find the nearest location of a specified source location with high precision values. Latitude and longitude are used to calculate geohash value: the higher the precision value, the more the length of geohash value. Instead of sequential processing of the database, a parallel search process must be done for a required destination using this geohash. To achieve this, MongoDB is integrated with Spark, which is an efficient distributed parallel processing.

1.1.1 Integration of Spark and MongoDB

Spark is a distributed system for processing variety of data, and MongoDB aims for high availability. By combining both Spark and MongoDB, spatial operations on geospatial data can be processed effectively [2]. The MongoDB Connector to Hadoop and Spark provides read/write data to MongoDB using Scala. When client submits a geospatial query, the master node initially chooses an executor. This executor selects sharding based on the shard key. This shard key is used to store and retrieve similar keys from the in-memory document collection with high precision and optimized processing time. Sharding techniques (horizontal scaling) distribute geospatial data into various nodes [3]. Figure 1.1 presents sharding at collection level.

A MongoDB shard cluster consists of shard, mongos, and config servers. Mongos acts as an interface where the application layer determines a particular sharded cluster for geospatial data. Hashed sharding computes a hash value for each of the shard keys.

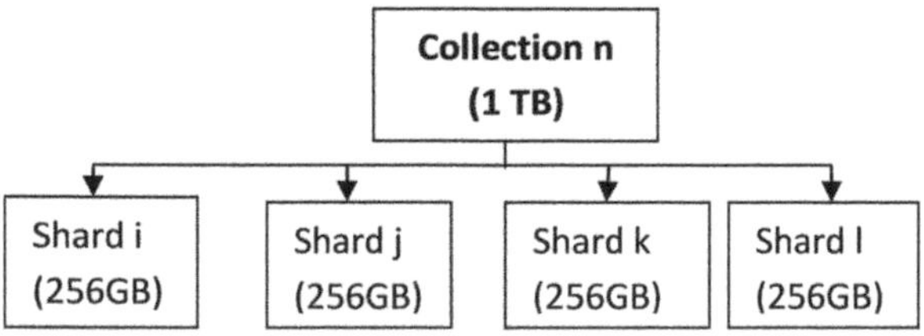

FIGURE 1.1 Sharding at collection level.

Based on this, each chunk is assigned a range. Ranged sharding divides data into ranges based on shard key values, and then each chunk is assigned a range [4].

The following paragraph presents how Range (Tag) Sharding and Zone Sharding compute the hash values. Figure 1.2 presents a sample geospatial input.

Figure 1.3 presents Geohash computed using MD5 algorithm that guarantees uniform distribution of writes across shards but is less optimal for range-based queries.

Figure 1.4 presents document after appending HashKey. Similar process is followed in both storage and retrieval operations for finding the corresponding shard.

The calculated MD5 value is the shard key and is appended to the document as shown in Figure 1.5.

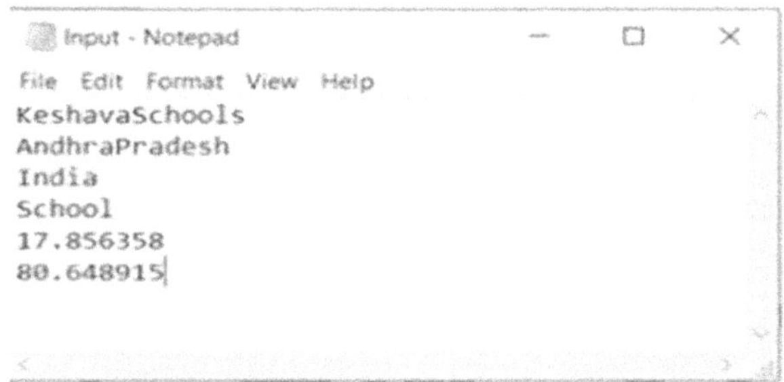

FIGURE 1.2 Input data given through a file.

17.856358

80.648915

d62fc425fed608500aa2f82d814ab7c2

FIGURE 1.3 Hashed sharding: generating geohash value.

{ "_id" : { "$oid" : "5a841114f3dc241dd41103f1"} , "Place" : "KeshavaSchools" , "State" : "Andhra Pradesh" , "Country" : "India" , "Place Type" : "School" , "Latitude" : 17.856358 , "Longitude" : 80.648915 , "MD5HashKey" : " d62fc425fed608500aa2f82d814ab7c2"}

FIGURE 1.4 Hashed sharding: appending the HashKey and storing in the database.

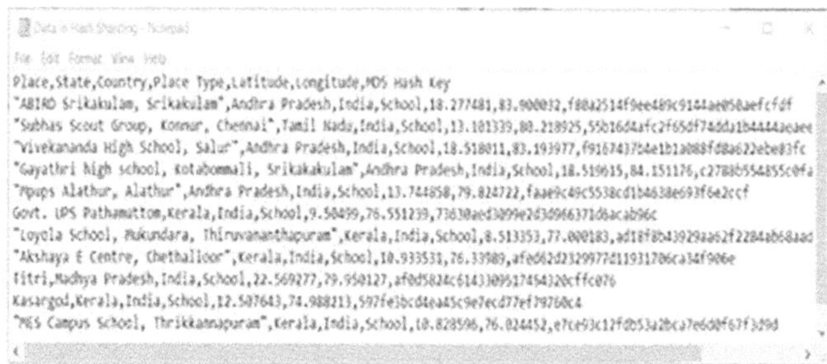

FIGURE 1.5 Data representation in hashed sharding.

The following Figure 1.6 presents the shard key value for range sharding.

Figure 1.7 presents document after appending shard key. Similar process is followed in both storage and retrieval operations for finding the corresponding shard.

The calculated shard value is the shard key and is appended to the document based on the latitude and longitude ranges as shown in Figure 1.8.

Figure 1.9 presents the shard key value for zone sharding.

```
17.856358
80.648915
SouthShard
```

FIGURE 1.6 Range sharding: generating shard key.

```
{ "_id" : { "$oid" : "59e1a21210d1801bec43c1ac"} , "Place" : "KeshavaSchools" ,
"State" : "AndhraPradesh" , "Country" : "India" , "Place Type" : "School" ,
"Latitude" : 17.856358 , "Longitude" : 80.648915 , "Tag" : "South"}
```

FIGURE 1.7 Range sharding: appending the shard key and storing in the database.

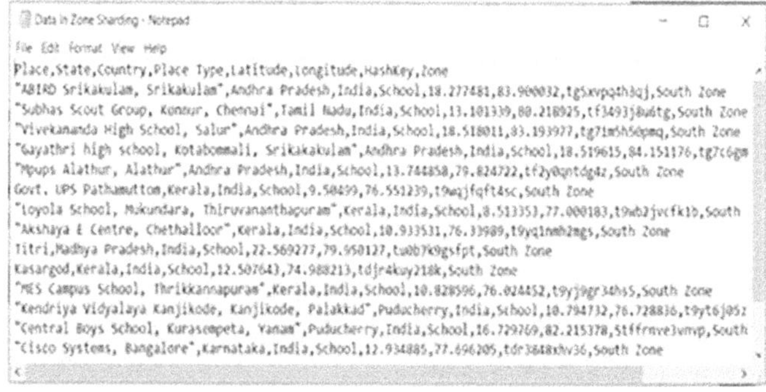

FIGURE 1.8 Data representation in range sharding.

16.49590337

80.67028580

tfcmgdyr9gkb

FIGURE 1.9 Zone sharding: generating shard zone based on geographic region.

Figure 1.10 presents document after appending the Shard Zone. Similar process is followed in both storage and retrieval operations for finding the corresponding shard. When this shard key range is modified, data is migrated to a new zone automatically.

The calculated geohash value is the shard key and is appended to the document based on the zone of input location along with the zone value, as shown in Figure 1.11.

LTRMP (Long Term Resource Monitoring Program Component) Spatial Data Query and Visualization Tool is developed Olsen [8]. It facilitates the display and querying of Upper Mississippi River Basin (UMRB) data sets. It requires an understanding of geographic information system (GIS) concepts and LTRMP component database structure. Data processing and visualization tools are reported in [9].

Refinement tools such as DataWrangler and Google Refine; conversion tools such as Mr. Data Convertor; statistical analysis tools such as R project for statistical computing; generic visualization applications such as Google fusion tables, Tableau, Many Eyes, CartoDB, and GeoCommons; wizards such as Google Chart Tools; JavaScript InfoVis Toolkit,D3.js, Protovis, and Recline.js; geospatial visualization tools such as OpenHeatMap, OpenLayers, and OpenStreetMap; temporal data visualization tools such as TimeFlow; tools for network analysis such as Gephi,NodeXL are discussed. In the category of web applications, Google Fusion tables, CArtoDB,GoogleChart Tools, Recline.js, and OpenLayers are widely used.

```
{ "_id" : { "$oid" : "5a7b2ba53767db96e54db709"} , "Place" : "Rohit School" ,
"State" : "Andhra Pradesh" , "Country" : "India" , "Place Type" : "School" ,
"Latitude" : 16.49590337 , "Longitude" : 80.67028580 , "HashKey" : "tfcmgdyr9gkb",
"Zone" : "SouthZone"}
```

FIGURE 1.10 Zone sharding: appending the shard zone.

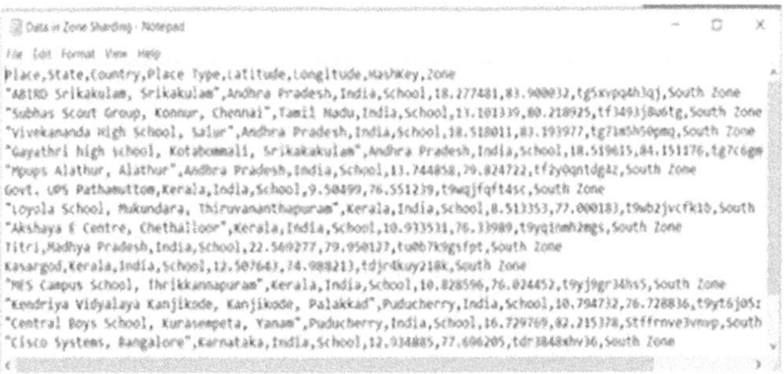

```
Data in Zone Sharding - Notepad
File Edit Format View Help
Place,State,Country,Place Type,Latitude,Longitude,HashKey,Zone
"ABIRD Srikakulam, Srikakulam",Andhra Pradesh,India,School,18.277481,83.900032,tg5xvpq4h3qj,South Zone
"Subhas Scout Group, Konnur, Chennai",Tamil Nadu,India,School,13.101339,80.218925,tf3493j8u6tg,South Zone
"Vivekananda High School, Salur",Andhra Pradesh,India,School,18.518011,83.193977,tg7im5h50pmq,South Zone
"Gayathri high school, Kotabommali, Srikakakulam",Andhra Pradesh,India,School,18.519615,84.151176,tg7c6gm
"Mpups Alathur, Alathur",Andhra Pradesh,India,School,13.744858,79.824722,tf2y0qntdg42,South Zone
Govt. UPS Pathamuttom,Kerala,India,School,9.50499,76.551239,t9wqjfqft4sc,South Zone
"Loyola School, Mukundara, Thiruvananthapuram",Kerala,India,School,8.513353,77.000183,t9wb2jvcfk10,South
"Akshaya E Centre, Chethalloor",Kerala,India,School,10.933531,76.33989,t9yq1nmh2mgs,South Zone
Titri,Madhya Pradesh,India,School,22.569277,79.950127,tu0b7k9gsfpt,South Zone
Kasargod,Kerala,India,School,12.507643,74.988213,tdjr4kuy218k,South Zone
"MES Campus School, Thrikkannapuram",Kerala,India,School,10.828596,76.024452,t9yj9gr34hs5,South Zone
"Kendriya Vidyalaya Kanjikode, Kanjikode, Palakkad",Puducherry,India,School,10.794732,76.728836,t9yt6j05:
"Central Boys School, Kurasempeta, Yanam",Puducherry,India,School,16.729769,82.215378,Stffrnve3vmvp,South
"Cisco Systems, Bangalore",Karnataka,India,School,12.934885,77.696205,tdr3848xhv36,South Zone
```

FIGURE 1.11 Data representation in zone sharding and storing in the database.

Our previous work [16,18] is on geospatial querying by integrating Cassandra with Hadoop. Geospatial querying within Hadoop environment faced significant data moments because of no proper scheduling mechanism for job execution. We extended the same framework for integrating MongoDB with Spark with visualization. The main objective of this chapter is to propose "GeoMongoSpark Architecture" for efficient processing of spatial queries in a distributed environment that takes advantage of integrating MongoDB with Spark for geospatial querying by using geospatial indexing and sharding techniques and visualizing the geospatial query processing. The main objectives are listed as follows:

1. To understand the need for integrating MongoDB and Spark and propose a new architecture for the integration.
2. Collect benchmark datasets for testing the proposed architecture.
3. Apply various sharding techniques to partition the data.
4. Compare the performance evaluation of various sharding techniques.
5. Use Tableau for visualization.

1.2 LITERATURE SURVEY

Query processing in a centralized database system has more response time and slower data access when compared with a decentralized database. Location-based services use voluminous spatial data to answer end user queries. Hence, scalable techniques that operate on distributed computing platforms and technologies are required for efficient processing of spatial queries. Spatial data processing based on Hadoop, and Spark comparison based on query language, spatial partitioning, indexing, and data analysis operators are reported in [5]. The STARK system provides a number of spatial predicates for filters and joins. The drawback with this framework is that filters can delay the processing time, irrespective of any kind of data, and persistent indexing causes memory wastage. The processing performance can be improved with the use of hashed indexing that uses less memory. The work reported in [6] presents a distributed query scheduler for spatial query processing and optimization. It generates query execution plans using spatial indexing techniques. Experiments are performed on real datasets and are compared with other works. Although the query execution performance is improved and communication cost is reduced, this architecture is costly to implement, and the use of filters can increase the processing time of a query. Spatial-Hadoop and GeoSpark are discussed in [7], where authors proved that GeoSpark is faster than Spatial-Hadoop for geospatial big data analytics. Spatial data processing system that schedules and executes range search is described in [10]. *k*-NN (k nearest neighbor), spatiotextual operation, spatial-join, and *k*-NN-join queries are described. Bloom filters are used to reduce network communication. Experiments are carried out using sample data. Hashed indexing can improve the query processing time by reducing the use of memory for storing global and local indexing, even though frequently accessed data is cached in Location Spark and less frequently used data is stored into disk. The usage of filters on the spatial data increases the implementation cost of this architecture. Authors in [11] described Panda architecture for spatial predictive queries, such as predictive range, *k*-NN, and aggregate queries. Panda System will display a system behavior statistics on the interface. Usage of grid and list data structures to store the data may cause memory wastage. Also, processing the grid data structures requires more amount of time. The identification of object moment is very important in system, and variations may lead to major differences. Distributed profitable-area query (DISPAQ) is described in [12]. It identifies profitable areas from raw taxi trip data using PQ-index. Z-skyline algorithm prunes multiple blocks during query processing. But usage of predictive function increases the computation overhead, though it reduces the processing time for executing frequent queries. Performance can be improved by dividing the places into zones. Data mining is used to increase marketing of an educational organization in [13]. Authors used *k*-means algorithm to cluster the locations. Student residential address is calculated using latitude and longitude and visualized minimum, maximum, and average distance. Such visualization helps to improve the admission rate. Haversine formula is used to calculate the distance between the two locations. Their application can be used for 1,000–2,000 students. Authors of [14] used open shape-based strategy (OSS) to eliminate the computation related to the world boundary. It also reduces I/O (Input-Output) and CPU (Central Processing Unit) costs when compared with range query strategies (RQS) and is scalable than RQS for large datasets. However, OSS is more expensive than the filter operations in RQS. Direction of object plays an important role in the evaluation of queries. Surrounding join query, that retrieves, for each point in Q, all its surrounding points in P, is described in [15]. Even though the surrounding join query enriches the semantics of the conventional distance-based spatial join query, creating a VD (Visitor Database) requires a lot of processing time. Also, the pruning of unnecessary nodes in hierarchical algorithm requires more time, which can be avoided by starting only with necessary nodes.

To summarize, the main drawback of the existing system is time complexity for processing large datasets. Our work mainly focuses on reducing the query response time by using hashed indexing and zones. Also, existing system has high time complexity for processing large geospatial datasets and usage of intermediate data. The usage of complex data structures to hold data leads to large processing times and requires complex infrastructures to handle. Visualization alone cannot help to perform analytics on the data that is handled by the system. Visualizing can be made fast and efficient using Tableau, as it handles different types of data and supports business analytics operations for generating automatic reports. Sharding techniques can be treated as an alternative to partitioning techniques, as such Range Sharding and Zone Sharding are used in our framework. We also made an attempt to determine which sharding performs well for geospatial query processing.

1.3 PROPOSED SYSTEM

This section presents a detailed architecture of GeoMongoSpark along with algorithms. Figure 1.12 presents the proposed GeoMongoSpark Architecture to perform geospatial querying, aiming at optimized query response time for spatial data analysis.

Figure 1.13 presents a detailed flow chart of the sequence of steps for the Visualization of Geospatial Data. Initially, data is stored by passing the data to Zone Sharding and Range Sharding. The output of a geospatial can be converted to JSON format and can be given as input to a tableau for visualization.

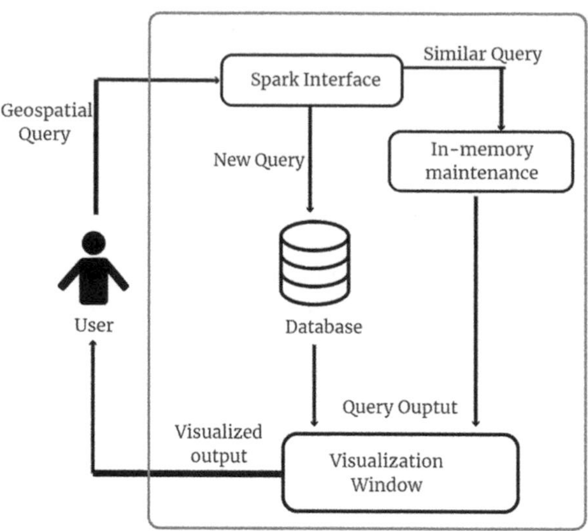

FIGURE 1.12 GeoMongoSpark architecture.

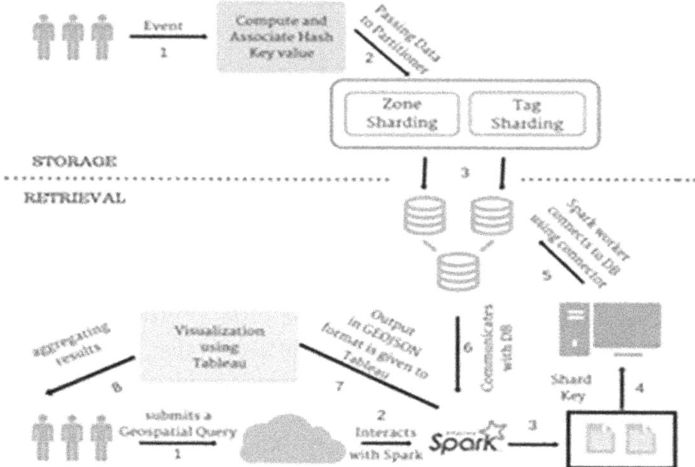

FIGURE 1.13 Visualization of geospatial data.

1.3.1 Methodology for Processing Spatial Queries

i. Spatial data storage
   a. Geospatial data is collected from various external devices or sensing systems. The collection of data can also be through event messages or current geo location of end users. This data is in the form of

<Latitude, Longitude> pair for different locations, where spatial queries are processed on this pair. Dataset is constructed with this pair of data.

b. Geohashing technique (that converts the longitude, latitude pair into a single value, represented in binary format) is applied for each tuple in the dataset, which takes <Latitude, Longitude> pair as input and produces "Geohash" value as output (with a precision of 12 bits). Thus, Geohash values are added as an additional column.

c. Hashed indexes support sharding using hashed shard keys. Hashed-based sharding uses a hashed index of a field as the shard key to partition data across the sharded cluster. Same data is sharded by applying Range and Zone Sharding and is stored in MongoDB.

ii. Spatial querying

a. User submits a spatial query.

b. Compute Geohash value for the queried area using Geohashing technique. Master-Slave framework is used for parallel implementation of the input query. The input query logic is executed in parallel on the in-memory chunks and over different mappers.

c. Spark Master-Slave framework performs Map and Reduce task.

d. The output of the geospatial query is given as input to the tableau for visualization.

e. Visualize the output.

1.3.2 Spark Master-Slave Framework

Data processing using Spark framework is divided into two daemons:

i. Master Daemon: Required attributes are mapped using `Map( )` function.

ii. Executor Daemon: The number of entries are reduced using `Reduce( )` function and returns the required tuples.

Here, both Master and Executor use two sharding techniques:

i. Range Sharding

Master communicates with the cluster manager and sends the task to the executor, as shown in Figure 1.14. Figure 1.15 presents the Executor Task in Range Sharding.

ii. Zone Sharding

The Master Task takes the user input location as the input, and the functions of it are represented as shown in Figure 1.16. Figure 1.17 presents Executor Task in Zone Sharding. The data can be retrieved from the spatial database based on the user location, and the query is specified as shown in Figure 1.18. During retrieval, we make a comparison of which sharding techniques give the best response time: Range Sharding or Zone Sharding.

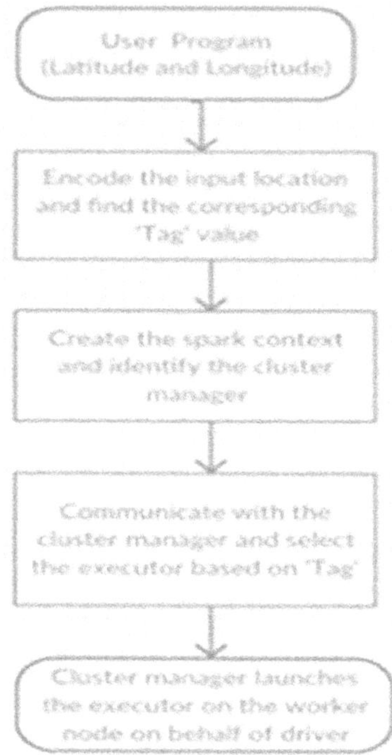

FIGURE 1.14 Master task in tag sharding.

For our experiments, we used a cluster of four nodes, where each node has an Intel Core i5 processor, 16 GB RAM, and a 1 TB disk. All experiments were executed in MongoDB Cluster mode with a single master that controls and manages each executor in cluster nodes. On our cluster, we run Windows 10 Pro with MongoDB 3.4.5, Spark 2.2.0, Scala 2.11.8, and Java 1.8.

1.3.3 Algorithms for Sharding

1.3.3.1 Algorithm for Range Sharding

1. Algorithm for `GetTag()`
   Input: User location, i.e., latitude and longitude
   Output: The "tag" value which is used as Shard key
   1. Find the range r in which given user location, i.e., latitude and longitude, belong.
   2. Identify the tag value t based on the input location range r
   3. Append the tag value t to document d
   4. Return Tag.
2. Algorithm—Range Query Using Range Sharding
   Input: The location is taken as input, i.e., the latitude and longitude.
   Output: Result List, R, i.e., locations in a particular range are retrieved.

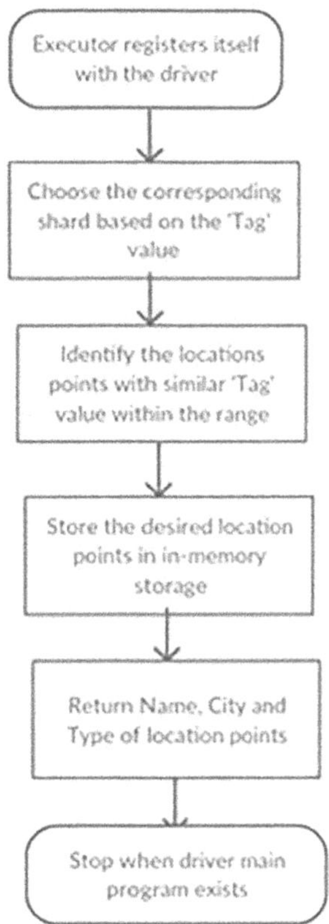

FIGURE 1.15 Executor task in tag sharding.

1. Find "Tag" variable using `GetTag()` //based on the tag of the input location//
2. Create "cursor" for input query //to access the desired documents from the database//
3. `if (cursor!=empty)`

```
Begin
                  While (cursor!=empty)
                    Begin  //retrieve the locations from
the database//
                R.append(document); //add the location to
Result  //
                  List
                  End
End
else
          Add null to R
```

4. Return result list, R

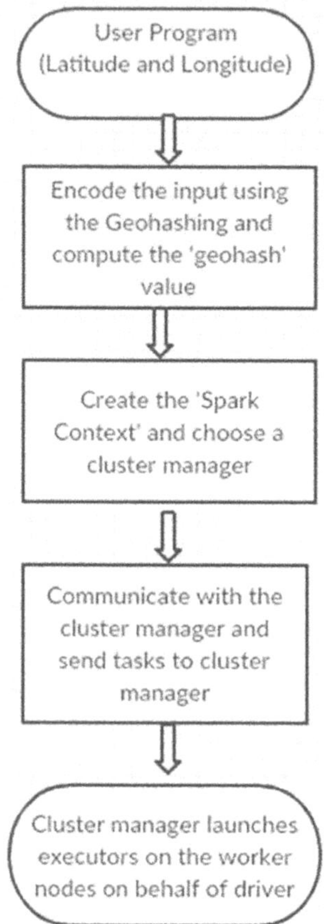

FIGURE 1.16 Master task in zone sharding.

3. Algorithm—Range Join Query Using Range Sharding
   Input: The location is taken as input, i.e., the latitude and longitude.
   Output: Result List, R, i.e., locations in a particular range are retrieved.
   1. Find "Tag" variable using `GetTag()` //based on the tag of the input location//
   2. Create "cursor A, B" for input query //access the desired documents from the database//
   3. `if (cursor A!= empty || cursor B!=empty)`

```
Begin
             While (cursor A!=empty)
             Begin
              //retrieve the locations from the database//
              R.append(document); //add the location to
Result    //
```

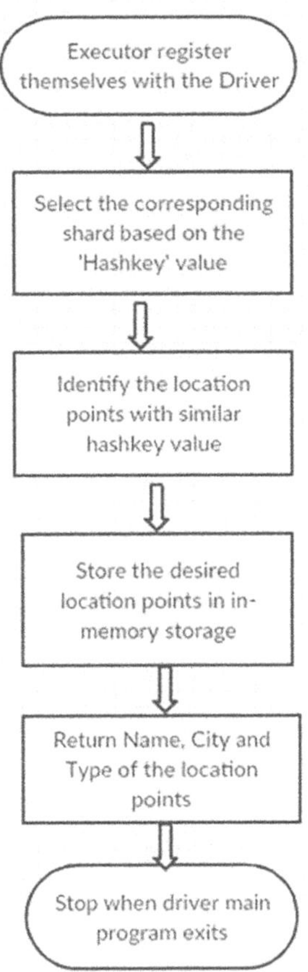

FIGURE 1.17 Executor task in zone sharding.

```
            List
            End
            While(cursor B!=empty)
            Begin
             //retrieve the locations from the database//
             R.append(document); //add the location to
Result  //
            List
            End
End
else
            Begin
          Add null to R
            End
```

4. Return result list, R

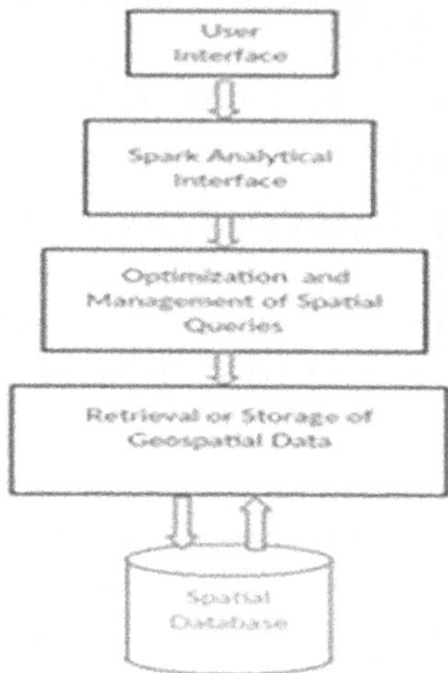

FIGURE 1.18 Retrieval of data from the spatial database.

4. Algorithm—*k*-NN Query Using Range Sharding

   Input: The location, i.e., the latitude and longitude and *k* value, is taken as input.

   Output: Result List, R, i.e., *k* nearest locations in a particular range are retrieved.

   1. Find "*k*" variable //number of nearest locations//
   2. Find "Tag" variable using `GetTag()` //based on the Tag of the input Location//
   3. Create "cursor" for input query //access the desired documents from database//
   4. `if (cursor!=empty)`

```
Begin
            While (cursor!=empty)
            Begin
          if (k==0) break;
                          //retrieve the locations from the
database//
          R.append(document); //add the location to Result
List //
              k--;
              End
End
else
         Add null to R
```

   5. Return result list, R

5. Algorithm—*k*-NN Join Query Using Range Sharding

   Input: The location, i.e., the latitude and longitude and *k* value, is taken as input.

Output: Result List, R, i.e., *k* nearest locations in a particular range, are retrieved.

1. Find "*k*" variable //number of nearest locations//
2. Find "Tag" variable using `GetTag()` //based on the Tag of the input Location//
3. Create "cursor A,B" for input query //to access the desired documents from the database//
4. `if (cursor A!=empty || cursor B!=empty)`

```
Begin
          While(cursor A!=empty)
          Begin
          if (k==0) break;
                    //retrieve the locations from the
database//
      R.append(document); //add the location to Result
List//
                    k--;
          End
          While (cursor B!=empty)
          Begin
          if (k==0) break;
                           //retrieve the locations from
the database//
         R.append(document); //add the location to Result
List//
                    k--;
          End
End
else
          Add null to R
```

5. Return result list, R

1.3.3.2 Algorithms for Zone Sharding

1. Algorithm for `GetZone()`

   Input: User location, i.e., latitude and longitude

   Output: The "zone" value is used as the Shard key

   1. Calculate the geohash value for a location, i.e., latitude and longitude
   2. Identify the tag value based on geohash, the input location, or shard key
   3. Append the zone value to document
   4. Return Zone.

2. Algorithm—Range Query Using Zone Sharding

   Input: The location is taken as input, i.e., the latitude and longitude.

   Output: Result List, R, i.e., locations in a particular range, are retrieved.

   1. Find "zone" variable //based on the zone of the input location//
   2. Create "cursor" for input query //to access the desired documents from the database//
   3. `if (cursor!=empty)`

```
Begin
                While (cursor!=empty)
                 Begin
             //retrieve the locations from the database//
             R.append(document); //add the location to
Result List//
                End
End
  Add null to R
```

4. Return result list, R

3. Algorithm—Range Join Query Using Zone Sharding
   Input: The location is taken as input, i.e., the latitude and longitude.
   Output: Result List, R, i.e., locations in a particular range, are retrieved.
   1. Find "zone" variable //based on the zone of the input location//
   2. Create cursor A, B for input query //to access the desired documents from the database//
   3. `if(cursor A!= empty || cursor B!=empty)`

```
Begin
While(cursor A!=empty)
                    Begin
                    //retrieve the locations from the
database//
                    R.append(document); //add the
location to Result List//
                  End
While(cursor B!=empty)
                    Begin
             //retrieve the locations from the database//
             R.append(document); //add the location to
Result List//
                  End
End
  Add null to R
```

   4. Return result list, R

4. Algorithm—*k*-NN Query Using Zone Sharding
   Input: The location, i.e., the latitude and longitude and *k* value, is taken as input.
   Output: Result List, R, i.e., *k* nearest locations in a particular range are retrieved.
   1. Find "*k*" variable //number of nearest locations//
   2. Find "zone" variable //based on the zone of the input location//
   3. Create "cursor" for input query //to access the desired documents from database//
   4. `if(cursor!=empty)`

```
Begin
               While(cursor!=empty)
                     Begin
              if(k==0) break;
```

```
//retrieve the locations from the database//
                R.append(document); //add the location to
Result List//
                       k--;
                       End
End
  Add null to R
```

5. Return result list, R

5. Algorithm—*k*-NN Join Query Using Zone Sharding

   Input: The location, i.e., the latitude and longitude and *k* value, is taken as input.

   Output: Result List, R, i.e., *k* nearest locations in a particular range, are retrieved.

   1. Find "*k*" variable //number of nearest locations//
   2. Find "zone" variable //based on the zone of the input location//
   3. Create "cursor A,B" for input query //to access the desired documents from the database//
   4. `if(cursor A!=empty || cursor B!=empty)`

```
                Begin
                     While(cursor A!=empty)
                            Begin
                     if(k==0) break;
//retrieve the locations from the database//
                     R.append(document); //add the
location to Result List//
                            k--;
                            End
                     While (cursor B!=empty)
                            Begin
                     if (k==0) break;
//retrieve the locations from the database//
                     R.append(document); //add the
location to Result List//
                            k--;
                            End
                End
else
       Add null to R
```

5. Return result list, R

1.3.4 Dataset and Statistics

The following datasets are considered for experimentation:

i. India Dataset—Geographical dataset in .csv format with 650 tuples of data is available at [19]. It contains the information about Place, State, Country along with its Latitude and Longitude values. The Place Type is used to summarize the type of location within the proximity of a location.

ii. Uber Taxi Trip Dataset—The dataset is in the form of .TSV and is available at [20]. It contains information about CarID, Time and Data, and Latitude and Longitude values for 1,048,500 tuples.
iii. Road Network Dataset—The dataset is in .csv format and is available at [21]. It contains information about NodeID and Latitude and Longitude values for about 21,000 tuples. Statistics of datasets is given in Table 1.1.

1.4 RESULTS AND PERFORMANCE EVALUATION

Results of query processing using two sharding techniques and four query types are presented here. Sample queries used in the implementation are shown as follows:

i. Range Query Using Zone Sharding and Range Sharding: Given latitude and longitude values, find the locations in a particular range.
ii. Range Join Query Using Zone Sharding and Range Sharding: Given latitude and longitude values, find the locations in a particular range of two place types.
iii. *k*-NN Query Using Zone Sharding and Range Sharding: Given latitude and longitude values, find the *k*-locations in a particular range.
iv. *k*-NN Join Query Using Zone Sharding and Range Sharding: Given latitude and longitude values, find the *k*-locations in a particular range of two place types.

Implementation of each query follows the following steps:

1. Initially, dataset containing places with their <latitude, longitude> is collected and stored in .csv format.
2. The dataset is further modified by adding the new fields in all the sharding techniques.
3. Spark—MongoDB server is connected.
4. Master-Slave executes Spatial data in parallel with each location <key, value> pair as <Place_Name, geohash>. Place Names will be given as output for each of the end user query. Type signifies the category of the place, like school, library, hospital, pubs, colleges, temples, museums, etc.

Table 1.2 presents response time for querying data of various queries using Zone and Range Sharding. It can be observed that for India Dataset and Taxi Trip Dataset,

TABLE 1.1
Statistics of Test Data Used

S. No.	Dataset Name with URL	Size in MB
1	India dataset [19]	0.06
2	Taxi trip dataset [20]	56.0
3	Road network dataset [21]	0.7

Zone sharding performed well when compared with Range sharding for all query types. For road network dataset, Zone sharding performed well for Range Join Query, k-NN Query, k-NN Join Query, and Range Sharding performed well for Range Query. In road network dataset, the latitude and longitude of each tuple are random in order.

Table 1.3 compares execution time for Join Query of GeoMongoSpark, GeoSpark [7], SpatialSpark [17], and Stark [5]. The time value taken for GeoMongoSpark is the average of Zone Sharding and Range Sharding techniques.

Figure 1.19 presents the processing time comparison for Zone Sharding, and Figure 1.20 presents the processing time comparison for Range Sharding for all the three datasets.

Table 1.4 presents the execution time for k-NN search when $k = 10, 20, 30$, and Table 1.5 presents the execution time (in seconds) of k-NN Join search when $k = 10, 20, 30$.

It can be observed that runtime increases as k value increases. Figure 1.21 presents the GUI for GeoMongoSpark. GUI presents various options to execute the query to the left window, and query visualization output is shown at the right window.

Figure 1.22 presents the visualization output of Range Join Query for Zone Sharding. Similar visualization is generated for all other queries. Different colors are used to plot so as to distinguish the type of place (museum, hospital, school, bank).

TABLE 1.2
Response Time (in ms) for Querying Data of Various Queries Using Zone Sharding (ZS) and Range Sharding (RS)

		Execution Time for Range Query		Execution Time for Range Join Query		Execution Time for k-NN Query		Execution Time for k-NN Join Query	
S. No.	**Dataset**	**ZS**	**RS**	**ZS**	**RS**	**ZS**	**RS**	**ZS**	**RS**
1	Dataset-1	86	92	95	97	83	90	101	108
2	Dataset-2	90	91	120	128	95	98	110	120
3	Dataset-3	93	90	115	120	92	100	115	122

TABLE 1.3
Execution Time for Join Queries in Various Frameworks

Geo Spark	Spatial Spark	STARK	GeoMongoSpark
92.5 s	40.07 s	57.5 s	130 ms

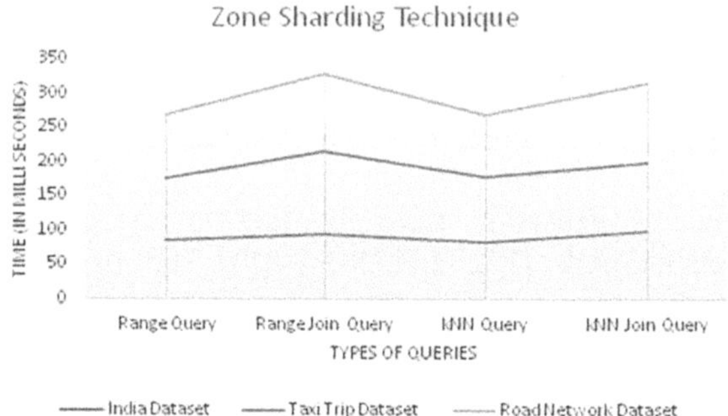

FIGURE 1.19 Processing time comparison for zone sharding.

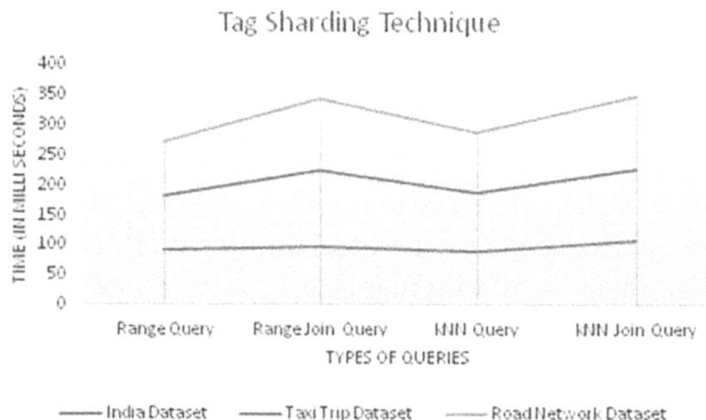

FIGURE 1.20 Processing time comparison for range sharding.

TABLE 1.4
Runtime (in µs) of *k*-NN Search for *k* = 10, 20, 30

k-NN Query	*k* = 10	*k* = 20	*k* = 30
Dataset-1	87,000	92,000	105,000
Dataset-2	98,000	108,000	113,000
Dataset-3	93,000	100,000	109,000

TABLE 1.5
Runtime (in s) of *k*-NN Join Search for *k* = 10, 20, 30

k-NN Query	*k* = 10	*k* = 20	*k* =30
Dataset-1	0.103	0.112	0.119
Dataset-2	0.110	0.116	0.125
Dataset-3	0.108	0.115	0.120

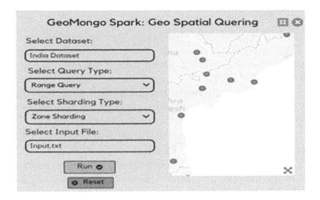

FIGURE 1.21 GeoMongoSpark main GUI page.

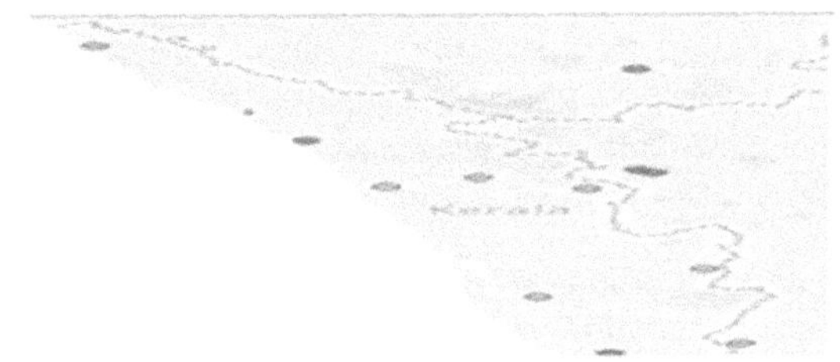

FIGURE 1.22 Visualization of range join query using zone sharding.

1.5 CONCLUSION

GeoMongoSpark added geospatial query processing that works as an effective storage and retrieval system. Using shard key during storage and retrieval helped in faster data access. Integrating Spark and MongoDB made us to process spatial query data stored in MongoDB without having to move data into Spark environment. Geohash helped to shard data, thereby improving query performance. When the shard key is appended with a geospatial query, mongos routes the query to a subset of shards in the cluster. Three benchmark datasets are used for experimenting on a variety of queries and two sharding techniques. GeoMongoSpark performance is compared with GeoSpark, SpatialSpark and Stark. Performance is also compared for different k values (10, 20, 30) for k-NN and k-NN join query. Results of geospatial queries are visualized using tableau. Zone sharding proved to be better than Range sharding.

The future work is to setup multinode configurations for Spark, which can produce the best results for processing large spatial datasets. An effective searching technique such as geoIntersect, geoWithin, and nearSphere is still required to perform join on fuzzy domains (when a search point belongs to more than one domain). Such searching technique can be used in applications such as Telecom industry to implant cell towers and also in tourism and development industries. Our future work also concentrates on building machine learning model using Scala. Sharding technique can be treated as an alternative to partitioning techniques, but it poses several drawbacks such as operational complexity, single point of failure, etc.

REFERENCES

1. Berry, J.K. (1987). Fundamental operations in computer-assisted map analysis. *International Journal of GIS*. 1. 119–136.
2. Apache Spark and MongoDB (2015), www.mongodb.com/products/spark-connector, Last accessed on [Feb. 11, 2018].
3. Alger, K.W. (2017), MongoDB Horizontal Scaling through Sharding, https://dev.to/kenwalger/mongodb-horizontal-scaling-through-sharding, Last accessed on [Feb. 07, 2018].
4. "Mongos" (2017), https://docs.mongodb.com/manual/core/sharded-cluster-query-router/, Last accessed on [Aug. 18, 2017].
5. Hagedorn, S., et al. (2017), The stark framework for spatial temporal data analytics on spark, *Proceedings of 20th International Conference on Extending Database Technology (EDBT)*, pp. 123–142, Bonn
6. Tang, M., et al. (2016), In-memory Distributed Spatial Query Processing and Optimization, pp. 1–23.
7. Lenka, R.K., et al. (2017), Comparative Analysis of SpatialHadoop and GeoSpark for Geospatial Big Data Analytics, arXiv:1612.07433v2 [cs.DC], pp. 484–488.
8. Olsen, D.A. (1997), Long Term Resource Monitoring Program Spatial Data Query and Visualization Tool, Environmental Management Technical Center, pp. 1–12.
9. Datos.gob.es (2013), Data Processing and Visualisation Tools, ePSIplatform Topic Report, pp. 1–23.
10. Tang, M., et al. (2016), Location spark: A distributed in memory data management system for big spatial data, *Proceedings of the VLDB Endowment*, Vol. 9, No. 13, pp. 1565–1586.
11. Hendawi, A.M., et al. (2017), Panda ∗: A generic and scalable framework for predictive spatio-temporal queries, *GeoInformatica*, Vol. 21, No. 2, pp. 175–208.
12. Putri, F.K., et al. (2017), DISPAQ: Distributed profitable-area query from big taxi trip data, *Sensors*, Vol. 17, p. 2201.
13. Hegde, V., et al. (2016), Student residential distance calculation using Haversine formulation and visualization through Googlemap for admission analysis, *Proceedings of IEEE International Conference on Computational Intelligence and Computing Research (ICCIC)*, pp. 1–5, Chennai, India
14. Liu, X., et al. (2003), Object-based directional query processing in spatial databases, *IEEE Transactions on Knowledge and Data Engineering*, Vol. 15, No. 2, pp. 295–304.
15. Li, L., et al. (2017), Surrounding join query processing in spatial databases, *Proceedings of ADC 2017*, Springer International Publishing, pp. 17–28, Brisbane, Australia
16. Vasavi, S., et al. (2018) Framework for geospatial query processing by integrating Cassandra with Hadoop, *Knowledge Computing and Its Applications*, Editors: S. Margret Anouncia Uffe Kock Wiil, Springer, Singapore, pp. 131–160.
17. You, S., Zhang, J., Gruenwald, L. (2015). Large-scale spatial join query processing in cloud. In *Proceedings of the 2015 31st IEEE International Conference on Data Engineering Workshops (ICDEW)*, pp. 34–41.
18. Vasavi, S., et al. (2018), Framework for geospatial query processing by integrating Cassandra with Hadoop, *GIS Applications in the Tourism and Hospitality Industry*, Editors: Somnath Chaudhuri, Nilanjan Ray, Chapter 1, IGI Global, pp. 1–41.
19. India dataset, www.latlong.net/country/india-102.html, Last accessed on [Feb. 11, 2018].
20. Taxitrip dataset, https://raw.githubusercontent.com/dima42/uber-gps-analysis/master/gpsdata/all.tsv, Last accessed on [Feb. 11, 2018].
21. Road network dataset, www.cs.utah.edu/~lifeifei/research/tpq/cal.cnode, Last accessed on [Feb. 11, 2018].

2 A Study on Metaheuristic-Based Neural Networks for Image Segmentation Purposes

Navid Razmjooy
Independent Researcher, Belgium

Vania V. Estrela
Universidade Federal Fluminence

Hermes J. Loschi
State University of Campinas – UNICAMP

CONTENTS

2.1 INTRODUCTION

Digital images that are generated from smaller elements are called pixels. By considering an image as function *f*, with two independent dimensional including *x*, and *y*, $f(x, y)$, where $-\infty < x, y < \infty$ can be illustrated on a page by taking a value for its intensity or brightness. Here, the point (x, y) is called the point of a pixel.

The process of applying different methods on these images is called image processing. The applications of image processing and machine vision in different industries and fields are increasing day by day. There are two principal limitations that make the digital image processing methods interesting:

1. Enhancing the visual information for human interpretation
2. Scene data processing for machine perception independently

By increasing the awareness of human life and by complicating their demand, the utilization of new machine vision systems in the real world is inevitable [1]. The visual perception of the human can be easily mistaken; this disadvantage increases the instability of the human, the inspecting cost, and changing of such assessments, which made it necessary to have a machine vision system in industrial and sensitive works.

One of the main parts of image processing is image segmentation. The main idea behind image segmentation is to classify a considered image into its main components. In other words, the main purpose of image segmentation is to simplify the next process steps by turning the image into a more meaningful data. Image segmentation uses assigning labels to all pixels of the image, where pixels with similar features are labeled in the same class [2]. This process includes an important part of most computer vision applications [3]. Image segmentation can be classified into four principal categories: histogram-based thresholding methods, methods based on classification and clustering techniques, texture analysis based methods, and region-based split and merge methods [4].

Among the different types of image segmentation, classification is a useful technique that separates the main components of an image, whereas the pixels with the same features include a special range of gray levels. In simple terms, classification-based segmentation has been given considerable attention in the last few years [5–10]. Furthermore, due to their ability to achieve near-optimal solutions for many applications, heuristic algorithms have been recently utilized to achieve proper solutions for the problem of image segmentation [5,10–12].

Recently, the neural network is turned into one of the most popular methods for image segmentation in different applications. Multilayer perceptron (MLP) is one of the most widely used neural network models in which the connection weight training is normally completed by a backpropagation (BP) learning algorithm [13–15].

BP learning algorithm uses a gradient descent method for minimizing the error between the desired and the output for classification. One of the main drawbacks of the gradient descent method is getting stuck in the local trap.

Using metaheuristic algorithms can minimize error in the neural network by optimizing the value of its parameters like weight training, architecture adaptation (for determining the number of hidden layers, number of hidden neurons, and node transfer functions), and learning rules [10,13,16].

This chapter focuses on a study about how metaheuristic algorithms can optimize the performance of a neural network and how they can be employed for image segmentation purposes. The main idea is to optimize the initial weights of a feedforward neural network.

2.2 SUPERVISED IMAGE SEGMENTATION

The main idea in the supervised classification in image processing is to divide all the pixels of the input image into two predefined classes. By using supervised classification, we categorize examples of the information classes of interest in the image.

Color is one of the considered cases that can become a classification issue. Furthermore, the purpose of color pixel classification is to decide what class a color pixel belongs to. Good image color pixel segmentation should give coverage for all complicated types. Such mentioned problem can be evaluated by artificial neural networks (ANNs), which have been proven as an efficient tool for pattern classification purposes, where decision rules are hidden in highly complex data and can be learned only from examples. The image is then classified by attempting to change the performance of each pixel and decide which of the signatures are the most similar.

ANNs are computational models that are inspired by natural neurons. Natural neurons receive signals through synapses located on the dendrites or membrane of the neuron. When the received signals get strong enough, the neuron is activated and emits a signal through the axon. This signal might be sent to another synapse and might activate other neurons.

From a practical point of view, ANNs are just parallel computational systems that include many simple processing elements connected together in a special way to perform a considered task. ANNs are strong computational devices that can learn and generalize from training data since there is no requirement for complicated feats of programming.

From a practical point of view, image segmentation can be performed within two different concepts: supervised (classification) methods and unsupervised (clustering) methods [1,17,18].

2.3 LITERATURE REVIEW

In the recent years, applications of automated computer-aided assessment systems increased from industrial quality inspection [1], agricultural systems [19], aerial imaging [20], to medical imaging [5]. These processes have been applied to the input of digital images from the camera.

The main purpose of supervised methods is to classify all the pixels of a digital image into one of several classes; the main idea behind supervised methods is that the main groups are first determined by a supervisor (expert), and then, the method classifies the data based on a given class [8,9,21,22].

In contrast, in unsupervised methods, the groups are not defined by en expert. So, these methods have two main works to do; first, identify some optimal groups, and second, classifying the given data among these groups.

One of the potent classification methods is neural networks. Nowadays, ANNs have been introduced as a pleasant potency for modeling the complicated systems in different applications [12,13,23,24]. Some of them are supervised [25] and some others are unsupervised [26].

Since using supervised methods are better for most of the image segmentation purposes, in this study, the supervised methods have been focused. There are different methods that have been introduced for image classification, from Knn (K nearest neighbor) classifiers to ANN-based classifiers.

This chapter describes the analysis and usage of a popular type of neural network known as an MLP network.

The main advantage of an MLP network is the generalization preference towards supervised problems because of its superabundant parallel interconnections and easiness to apply on any complicated problem using just training samples.

However, it is important to know that the performance of this network is directly related to the training sample; in other words, if wrong samples are used for training, the system will have wrong results. Another point is that always using more training samples is not a correct way; the main idea is to have enough sampling data, not extra data. The extra, on the other hand, can cause overfitting and result in incorrect output [27,28].

There are a lot of research studies that have proposed to cover this drawback [24,29]. For instance, Genetic Algorithm (GA) [9,30,31], Particle Swarm Optimization (PSO) algorithm [12,32–34], Quantum Invasive Weed Optimization (QIWO) algorithm [35], and World Cup Optimization (WCO) algorithm are some of the examples [36].

2.4 ARTIFICIAL NEURAL NETWORKS

Recently, ANNs have been turned into a popular and high-performance method in machine computing, especially machine learning that can be used for describing the knowledge and implementing the obtained knowledge for output overestimation of complicated systems [37].

The human brain, according to many scientists, is the most sophisticated system that is ever seen and studied in the universe. But this complex system has neither its dimensions nor its number of components, but more than today's supercomputers [38]. The mysterious complexity of this unique system returns to the many connections among its components. This is what distinguishes 1,400 grams of the human brain from all other systems. However, research has shown that human brain units are about 1 million times slower than the transistors used in CPU (Central Processing Unit) silicon chips in terms of performance [39,40]. Apart from this, the high capability of the brain performance in solving a variety of problems has made the mathematicians and engineers to work on the simulation of hardware and software architectures. Two main parts in simulating the structure of neural networks are nodes (neurons) and weighted communications (synapses) [19]. A simple structure of a neural network is shown in Figure 2.1.

Different types of computational models have been introduced under the name of ANN, each of which is usable for a range of applications and are inspired based on a certain aspect of capabilities and characteristics of the human brain [13,41–44].

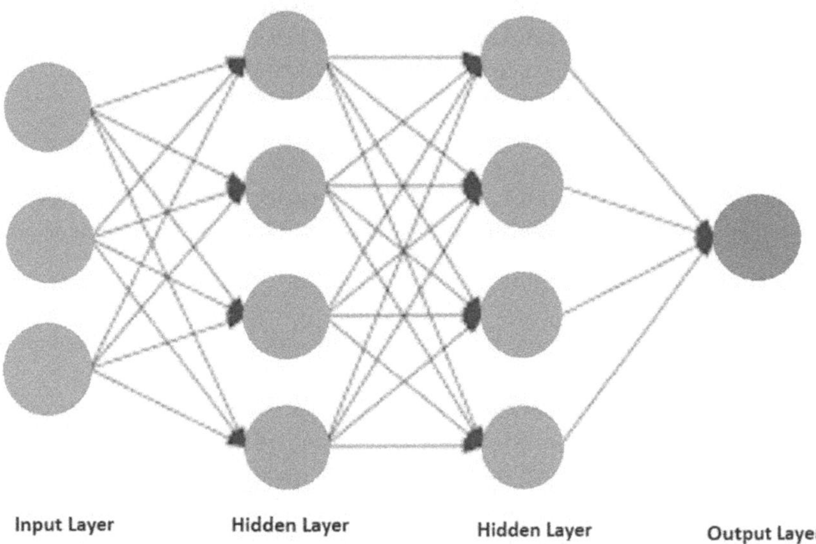

FIGURE 2.1 Schematic of the neural network.

Recently, a large number of different types of ANNs have been presented. For instance, all of the introduced networks, such as *MLP networks*, *Kohnen networks*, *Hopfield networks*, etc., are trained based on different conceptions, like *Error BP method* [45–46]. From the point of solution nature, ANNs can be divided into two different groups of postpropagation and recurrent networks (in which they use the output feedback). The style of how to address the aforeexplained cases is one of the most significant differences between human memory and computer memory. For a computer memory, addressing is based on the memory blocks or the information on the permanent memory. For instance, to get a specific text or image, the memory address or file associated with that image or text should exist.

In this type of addressing, by having a picture or text, the structure cannot recognize the considered target, and it needs a lot of learning samples to compare the target with them. This is so clear that doing such a thing is so time-consuming and costly. Unlike the computer memory, the human brain does this work so easily and fast. For instance, if you see an image incomplete, you can imagine the total image in your brain, if you see a letter in your notebook, you can remember all the things that you should do, etc.

This kind of memory is the so-called *Content Addressable Memory* [47,48]. In this type of memory, after receiving the contents of a memory block, the address is immediately given as output.

One of the most basic neural models available is the MLP model, which simulates the transfusion function of the human brain. In this kind of neural network, most of the human brain's network behavior and signal emission have been considered, and hence is sometimes referred to as feedforward networks [49,50].

Each of the neurotic cells, called neuron, receives an input (from a neuronal or nonneuronal cell), and then transmits it to another cell (neuronal or nonneuronal). This behavior continues until a certain outcome, which will eventually lead to decision, process, thinking, or movement. The structure of MLPs comprises sets of sensory units, which includes an input layer, one or more hidden layers, and an output layer. Figure 2.1, indeed, is a schematic of an MLP [51]. The error BP algorithm in MLPs changes the weights and bias values of the network to decrease the error value more quickly. The error BP algorithm can be described using the following formula:

$$x_{k+1} = x_k - \alpha_k g_k \tag{2.1}$$

where x_k describes the weight and bias vectors in the kth iteration, α_k is the learning rate in the kth iteration, and g_k describes the gradient in the kth iteration.

Another commonly used network is the Radial Basis Function (RBF) based networks; similar to the pattern of MLP neural networks, there are other types of neural networks in which processor units are centrally focused on processing [52,53]. This focus is modeled using an RBF. In terms of overall structure, RBF neuronal networks do not differ much from MLP networks, and they differ only in the type of processing that neurons perform on their cores. However, RBF networks often have a faster process of learning and preparation. In fact, due to the concentration of neurons on a specific functional range, it would be easier to adjust them [54,55].

RBF networks need more neurons than the standard MLP networks, but most of these kinds of networks need less time than the MLPs for learning. When abundant data exists, RBF networks result in better performance than MLPs. RBF networks comprise a kind of feedforward neural networks with a similar structure to MLP networks. The schematic of an RBF network is shown in Figure 2.2.

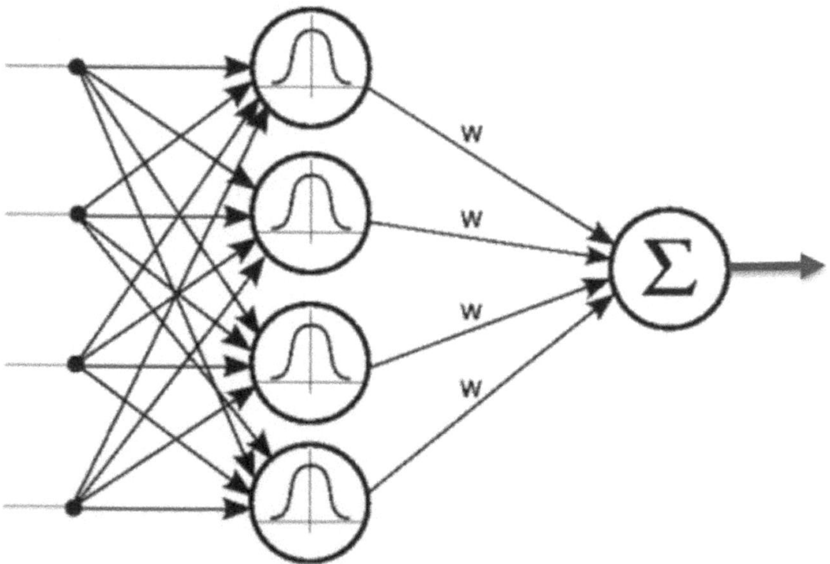

FIGURE 2.2 Schematic of the RBF networks.

One of the differences between RBF and other ordinary network is their inputs. In the RBF network, the inputs of the transfer function are equal to the distance vector between weights and inputs multiplied by bias. The input values are sent to the hidden layer nodes by the input layers. Each node in the hidden layer is described by a transmission function *f* that transmits the input signals; in Figure 2.3, the radial transfer function example diagram is shown.

For the *p*th input pattern, i.e., X^p, the *J*th hidden node's response, which means y_j, is equal to

$$y_j = f\left\{\frac{\left\|X^p - U_j\right\|}{2\sigma_j^{\,2}}\right\} \tag{2.2}$$

where $\|\cdots\|$ describes the Euclidean norm, σ is the RBF range, and U_j is the center of the *j*th RBF.

The network output is achieved using the weighted linear sum of the hidden layer responses in each of the output nodes. The following equation illustrates how to obtain the output of the *k*th node in the output layer:

$$Z_{pk} = \sum_{j=1}^{L} y_i w_{kj} \tag{2.3}$$

In all of these models, there is a mathematical structure that can be graphically displayed and has a set of parameters and adjusting parameters. This general structure is adjusted and optimized by an algorithm of learning algorithms, which can display a proper behavior.

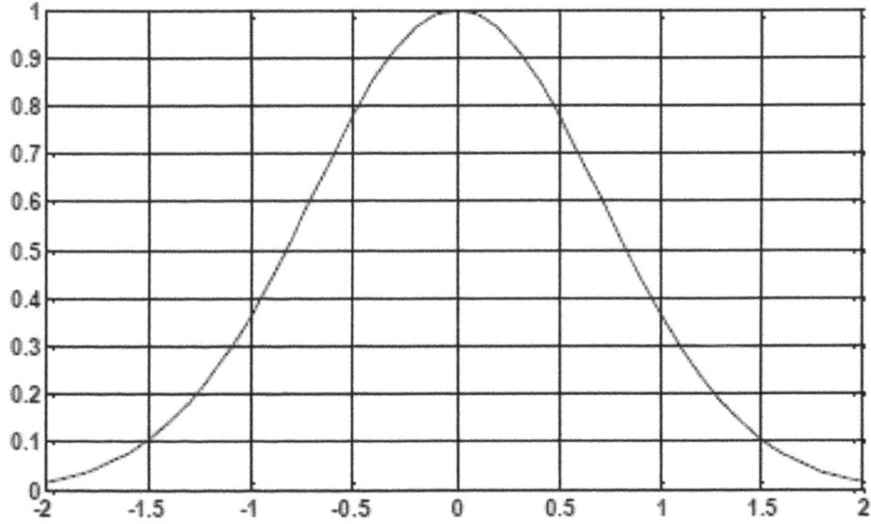

FIGURE 2.3 A radial basis transfer function.

A look at the learning process in the human brain also shows that in fact we also experience the same process in our brains, and all our skills, thoughts, and memories are shaped by the weakening or strengthening of the relationship between the neural cells of the brain. This amplification and weakening in mathematical language models describe itself as the setting of a parameter (known as Weight) [5,10,12,24,56,57].

2.5 OPTIMIZATION

The main purpose of optimization is to find the best acceptable solution, given the constraints and requirements of the problem [58,59]. For a problem, there may be several different answers that are defined to compare them and select an optimal answer, a function called a fitness (cost) function.

How to choose a fitness function depends on the nature of the problem. For instance, consumed time or cost is one of the common goals of optimizing transportation networks. However, selecting an appropriate fitness function is one of the most important steps in optimization.

Sometimes, in optimization, multiple objective functions are considered simultaneously, which are called multiobjective optimization problems [60,61].

The simplest way to deal with these problems is to form a new objective (fitness) function as a linear combination of main objective functions, in which the effect of each function is determined by the weight assigned to it. Each optimization problem has a number of independent variables, which are called design variables, which are represented by n dimensions of vector x. The purpose of optimization is to determine the design variables in such a way that the objective function is minimized or maximized.

Optimization problems are divided into two main categories:

1. Unconstrained optimization problems: In these problems, the maximization or minimization of objective function is without any constraints on the design variables.
2. Constrained optimization problems: Optimization in most applications is subject to limitations; constraints on the behavior and function of a system, and the behavioral limitations and constraints that exist in the physics and geometry of the problem are called geometric or lateral constraints.

Equations representing constraints may be equal or unequal, and in each case, the optimization method is different. However, constraints determine the acceptable design area. The optimization process can be considered in the following stages:

Problem formulation: At this stage, a decision problem is defined along with a general structure of it. This general structure may not be very accurate but states the general state of the problem, which includes the input and the output factors and the objectives of the problem. Clarifying and structuring the problem may be complicated for many optimization problems [62–64].

Problem modeling: At this stage, a general mathematical model for the problem is constructed. Modeling may help with similar models in the subject's history. This step resolves the problem to one or more optimization models.

Problem optimization: After modeling the problem, the solution routine produces a good solution to the problem. This solution may be optimal or almost optimal. The point to be taken into consideration is that the obtained result is a solution to the model, not to the real problem. During formulation and modeling, there may be some changes in the real problem, and the new problem is far from the actual one.

Problem deployment: The solution obtained by the decision maker is examined, and, if acceptable, is used, and, if the solution is not acceptable, the optimization model or the algorithm should be developed and the optimization process repeated.

A general description of the optimization problem is given in the following:

$$\begin{aligned} &\underset{x \in \Re^n}{\text{minimize}}\, f_i(x), (i = 1, 2, \ldots, M). \\ &\text{s.t. } h_j(x) = 0, (j = 1, 2, \ldots, J) \\ &g_k(x) \leq 0, (k = 1, 2, \ldots, K) \end{aligned} \tag{2.4}$$

where $f_i(x)$, $h_j(x)$, and $g_k(x)$ are functions of the vector. Each x_i member of x is called the decision variable that can be continuous, real, discrete, or a combination of them.

While using the optimization algorithms, a few cases should be considered:

- The functions $f_i(x)$ in which $i = 1, 2, \ldots, M$ are called fitness, objective, or cost function.
- If the optimization is a single objective, then $M = 1$.
- Search space ($\Re^n$) is divided by the decision variables.
- Solution (response) space is a space that can be formed by the values of the objective function.
- Problem constraints can be considered as equations (h_j) and inequalities (g_k).
- All inequalities can be rewritten as more than zero.
- The problem function can be rewritten as the maximization problem.

2.6 METAHEURISTIC ALGORITHMS

The goal of metaheuristic algorithms is to provide a solution within a reasonable time frame that is appropriate to solve the problem, and the exploratory algorithm may not be the best solution to solve the problem, but can be the closest to the solution.

Metaheuristic algorithms can be combined with optimization algorithms to improve the efficiency of the algorithm. A metaheuristic algorithm is a combination of heuristic algorithms that are designed to find, generate, or select each exploration in each step, and provide a good solution to problems that are optimizing. Metaheuristic algorithms consider some of the hypotheses of the optimization problems to be solved.

Metaheuristic algorithms are some kinds of algorithms that are generally inspired by nature, physics, and human life and are used to solve many of the optimization problems. Metaheuristic algorithms are usually used in combination with other algorithms to reach the optimal solution or escape from the local optimal solution [65].

In recent years, one of the most important and promising researches has been "Inventive methods derived from nature," which have similarities with social or natural systems. Their application is based on continuous inventive techniques that have interesting results in solving the Nondeterministic Polynomial-time hard (NP-Hard) problems.

First, we begin by defining the nature and the naturalness of the methods; the methods are derived from physics, biology, and sociology, and are as follows:

- Using a certain number of attempts and repetitive efforts
- Using one or more agents (neuron, pepper, chromosome, ant, etc.)
- Operations (in multiagent mode) with a competitive cooperation mechanism
- Creating self-transformation methods

Nature has two great measures:

- Selection of rewards for strong personal qualities and penalties for the weaker person
- A mutation that introduces random members and the possibility of a new person's generation

Generally, there are two situations that are found in the original methods derived from nature, selection, and mutation: choosing the basics for optimizing and jumping the basics of ideas for online searching.

Among the characteristics of the innovative methods derived from nature, one can mention the following:

- Model real phenomena in nature
- With no interruptions
- They often introduce the same combination of factors (multiple agents)
- They are compatible

The earlier characteristics provide sensible behavior to provide intelligence. The intelligent definition also includes the power to solve the problems; therefore, intelligence leads to a proper solution to complicated optimization problems. In the following, some population-based methods that are used in this paper are explained.

2.6.1 Genetic Algorithm

GA is one of the most important algorithm that is used to optimize for defined functions on a limited domain. In this algorithm, the past information is extracted according to the inheritance of the algorithm and is used in the search process. The concept of GA was developed in 1989 by John Holland [66,67].

GAs are usually implemented as a computer simulator in which the population of an abstract sample (chromosomes) of the candidates for the solution of an optimization problem leads to a better solution.

Traditionally, solutions were in the form of strings of 0 and 1, but today they are implemented in other ways. The hypothesis begins with a completely randomized unique population and continues in generations. In each generation, the capacity of the whole population is evaluated; several individuals are selected in the random process of the current generation (based on competencies) and modified to form the new generation (fraction or recombination), and in the next repetition of the algorithm is converted to the current generation. The structure of the GA is as follows:

A. Chromosome: A string or sequence of bits, encoded as a form, is a possible (appropriate or inappropriate) solution to the problem [9]. If binary coding is used, each bit can be one of the values of zero and one each of the bits of the chromosome of the recent problem is a potential solution to the problem variables. The basis of GA is to encode each set of solutions. This coding is called a chromosome. In fact, the encrypted form is a probable solution to the problem. In this case, each chromosome is a solution to a problem that can be justified or unjustified.

B. Objective (fitness) function: A function whose value of the problem variable is placed, thus, the desirability of each solution is determined. In optimization problems, the objective function is used as a fitness function. The objective function is used to determine how people act in the problem boundary, and the fitness function is usually used to convert the value of the objective function into a dependent fitness value. In other words:

$$F(n) = G(F(x)) \tag{2.5}$$

where *f* is an objective function, and the function *G* converts the value of the objective function to a nonnegative number, and *F* is the amount of fitness it relates to.

The suitability or not of the solution is measured by the amount of the obtained objective function.

C. The population size and the number of production: the number of chromosomes is called population size. One of the advantages of GA is that it uses parallel searches rather than traditional search methods. By definition, population size is the size of parallel searches [68,69].

Genetic operators should be used to find a point in the search space. Two of these operators to generate children are *Crossover* and *Mutation*.

Crossover operator is the main operator for generating new chromosomes in GA [70]. Like its counterpart in nature, this operator produces new people whose components (genes) are composed of their parents. Some types of intersecting operators are one-point crossover, two-point crossover, uniform crossover, etc.

Mutation operator is a random process in which the content of a gene is replaced by another gene to produce a new genetic structure. Note that each repetition of the algorithm that leads to the creation of a new population is called a generation [71]. The flow chart of the GA is given in Figure 2.4:

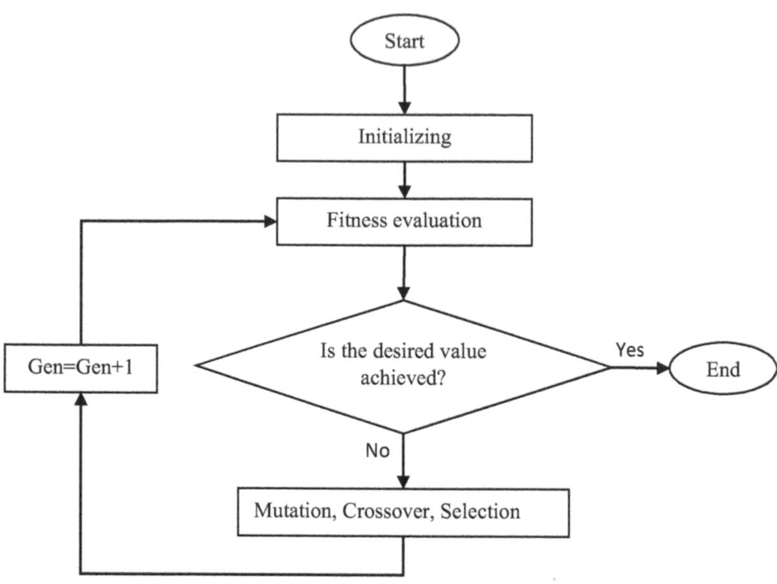

FIGURE 2.4 The flow chart of the GA.

2.6.2 Particle Swarm Optimization Algorithm

The PSO Algorithm is first introduced by James C. Kennedy and Russell Eberhart. The PSO algorithm is a group of optimization algorithms that operate on the basis of random population generation [12,72]. This algorithm is based on the modeling and simulation of the collective flying (grouping) of birds or collective movement (grouping) of fish.

Each member in this group is defined by the vector of velocity and position vector in the search space. In each repetition, the new particle position is defined due to the velocity vector and position vector in the search space [73,74].

At each time interval, the new particle position is updated according to the current velocity vector, the best position found by that particle, and the best position found by the best particle in the group.

Sometimes, differences can be seen in different ways of implementing the PSO algorithm. But this does not cause a problem in programming, because what matters is the execution of program steps in the sequence that follows, and how these steps are separated. For example, in some references, they combined the update phase of the particle velocity and the transfer of particles to new locations as a stage. This change will not cause a problem in the implementation of the algorithm.

General steps of applying the PSO algorithm are given later:

Step 1: Generate a random initial particle population: Random generation of the initial population is simply a random determination of the initial site of the particles with uniform distribution in the space of space (search space). The randomized generation of the initial population exists in almost all metaheuristic optimization algorithms. But in this algorithm, in addition to the initial random location of particles, a value is also allocated for the initial particle velocity. The initial propagation range for particle velocity can be extracted from the following equation.

$$\frac{X_{\min} - X_{\max}}{2} \leq V \leq \frac{X_{\max} - X_{\min}}{2} \tag{2.6}$$

Step 2: Selecting the number of primary particles: increasing the number of primary particles reduces the number of iterations needed to converge the algorithm. But, sometimes, it is observed that users of the optimization algorithms assume that this reduction in the number of iterations means reducing the program execution time to achieve convergence, while such an idea is completely wrong.

However, an increase in the primary number reduces the number of iterations. However, an increase in the number of particles makes the algorithm to spend more time in the particle evaluation stage, which increases the time of evaluation so that the algorithm's execution time does not decrease, until the convergence progresses despite a decrease in the number of iterations.

Step 3: Evaluate the objective function of particles: at this stage, we must evaluate each particle, which represents a solution to the problem under consideration. Depending on the problem under consideration, the evaluation method will be different. For example, if it is possible to define a mathematical function for a target, by simply inserting the input parameters (extracted from the particle position vector) in this mathematical function, the cost of this particle will be easily calculated. Note that each particle contains complete information about the input parameters of the problem, which is extracted from this information and is in the objective function.

Step 4: Convergence test: the convergence test in this algorithm is similar to other optimization algorithms. There are different methods for examining the algorithm. For instance, it is possible to determine the exact number of repetitions from the beginning, and at each step, it was checked whether the number of iterations reached the specified value or not. If the number of iteration is smaller than the initial value, then you must go back to Step 2. Otherwise, the algorithm ends.

Another method that is often used in the convergence test of the algorithm is that, if in the successive iteration, for example, 15 or 20 iterations, no change in the cost of the best particle is created, then the algorithm ends; otherwise, you need to return to Step 2.

The flow chart diagram of the PSO algorithm is shown in Figure 2.5.

2.6.3 Imperialist Competitive Algorithm

The Imperialist Competitive Algorithm (ICA) algorithm is one of the newest intelligent optimization algorithms that have been introduced in the field of artificial intelligence and metaheuristic algorithms [75–77]. The main reason for this algorithm

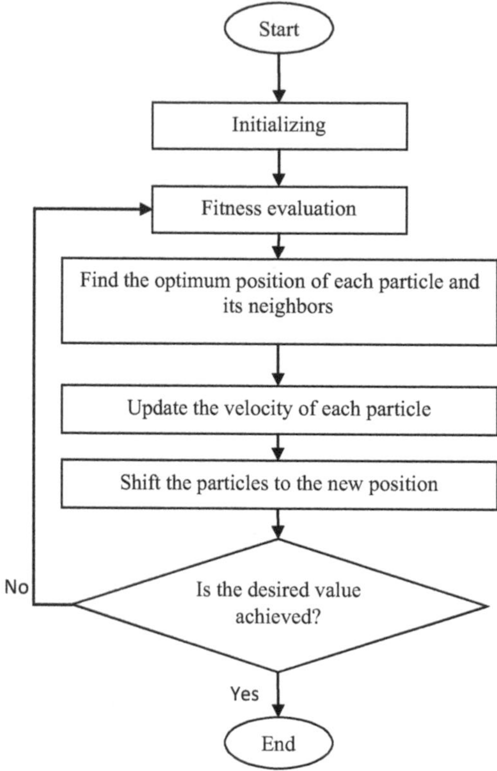

FIGURE 2.5 The flow chart of the PSO algorithm.

is to simulate the colonial political process. In the same way that the GA simulates biological evolution, in the imperialist competition algorithm, political evolution has been used. This algorithm was introduced in 2007 by Atashpaz-Gargari and Lucas and has been ever since used as a tool for many applications and research fields. The high power of this algorithm, especially in dealing with continuous issues, has led the imperialist competition algorithm to be considered as one of the major tools for optimization [78].

Like other metaheuristic algorithms, ICA has an initialization to random generation of a set of possible solutions. These initial solutions are called country.

ICA with the following process will gradually improve these initial solutions (countries) and ultimately provide an optimal solution to the optimization problem.

The main foundations of this algorithm are assimilation, imperialistic competition, and revolution. This algorithm, by imitating the social, economic, and political evolution of countries and by mathematical modeling of them, provides parts of this process with regular operators in the form of algorithms that can help to solve complicated optimization problems. In fact, this algorithm looks at the optimization problems in the form of countries and tries to improve these responses during the iterative process and ultimately to the optimal solution for the problem.

Like other evolutionary algorithms, this algorithm also begins with a number of random initial populations; each member is called a “country.” Some of the best members of the population (the equivalent of elites in the GA) are selected as imperialists.

The remaining population is considered as colonies. Imperialists, depending on their strength, are pulling their colonies with a specific process. The power of all empires depends on both the imperialist country (as the core) and its colonies. In mathematical terms, this dependence is modeled on the definition of imperial power, in particular, the power of the imperialist country, in addition to its average colonial power.

With the formation of early empires, colonial competition begins among them. Any empire that fails to succeed in colonial competition and adds to its power (or at least minimizes its influence) will be eliminated from the imperialist competition, so the survival of an empire is dependent on its power to capture the colonies of competing empires and to control them.

Consequently, during colonial struggles, the power of the empires will be gradually increased, and the weaker empires will be eliminated. Empires will have to advance their own colonies to boost their power.

In brief, this algorithm considered colonialism as an inseparable part of the evolution of human history, and its impact on imperial and colonial countries and the entire history has been used as the source of inspiration for an efficient and innovative algorithm for evolutionary computing.

In this study, two parameters of ICA, including colony assimilation coefficient and revolution probability, are defined by *Beta* and *pRevolution.*

The flow chart of the ICA is shown in Figure 2.6.

2.7 OPTIMIZATION OF THE NEURAL NETWORKS WEIGHTS USING OPTIMIZATION ALGORITHMS

One of the proper methodologies for optimizing of ANNs is to find the optimal values for networks’ weights. This technique can reduce the drawback of the BP algorithm based on metaheuristic algorithms’ exploration search technique and its improvements on the neural network architecture for reconstructing and fixing it during the evolution.

The numerical value of the weights can be determined based on minimizing an error function for training the network from the best position [5,10,24,29].

Mathematically, a network function $f(x)$ of a neuron can be considered as a combination of other functions $g_i(x)$, which can be easily considered as a structure for a network, with arrows representing the dependencies among variables.

A traditional model for forming the nonlinear weighted sum is given in the following:

$$f(x) = K\left(\sum_i \omega_i g_i(x) + b_i\right) \tag{2.7}$$

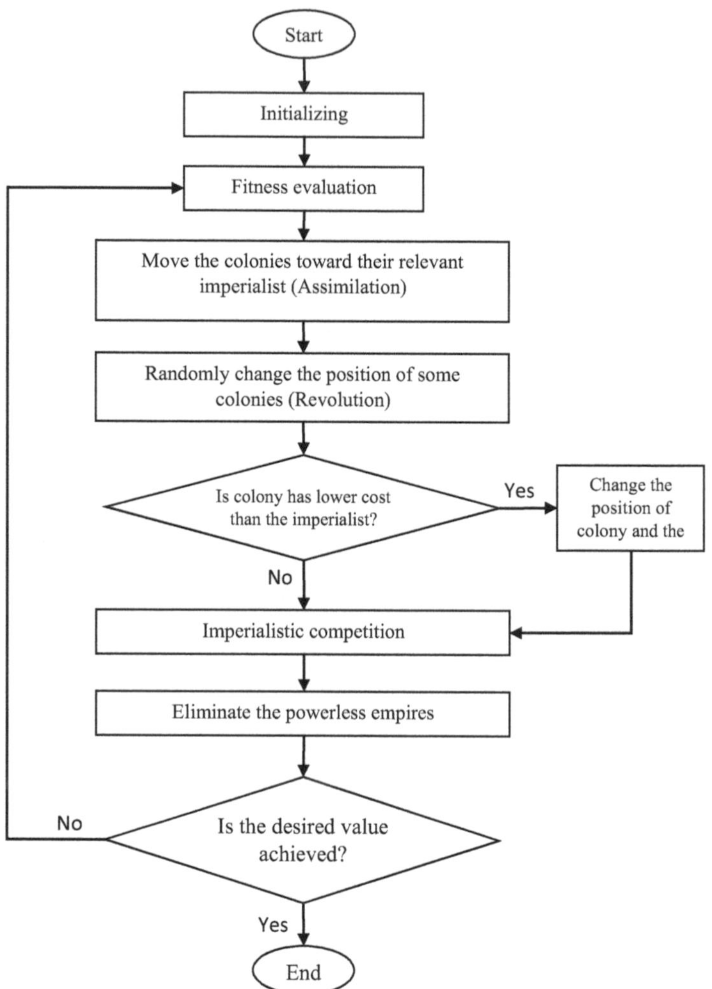

FIGURE 2.6 Flow chart of the ICA.

where K is a predefined function like the sigmoid function. A good vision for simplifying the collection of functions g_i is to assign them as a vector of $g = (g_1 \ldots g_n)$.

As explained earlier, BP method is one of the widely used methods for solving feedforward networks. It calculates the error on all training pairs and adjusts its weights to achieve an optimal solution for the desired output. It will be performed in several iterations to achieve the optimal solution and the minimum value of error for the training set.

After the training step, the network with optimal values of weights is prepared for validation test and utilization. A schematic diagram of how to adjust an MLP by metaheuristics is shown in Figure 2.7.

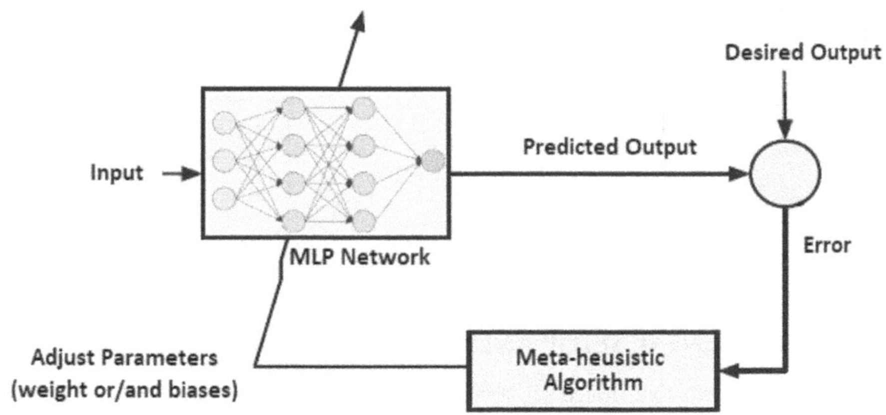

FIGURE 2.7 Schematic diagram of how to adjust an MLP by metaheuristics.

Generally, the gradient descent algorithm is utilized for obtaining the minimum value for the BP network. This algorithm includes a very important problem: trapping into the local minimum. This drawback is totally dependent on initial (weight) settings. One of the advantages of the metaheuristic algorithms is to escape from the local minimum based on its *exploration* characteristics.

For optimizing the weights of the network, ANN is first trained using a metaheuristic algorithm to find the optimal initial weights. Afterward, the neural network is trained by using a BP algorithm that includes an optimal BP network.

The mean squared error of the network (MSE) that should be minimized using metaheuristics is given later:

$$\text{MSE} = \frac{1}{2}\sum_{k=1}^{g}\sum_{j=1}^{m}\left(Y_j(k) - T_j(k)\right)^2 \tag{2.8}$$

where m describes the number of nodes in the output, g is the number of training samples, $Y_j(k)$ is the desired output, and $T_j(k)$ defines the real output.

The main purpose of image segmentation based on the optimized neural network in different input images is to categorize all of their pixels into some considered classes for simplifying the feature extraction, classification, etc. into the next level.

In the image segmentation based on supervised techniques, we have to first network based on several different samples with their solutions (subject with its label). After this learning, the neural network starts the optimal learning process based on adjusting its weights according to the illustrated methodology. Afterward, the neural network has the ability to classify an input image based on its learning. Figure 2.8 shows the steps of image segmentation based on a supervised technique.

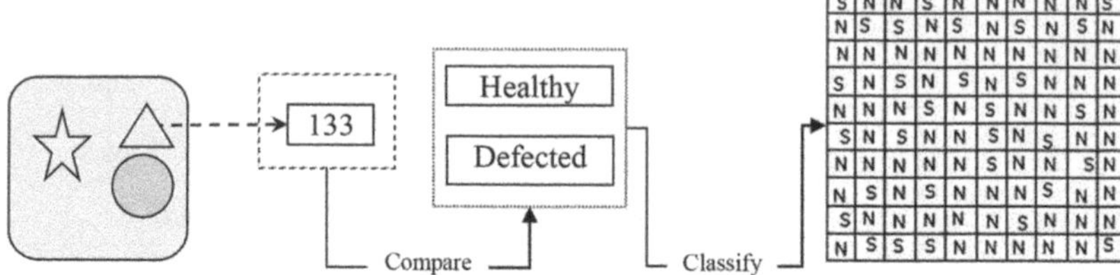

FIGURE 2.8 Steps in supervised image segmentation.

2.8 EXPERIMENTAL SETUP AND METHOD ANALYSIS

In this research, two-fold image segmentation (binarization) is considered for testifying. For analyzing this technique, based on pixel classification, two different databases are analyzed:

1. Australian Cancer Database (ACD), a well-known and widely used skin cancer database [79].
2. United State Department of Agriculture (USDA) [80] and the Canadian Food Inspection Agency (CFIA) [81] are the two more popular datasets used for the quality inspection of potatoes.

The purpose of the proposed method for the first database is to detect cancer-like pixels from its healthy pixels, and the purpose of the proposed method for the second and third database is to separate the defected parts of potatoes from its healthy parts.

The input layer of the network considered three neurons from each image as either cancer (defected) or noncancer (healthy). In this analysis, a sigmoid function is utilized as an activation function for the optimized MLP network. The output generated a pixel with intensity between 0 and 255. Because of using sigmoid function and for achieving a two-fold output, a single threshold value is applied to the image. In the following, some examples of the proposed technique based on ICA algorithm are shown in Figure 2.9.

As it is clear, using the proposed technique can generate a strong method for image segmentation. It is important to know that, in this part, there is no postprocessing (like morphology) for it.

For analyzing the accuracy of three optimization algorithms, CDR, FAR, and FRR are evaluated.

CDR defines the correct detection rate, and FAR defines the false acceptance rate which is the percentage of identification moments in which false acceptance happens, and FRR is the false rejection rate that describes the percentage of identification moments in which false rejection happens. In the following, the formula for achieving each of this index is given.

$$CDR = \frac{\text{No. of pixels correctly classified}}{\text{Total pixels in the test dataset}} \tag{2.9}$$

$$\text{FAR} = \frac{\text{No. of nonhealthy pixels classified as cancer pixels classified}}{\text{Total pixels in the test dataset}} \quad (2.10)$$

$$\text{FRR} = \frac{\text{No. of defected pixels classified as nondefected pixels classified}}{\text{Total pixels in the test dataset}} \quad (2.11)$$

The value of the parameters for the analyzed metaheuristic algorithms is given in Table 2.1.

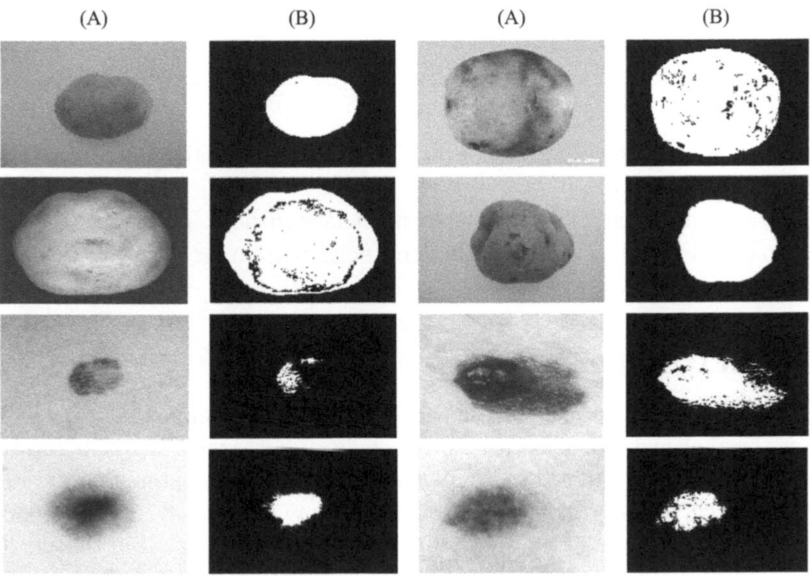

FIGURE 2.9 Some of the results for the algorithms: (A) original image and (B) segmented image based on optimized MLP network.

TABLE 2.1
The Value for the Parameters in the Analyzed Metaheuristic Algorithms

Algorithm	Parameter	Value
GA	Crossover probability	0.9
	Mutation probability 0.01	0.1
	Selection mechanism Roulette wheel	Roulette wheel
PSO	Acceleration constants	[2,2]
	Inertia weights	0.1
ICA	Number of countries	500
	Number of imperialists	10
	Beta	5
	pRevolution	0.1
	Zeta	0.1

TABLE 2.2
Classification Comparison of Performance in the Proposed Technique with Three Different Metaheuristics

	Metric	Ordinary MLP	MLP-GA	MLP-PSO	MLP-ICA
CFIA + USDA	CDR (%)	88	90	89	90
	FAR (%)	7.5	8	7	6.5
	FRR (%)	4.5	4	4	3.5
ACD	CDR (%)	90	90	93	94
	FAR (%)	6	7	4	3
	FRR (%)	4	3	3	3

Table 2.2 presents the efficiency of the presented segmentation algorithms based on CDR, FAR, and FRR.

It is clear from the earlier results that using metaheuristics can improve the system accuracy. It is also clear that ICA has better efficiency from the other metaheuristics for this purpose.

2.9 CONCLUSIONS

Image segmentation is a part of data science that works on dividing the image as data to its principal elements. In this chapter, the application of metaheuristic algorithms, including GA, PSO algorithm, and ICA, has been studied for optimizing the structure of neural networks for image segmentation purposes. For analyzing the proposed technique, two different databases including potato and melanoma skin cancers have been employed, where the main purpose is to separate the defected parts from the healthy parts based on intensity pixel classification by training the neural network. The weights of the neural network are optimized based on a metaheuristic algorithm to achieve a suitable accuracy. Simulation results showed that using metaheuristic algorithms for optimizing the neural network speeds up the convergence speed and reduces the root-mean-square error. To compare the performance of the three introduced metaheuristics, three metrics including CDR, FAR, and FRR have been employed, and the results show good efficiency for all of them, especially for world cup optimization algorithm. It is important to know that, since using these types of approaches increases the designing cost and the complexity of the system, it is better to utilize them for the applications that classic methods like this don't give proper results.

REFERENCES

1. Razmjooy, N., Mousavi, B. S., & Soleymani, F. (2012). A real-time mathematical computer method for potato inspection using machine vision. *Computers & Mathematics with Applications, 63*(1): 268–279.
2. Ghamisi, P., Couceiro, M. S., Benediktsson, J. A., & Ferreira, N. M. (2012). An efficient method for segmentation of images based on fractional calculus and natural selection. *Expert Systems with Applications, 39*(16): 12407–12417.

3. Razmjooy, N., Mousavi, B. S., Sargolzaei, P., & Soleymani, F. (2011). Image thresholding based on evolutionary algorithms. *International Journal of Physical Sciences, 6*(31): 7203–7211.
4. Brink, A. (1995). Minimum spatial entropy threshold selection. *IEE Proceedings-Vision, Image and Signal Processing, 142*(3): 128–132.
5. Razmjooy, N., Mousavi, B. S., & Soleymani, F. (2013). A hybrid neural network Imperialist Competitive Algorithm for skin color segmentation. *Mathematical and Computer Modelling, 57*(3): 848–856.
6. Banimelhem, O., & Yahya, Y. A. (2011). Multi-thresholding image segmentation using genetic algorithm. Jordan University of Science and Technology, Irbid, Jordan: 1–6.
7. Esteva, A., Kuprel, B., Novoa, R. A., Ko, J., Swetter, S. M., Blau, H. M., & Thrun, S. (2017). Dermatologist-level classification of skin cancer with deep neural networks. *Nature, 542*(7639): 115–118.
8. Gautam, D., & Ahmed, M. (2015). Melanoma detection and classification using SVM based decision support system. *Paper presented at the India Conference (INDICON)*, 2015, Annual IEEE, New Delhi.
9. Mousavi, B. S., & Soleymani, F. (2014). Semantic image classification by genetic algorithm using optimised fuzzy system based on Zernike moments. *Signal, Image and Video Processing, 8*(5): 831–842.
10. Razmjooy, N., Sheykhahmad, F. R., & Ghadimi, N. (2018). A hybrid neural network–world cup optimization algorithm for melanoma detection. *Open Medicine, 13*(1): 9–16.
11. Petroski Such, F., Madhavan, V., Conti, E., Lehman, J., Stanley, K. O., & Clune, J. (2017). Deep neuroevolution: Genetic algorithms are a competitive alternative for training deep neural networks for reinforcement learning. arXiv preprint arXiv:1712.06567.
12. Razmjooy, N., & Ramezani, M. (2016). Training wavelet neural networks using hybrid particle swarm optimization and gravitational search algorithm for system identification, *International Journal of Mechatronics, Electrical and Computer Technology, 6*(21), 2987–2997.
13. Zhang, Z. (2018). Artificial neural network. *Multivariate Time Series Analysis in Climate and Environmental Research*, Editor: Zhihua Zhang (pp. 1–35): Springer, Switzerland.
14. Mastorakis, N. (2018). Neural network methods for image segmentation. *Paper presented at the Applied Physics, System Science and Computers II: Proceedings of the 2nd International Conference on Applied Physics, System Science and Computers (APSAC2017)*, September 27–29, 2017, Dubrovnik, Croatia.
15. Roffman, D., Hart, G., Girardi, M., Ko, C. J., & Deng, J. (2018). Predicting non-melanoma skin cancer via a multi-parameterized artificial neural network. *Scientific reports, 8*(1): 1701.
16. DeGroff, D., & Neelakanta, P. S. (2018). *Neural Network Modeling: Statistical Mechanics and Cybernetic Perspectives*: CRC Press, Boca Raton.
17. Hemanth, D. J., & Estrela, V. V. (2017). *Deep Learning for Image Processing Applications* (Vol. 31): IOS Press.
18. Nascimento, J. D., da Silva Tavares, R., Estrela, V. V., de Assis, J. T., & de Almeida, J. C. H. *Image Processing Techniques Applied to Microtectonics*, Santo Amaro, Brazil.
19. Moallem, P., Razmjooy, N., & Ashourian, M. (2013). Computer vision-based potato defect detection using neural networks and support vector machine. *International Journal of Robotics and Automation, 28*(2): 137–145.
20. Estrela, V. V., & Coelho, A. M. (2013). State-of-the art motion estimation in the context of 3D TV. *Multimedia Networking and Coding,* Editors: Reuben A. Farrugia, Carl J. Debono, (pp. 148–173): IGI Global.

21. Zhang, C., Pan, X., Li, H., Gardiner, A., Sargent, I., Hare, J., & Atkinson, P. M. (2018). A hybrid MLP-CNN classifier for very fine resolution remotely sensed image classification. *ISPRS Journal of Photogrammetry and Remote Sensing, 140*: 133–144.
22. Özdoğan, M. (2019). Image classification methods in land cover and land use. *Remote Sensing Handbook-Three Volume Set*, Editor: Prasad Thenkabail (pp. 265–280): CRC Press, Boca Raton.
23. Irani, R., & Nasimi, R. (2011). Application of artificial bee colony-based neural network in bottom hole pressure prediction in underbalanced drilling. *Journal of Petroleum Science and Engineering, 78*(1): 6–12.
24. Moallem, P., & Razmjooy, N. (2012). A multi layer perceptron neural network trained by invasive weed optimization for potato color image segmentation. *Trends in Applied Sciences Research, 7*(6): 445.
25. Fadaeddini, A., Eshghi, M., & Majidi, B. (2018). A deep residual neural network for low altitude remote sensing image classification. *Paper presented at the Fuzzy and Intelligent Systems (CFIS), 2018 6th Iranian Joint Congress on*, Kerman, Iran.
26. Mikaeil, R., Haghshenas, S. S., Haghshenas, S. S., & Ataei, M. (2018). Performance prediction of circular saw machine using imperialist competitive algorithm and fuzzy clustering technique. *Neural Computing and Applications, 29*(6): 283–292.
27. Moghaddam, M. H. R., Sedighi, A., Fasihi, S., & Firozjaei, M. K. (2018). Effect of environmental policies in combating aeolian desertification over Sejzy Plain of Iran. *Aeolian Research, 35*: 19–28.
28. Azadi, S., & Karimi-Jashni, A. (2016). Verifying the performance of artificial neural network and multiple linear regression in predicting the mean seasonal municipal solid waste generation rate: A case study of Fars province, Iran. *Waste management, 48*: 14–23.
29. Montana, D. J., & Davis, L. (1989). Training feedforward neural networks using genetic algorithms, *Proceedings of the 11th international joint conference on Artificial intelligence*, pp: 762–767, vol 1, 1989, San Francisco.
30. Sharma, C., Sabharwal, S., & Sibal, R. (2014). A survey on software testing techniques using genetic algorithm. arXiv preprint arXiv:1411.1154.
31. Hosseini, H., Farsadi, M., Khalilpour, M., & Razmjooy, N. (2012). Hybrid Energy Production System with PV Array and Wind Turbine and Pitch Angle Optimal Control by Genetic Algorithm (GA), *Journal of World's Electrical Engineering and Technology, 1*(1): 1–4.
32. Trelea, I. C. (2003). The particle swarm optimization algorithm: convergence analysis and parameter selection. *Information processing letters, 85*(6): 317–325.
33. Moallem, P., & Razmjooy, N. (2012). Optimal threshold computing in automatic image thresholding using adaptive particle swarm optimization. *Journal of applied research and technology, 10*(5): 703–712.
34. de Jesus, M. A., Estrela, V. V., Saotome, O., & Stutz, D. (2018). Super-resolution via particle swarm optimization variants. *Biologically Rationalized Computing Techniques For Image Processing Applications*, Editors: Jude Hemanth Valentina Emilia Balas, (pp. 317–337): Springer.
35. Razmjooy, N., & Ramezani, M. (2014). An Improved Quantum Evolutionary Algorithm Based on Invasive Weed Optimization. *Indian J. Sci. Res, 4*(2): 413–422.
36. Razmjooy, N., Khalilpour, M., & Ramezani, M. (2016). A new meta-heuristic optimization algorithm inspired by FIFA World Cup competitions: Theory and its application in PID designing for AVR system. *Journal of Control, Automation and Electrical Systems, 27*(4): 419–440.
37. Razmjooy, N., Mousavi, B. S., Soleymani, F., & Khotbesara, M. H. (2013). A computer-aided diagnosis system for malignant melanomas. *Neural Computing and Applications, 23*(7–8): 2059–2071.

38. Vijayalakshmi, Y., Jose, T., Babu, S. S., Jose, S. R. G., & Manimegalai, P. (2017). Blue brain - A massive storage space. *Advances in Computational Sciences and Technology, 10*(7): 2125–2136.
39. Cao, M., He, Y., Dai, Z., Liao, X., Jeon, T., Ouyang, M., …, Dong, Q. (2016). Early development of functional network segregation revealed by connectomic analysis of the preterm human brain. *Cerebral Cortex, 27*(3): 1949–1963.
40. Dumoulin, S. O., Fracasso, A., van der Zwaag, W., Siero, J. C., & Petridou, N. (2018). Ultra-high field MRI: advancing systems neuroscience towards mesoscopic human brain function. *Neuroimage, 168*: 345–357.
41. Behrang, M., Assareh, E., Ghanbarzadeh, A., & Noghrehabadi, A. (2010). The potential of different artificial neural network (ANN) techniques in daily global solar radiation modeling based on meteorological data. *Solar Energy, 84*(8): 1468–1480.
42. Ostad-Ali-Askari, K., Shayannejad, M., & Ghorbanizadeh-Kharazi, H. (2017). Artificial neural network for modeling nitrate pollution of groundwater in marginal area of Zayandeh-rood River, Isfahan, Iran. *KSCE Journal of Civil Engineering, 21*(1): 134–140.
43. Khoshroo, A., Emrouznejad, A., Ghaffarizadeh, A., Kasraei, M., & Omid, M. (2018). Topology of a simple artificial neural network Sensitivity analysis of energy inputs in crop production using artificial neural networks. *Journal of Cleaner Production, 197*(1), 992–998.
44. Rafiei, M., Niknam, T., Aghaei, J., Shafie-khah, M., & Catalão, J. P. (2018). Probabilistic load forecasting using an improved wavelet neural network trained by generalized extreme learning machine. *IEEE Transactions on Smart Grid.*
45. Li, J., Cheng, J.-h., Shi, J.-y., & Huang, F. (2012). Brief introduction of back propagation (BP) neural network algorithm and its improvement. *Advances in Computer Science and Information Engineering*, Editors: D. Jin, S. Lin, (pp. 553–558): Springer, Berlin.
46. Ding, S., Li, H., Su, C., Yu, J., & Jin, F. (2013). Evolutionary artificial neural networks: a review. *Artificial Intelligence Review, 39*(3): 251–260.
47. Jarollahi, H., Gripon, V., Onizawa, N., & Gross, W. J. (2013). A low-power content-addressable memory based on clustered-sparse networks. *Paper presented at the Application-Specific Systems, Architectures and Processors (ASAP), 2013 IEEE 24th International Conference on*, Washington, DC.
48. Santoro, A., Bartunov, S., Botvinick, M., Wierstra, D., & Lillicrap, T. (2016). Meta-learning with memory-augmented neural networks. *Paper presented at the International Conference on Machine Learning.*
49. Ansari, H., Zarei, M., Sabbaghi, S., & Keshavarz, P. (2018). A new comprehensive model for relative viscosity of various nanofluids using feed-forward back-propagation MLP neural networks. *International Communications in Heat and Mass Transfer, 91*: 158–164.
50. Mohammadi, J., Ataei, M., Kakaei, R. K., Mikaeil, R., & Haghshenas, S. S. (2018). Prediction of the production rate of chain saw machine using the multilayer perceptron (MLP) neural network. *Civil Engineering Journal, 4*(7): 1575–1583.
51. Rezaee, M. J., Jozmaleki, M., & Valipour, M. (2018). Integrating dynamic fuzzy C-means, data envelopment analysis and artificial neural network to online prediction performance of companies in stock exchange. *Physica A: Statistical Mechanics and its Applications, 489*: 78–93.
52. Han, H.-G., Lu, W., Hou, Y., & Qiao, J.-F. (2018). An adaptive-PSO-based self-organizing RBF neural network. *IEEE Transactions on Neural Networks and Learning Systems, 29*(1): 104–117.
53. Ahangarpour, A., Farbod, M., Ghanbarzadeh, A., Moradi, A., & MirzakhaniNafchi, A. (2018). Optimization of continual production of CNTs by CVD method using Radial Basic Function (RBF) neural network and the Bees Algorithm. *Journal of Nanostructures, 8*(3): 225–231.

54. Tafarroj, M. M., & Kolahan, F. (2019). Using an optimized RBF neural network to predict the out-of-plane welding distortions based on the 3-2-1 locating scheme. *Scientia Iranica, 26*(2), 869–878.
55. Xie, S., Xie, Y., Huang, T., Gui, W., & Yang, C. (2019). Generalized predictive control for industrial processes based on neuron adaptive splitting and merging RBF neural network. *IEEE Transactions on Industrial Electronics, 66*(2): 1192–1202.
56. Nur, A. S., Radzi, N. H. M., & Ibrahim, A. O. (2014). Artificial neural network weight optimization: A review. *Indonesian Journal of Electrical Engineering and Computer Science, 12*(9): 6897–6902.
57. Aljarah, I., Faris, H., & Mirjalili, S. (2018). Optimizing connection weights in neural networks using the whale optimization algorithm. *Soft Computing, 22*(1): 1–15.
58. Fahimnia, B., Davarzani, H., & Eshragh, A. (2018). Planning of complex supply chains: A performance comparison of three meta-heuristic algorithms. *Computers & Operations Research, 89*: 241–252.
59. Akbari, M., Gheysari, M., Mostafazadeh-Fard, B., & Shayannejad, M. (2018). Surface irrigation simulation-optimization model based on meta-heuristic algorithms. *Agricultural Water Management, 201*: 46–57.
60. Donoso, Y., & Fabregat, R. (2016). *Multi-objective Optimization in Computer Networks using Metaheuristics*: Auerbach Publications, Boca Raton.
61. Sadeghi, M., Nemati, A., & Yari, M. (2016). Thermodynamic analysis and multi-objective optimization of various ORC (organic Rankine cycle) configurations using zeotropic mixtures. *Energy, 109*: 791–802.
62. Wang, G.-G., Gandomi, A. H., Alavi, A. H., & Dong, Y.-Q. (2016). A hybrid meta-heuristic method based on firefly algorithm and krill herd. *Handbook of Research on Advanced Computational Techniques for Simulation-Based Engineering,* Editor: Pijush Samui, (pp. 505–524): IGI Global.
63. Stützle, T., & López-Ibáñez, M. (2019). Automated design of metaheuristic algorithms. *Handbook of Metaheuristics*, Editors: Michel Gendreau, Jean-Yves Potvin (pp. 541–579): Springer, Berlin.
64. Memon, Q. (2019). On assisted living of paralyzed persons through real-time eye features tracking and classification using support vector machines. *Medical Technologies Journal*, 3(1): 316–333.
65. Khalilpuor, M., Razmjooy, N., Hosseini, H., & Moallem, P. (2011). Optimal control of DC motor using invasive weed optimization (IWO) algorithm. *Paper presented at the Majlesi Conference on Electrical Engineering*, Majlesi town, Isfahan, Iran.
66. Holland, J. H. (1992). Genetic algorithms. *Scientific American, 267*(1): 66–73.
67. Khan, G. M. (2018). Evolutionary computation. *Evolution of Artificial Neural Development,* Editor: Khan, G. M. (pp. 29–37): Springer, Berlin.
68. Askarzadeh, A. (2018). A memory-based genetic algorithm for optimization of power generation in a microgrid. *IEEE Transactions on Sustainable Energy, 9*(3): 1081–1089.
69. Mirjalili, S. (2019). Genetic algorithm. *Evolutionary Algorithms and Neural Networks*, Editor: Mirjalili, S., (pp. 43–55): Springer, Berlin.
70. Hosseinabadi, A. A. R., Vahidi, J., Saemi, B., Sangaiah, A. K., & Elhoseny, M. (2018). Extended genetic algorithm for solving open-shop scheduling problem. *Soft Computing, 23*: 1–18.
71. Pal, S. K., & Wang, P. P. (2017). *Genetic Algorithms for Pattern Recognition*: CRC press, Boca Raton.
72. Bansal, J. C. (2019). Particle swarm optimization. *Evolutionary and Swarm Intelligence Algorithms*, Editors: Bansal, Jagdish Chand, Singh, Pramod Kumar, Pal, Nikhil R. (pp. 11–23): Springer, Berlin.

73. AminShokravi, A., Eskandar, H., Derakhsh, A. M., Rad, H. N., & Ghanadi, A. (2018). The potential application of particle swarm optimization algorithm for forecasting the air-overpressure induced by mine blasting. *Engineering with Computers, 34*(2): 277–285.
74. Mahi, M., Baykan, O. K., & Kodaz, H. (2018). A new approach based on particle swarm optimization algorithm for solving data allocation problem. *Applied Soft Computing, 62*: 571–578.
75. Atashpaz-Gargari, E., & Lucas, C. (2007). Imperialist competitive algorithm: An algorithm for optimization inspired by imperialistic competition, *Proceedings of Evolutionary Computation, 2007. CEC 2007. IEEE Congress on*, Singapore.
76. Kaveh, A., & Talatahari, S. (2010). Optimum design of skeletal structures using imperialist competitive algorithm. *Computers & Structures, 88*(21–22): 1220–1229.
77. Aliniya, Z., & Mirroshandel, S. A. (2019). A novel combinatorial merge-split approach for automatic clustering using imperialist competitive algorithm. *Expert Systems with Applications, 117*: 243–266.
78. Mirhosseini, M., & Nezamabadi-pour, H. (2018). BICA: A binary imperialist competitive algorithm and its application in CBIR systems. *International Journal of Machine Learning and Cybernetics, 9*(12): 2043–2057.
79. Database, A. C., from www.aihw.gov.au/australian-cancer-database/.
80. United State Department of Agriculture, from www.usda.gov/.
81. Agency, C. F. I., from http://inspection.gc.ca/eng/1297964599443/1297965645317.

3 A Study and Analysis of a Feature Subset Selection Technique Using Penguin Search Optimization Algorithm

Agnip Dasgupta, Ardhendu Banerjee, Aniket Ghosh Dastidar, Antara Barman, and Sanjay Chakraborty
TechnoIndia

CONTENTS

3.1 INTRODUCTION

Machine learning is a branch of artificial intelligence (AI), which allows applications to become more authentic in anticipating to which class a particular data resides. Various applications of machine learning are spread over the areas like healthcare, finance, retail, travel, social media, advertisements, and most importantly, data mining [1]. There are two types of learning: supervised learning and unsupervised learning. In machine learning, feature alone is a measurable property or trait of a phenomenon being observed. Feature subset selection is one of the important tasks in data mining. To perform effective data mining and pattern recognition tasks, we need the help of an efficient feature selection technique. However, due to a large number of features present in high-dimensional datasets, there is a chance of unnecessary overfitting, which increases the overall computational complexity and reduces the prediction accuracy of the procedure. This feature selection problem belongs to a set of NP (non-deterministic polynomial)-hard problems, where the complexity increases exponentially if the number of features along with the size of datasets is increasing. Feature subset selection not only helps us to get rid of the curse of dimensionality but also helps us to shorten the training time and simplifying the model, making it easier for the analysts to interpret it. There are various approaches that deal with both supervised and unsupervised ways of feature subset selection [2–5]. There are two classes of feature selection methods, such as (i) filter-based feature selection and (ii) wrapper-based feature selection.

Filter methods are generally a part of the preprocessing step. Here, each feature is selected on the basis of their scores in various statistical tests, some of which are Pearson's correlation (PCA), linear discriminant analysis (LDA), analysis of variance (ANOVA), and chi-square. The other methods are wrapper methods, where a set of a subset is randomly chosen and then the efficiency is checked, and after that, other features apart from the subset are chosen, and the results are checked again, or some irrelevant or less important features are removed from the subset. This continues until we find an ideal subset. In wrapper methods, the problem is reduced to a search problem. Using wrapper methods are quite expensive [6]. Some of the examples of wrapper methods are forward selection, backward elimination, and recursive feature elimination. Apart from these two methods, we can also use an embedded method that includes the qualities of both wrapper methods and filter methods. It is used by those algorithms that have built-in feature subset selection methods [7].

After feature subset selection, we use classifiers to classify which class the particular data belongs to. However, we have tested our algorithm using K-nearest neighbors (KNN), support vector machine (SVM), and Random Forest classifiers [8]. KNN is a very simple nonparametric decision procedure that is used to assign unclassified observations a class with the use of a training dataset [9].

In this chapter, we have worked upon feature subset selection using "penguin search algorithm," inspired by the hunting strategy of the penguins [10]. Also, we have tested our algorithm on some popular UCI (unique client identifier) datasets, including Iris, Pima, Wisconsin, etc. Then, we have compared our work with the existing feature subset algorithm. We have also used three different types of classifiers and checked how good our algorithm works in terms of parameters such as accuracy, precision, recall, and F1 score. There are several existing algorithms or approaches for "feature subset selection," which are inspired by nature like ant, bee colony, whale, etc., and so we are using "penguin search optimization algorithm (PeSOA)" in our work, which is based on the way penguins hunt for food [10]. Penguins take random jumps in different places and random depths to find out fish, and after they find fish, they come back and communicate about the food availability with the other penguins, and this continues till the penguins find the best place or the place where maximum fish are present or the global maxima. The goal of finding the global maxima continues until the oxygen level of penguins does not get depleted. Each penguin has to return back to the surface after each trip. The duration of the trip is measured by the amount of oxygen reserves of the penguins, the speed at which they use it up, or their metabolism rate. This behavior of the penguins has given us motivation for the development of a new optimization method based on this strategy of penguins. Penguins are sea birds, and they are unable to fly [10]. Metaheuristics is mainly used for the development of new artificial systems, and it is effective in solving NP-hard problems. It can be classified in various ways. The first work in the field of optimization commenced in 1952 based on the use of a stochastic manner. Rechenberg diagrammed the first algorithm using evolution strategies in the optimization in 1965 [10]. Most of the methods utilize the thought of population, in which a set of solutions are calculated in parallel at each iteration, such as genetic algorithms (GAs), particle swarm optimization (PSO) algorithm, and ant colony optimization algorithm (ACO) [11–12]. Other metaheuristic algorithms use the search results based on their past experiences to guide the optimization in following the iterations by putting a learning stage of intermediate results that will conduct to an optimal solution. The work reported in this chapter is an example of it. It mainly depends on the collaborative hunting strategy of penguins. The catholic optimization process starts based on the individual hunting process of each penguin, who must share information to his group related to the number of fish found of their individual hunting area. The main objective of the group conversation is to achieve a global solution (the area having abundant amount of food). The universal solution is chosen by a selection of the best group of penguins who ate the maximum number of fish. Comparative studies based on other metaheuristics have proved that PeSOA accomplishes better answers related to new optimization strategy of collaborative and dynamic research of space solutions.

This rest of the book chapter is organized as follows. A brief literature review has been done in Section 3.2. In Section 3.3, we have described our proposed work with a suitable flow chart diagram. Then, we have described a detailed performance analysis of our proposed approach in Section 3.4. We have also compared our proposed work with some previous studies related to different parameters of classification in Section 3.4, and finally, Section 3.5 describes the conclusion of this chapter.

3.2 LITERATURE REVIEW

In the last decade, various researchers adopted various optimization methods for solving the feature selection problem. In one of the earliest works, a novel marker gene feature selection approach was introduced. In this approach, a few high-graded informative genes were elected by the signal–noise ratio estimation process. Then, a novel discrete PSO algorithm was used to choose a few marker genes, and SVM was used as an evaluator for getting excellent prediction performance on colon tumor dataset. The authors have introduced an algorithm called swarm intelligence feature selection algorithm, which is mainly based on the initialization and update of the swarm particles. In their learning process, they had tested the algorithm in 11 microarray datasets for brain, leukemia, lung, prostate, etc. And they have noticed that the proposed algorithm was successfully increasing the classification accuracy and reducing the number of chosen features compared with other swarm intelligence process. The authors have compared the utilization of PSO and GA (both illuminated with SVM) for the classification of high-dimensional microarray data. Those algorithms are mainly used for finding small samples of informative genes amongst thousands of them. An SVM classifier with tenfold cross-validation was applied to validate and assess the provided solutions [13]. There is one research work, where whale optimization algorithm (WOA) was introduced, through which a new wrapper feature selection approach is proposed. It is a recently proposed algorithm that has not been systematically used to feature selection problems. Two binary variants of the WOA algorithm were mainly introduced to find the subsets of the optimal feature for classification purposes. In the first case, the aim was mainly to study the impact of using the Tournament and Roulette Wheel selection mechanisms instead of using a random operator in the searching procedure. In the second case, crossover and mutation operators are used to prolong the exploitation of the WOA algorithm. The proposed methods are tested based on standard benchmark datasets. That paper also considers a comprehensive study related to the parameter setting for the proposed technique. The results display the ability of the proposed approaches in searching for the optimal feature subsets [14]. In a paper, a noble version of the binary gravitational search algorithm (BGSA) was proposed and used as a tool to choose the best subset of features with the objective of improving classification accuracy. The BGSA has the ability to overcome the stagnation situation by enhancing the transfer function. And then the search algorithm was used to explore a larger group of possibilities and avoid stagnation. To assess the proposed improved BGSA (IBGSA), classification of some familiar datasets and improvement in the accuracy of CBIR systems were dealt. And the results were compared with those of the original BGSA, GA, binary PSO (BPSO), and electromagnetic-like mechanism. Relative results ensured the effectiveness of the proposed IBGSA in feature selection [15]. Various optimization techniques (hill-climbing, PSO, etc.) are used to do an efficient and effective unsupervised feature subset selection [7].

It can be seen from a survey of existing work that a few researchers in the last decade have tried to solve the problem of feature selection using optimization techniques. There have been a couple of attempts by researchers to unify diverse algorithms for supervised and unsupervised feature selections. However, we think PeSOA can provide a better selection strategy in case of wrapper-based feature selection process. In our recent work, we discuss this area of feature selection.

3.3 PROPOSED WORK

In our algorithm, each location where the penguin jumps into is considered a feature or a dimension, and as the penguins dive to find the food, the penguins dive deep into the feature to check whether that particular feature is important or not or the amount of food or fish is ample or not. The results are shared among the other penguins once they come to the surface, and this search for finding the optimum result or the minimum number of features by which classification gives good results continues till the oxygen level depletes or the best features or best positions are found out by the penguins, i.e. [16], in terms of penguins places with the highest amount of fish or food. On the basis of these things, we have designed our algorithm.

3.3.1 Pseudocode of the Proposed FS-PeSOA Algorithm

Feature Selection Using PeSOA (FS-PeSOA)

```
Start;
Obtain Dataset;
Split the dataset in the ratio of 80:20 as training set and
test set;
Generate Random population of P penguins;
Initialize Oxygen reserve for Penguins
Initialize the first location of Penguins
While (iterations<Oxygen) do
                For each penguin j do
                       Look for Fish available (Calculate
fitness of available data for
the current Penguin with the help of Fitness Function 4.3.2).
                       Determine the quantity of fish available.
                       Update the Position of the Penguin
(based on the Position Update Logic 4.3.3)
                 End for;
                 Update the best solution;
                 Get the food quantity (fitness) data from the
penguins to update the group.
                 Scale the food quantity (fitness) data for
position update in next iteration.
                 Update the Oxygen reserve for the penguins
(using Oxygen Update Logic 4.3.4).
End While;
Find out the features that qualify the fitness cutoff
Save the obtained feature subset;
Use this subset of features to undergo classification using
SVM, KNN and Random Forest;
Performance analysis using precision, recall, f-score, and
accuracy;
End;
```

The first step of the Machine Learning is choosing the dataset we intend to work upon. These datasets have been explained in Table 3.1. After the dataset has been chosen, the data needs to be normalized or scaled to a particular range. This is done because it might be possible that one attribute ranges between 1 and 100 and the other attribute ranges from 10,000 to 50,000. This type of variance in the dataset will be affecting our results. Hence, we need to scale it down to a particular range, setting the upper limit and the lower limit. After the scaling has been done, the dataset is divided into two parts—training dataset and testing dataset. This is generally in the ratios of 70:30 or 80:20. The training dataset, as the name suggests, is used to train our machine, and the testing dataset is used to test our algorithm and how efficiently it works. This step is also called as data splitting. The overall workflow of our proposed approach is shown in Figure 3.1.

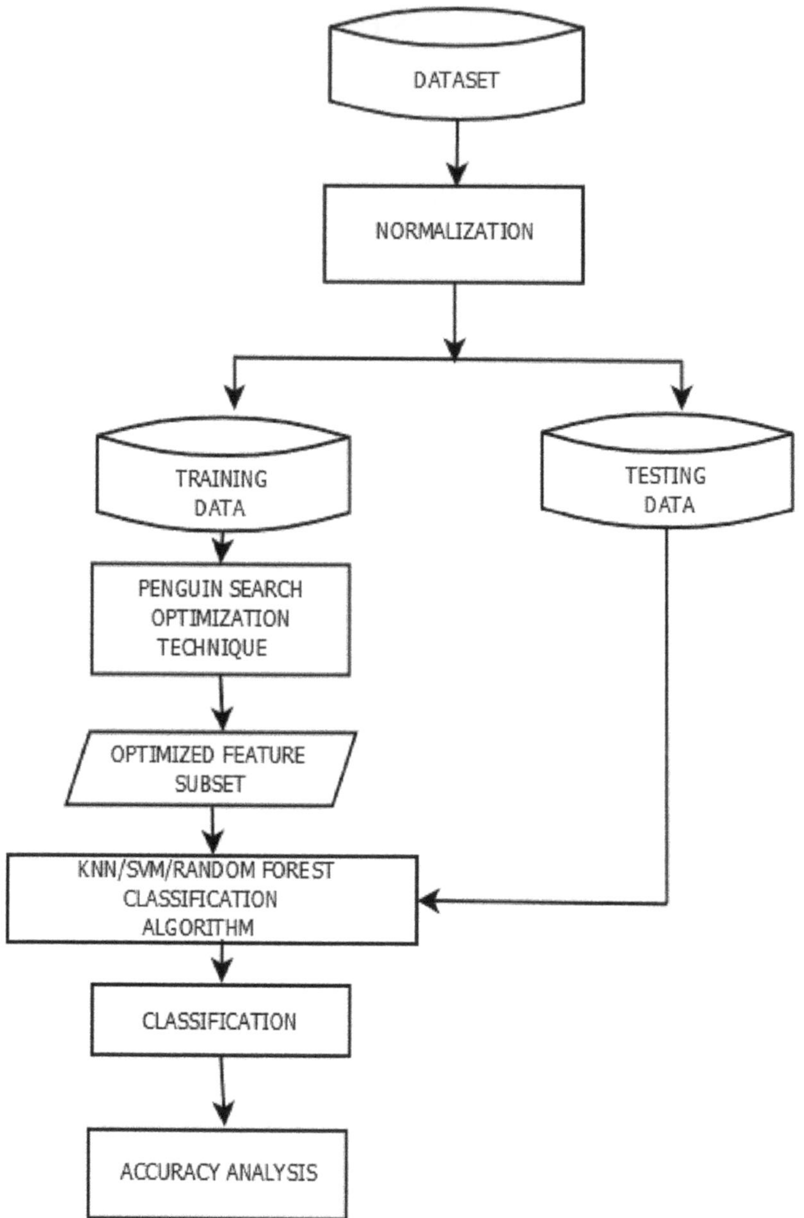

FIGURE 3.1 The overall flow chart of the proposed FS-PeSOA algorithm.

Now, after the data gets split into two parts, we will execute our proposed algorithm for feature subset selection, and the best features are selected out. This step is known as feature subset selection, which has been explained in the Introduction part. Briefly, it means selecting the minimum number of features by which our machine can identify which class a particular data belongs to. After feature subset selection has been done, several classifiers are used like SVM, KNN, Random Forest, etc. for further classification process. However, after the completion of the training phase, we would like to test our algorithm with the testing dataset, which we kept aside and will be used for performing the same steps again for classification.

After the classification is done, we are checking how accurate our results are and how well it performs when compared with other benchmark feature selection algorithms.

In the "Result Analysis" section, the comparison factors are accuracy, precision, recall, and F1 score, and the benchmark feature selection algorithms which we have compared are LDA and PCA.

3.3.2 Discussion

3.3.2.1 Hunting Strategy of Penguins

The hunting strategy of penguins is theorized in [10]. This hypothesis theorized that their hunting strategy may be explained in an economical manner. Penguins are a biological being that have a definite amount of capacity; by capacity it means that they have a definite amount of oxygen and strength to fulfill their search for food. During the process of finding the food for their survival, they tend to dive in water for finding fish. They go economical where they need to find food, and the point is they go for a search in food where the amount of food found is on par with the energy spent to find the food. Moreover, they have a limited amount of oxygen left with them that limits the amount of time they can continue with the total hunting procedure. They do dive in the water and look for the available food and consume them, and when they come back to the surface, they again communicate with the whole group about the amount of food found and at which location, and this communication among the penguins plays a vital role in this whole searching of food by the penguins. Whenever they communicate among them about the data of the location and quantity of food found, the penguins that have found a lesser amount of food tend to travel to the location that has been reported to have more amount of food in comparison to others. However, we have tried to visualize this hunting strategy of penguins to optimize our searching technique for an optimal subset of all features of the whole dataset in this chapter. We have tried to visualize the amount of food found by the penguins as the goodness or the fitness of features the penguins travel to. All the penguins work as a whole unit to deliver a single objective, to find the maximum amount of food that they can find with a limited number of resources. There is a cycle where they go in search of food and come back with food and communicate with the other penguins with the data of the food and location, and then the penguins travel to other locations, and then they again go in search of food. This whole cycle goes on and on, until and unless they have found the required amount of food.

3.3.2.2 Fitness Function Evaluation

In contrary to the actual world phenomenon of finding food by the penguins, they look for real-time quantity of food, and based on this quantity of food, they compare the quality of a particular location. For our datasets to find the quality or the fitness of features with their set of observations, we have used the Eigenvector Function as described in [17]. The eigenvalue works on the principle of calculating the variance of the particular feature with the total number of records for the particular feature. The basic idea is that the more the values of a particular feature scattered in a more varied range, the feature is fit for being used for the purpose of classification. Whenever a penguin goes to a particular feature, the eigenvalue function is used to get the variance of the feature, and this is the fitness. Based on this fitness value, the features are selected based on the cut-off criteria. Now, suppose we have a dataset of $\{x(i), i = 1, 2, 3, \ldots, m\}$ of m different features. So the data lies in an m-dimensional subspace and the data basically lies between the diagonal of this m dimensional subspace. Now, we find the mean of the data using Equation (3.1) as follows [17].

$$\mu = \frac{1}{m} \sum_{i=1}^{m} x^{(i)} \tag{3.1}$$

Replace the $x^{(i)}$ data with $x^{(i)} - \mu$, and from this, we get the mean of the data normalized. However, the data with zero and no mean is omitted and normalized. Now, we find the variance of the data using Equation (3.2).

$$\sigma_j{}^2 = \frac{1}{m} \sum_{i} \left(x_j^{(i)}\right)^2 \tag{3.2}$$

Replace $x^{(i)}$ with $X^{(i)}/\sigma$, and from this, we get the data scaled in a particular range, and this normalizes the data with respect to the whole dataset and all features. This is the final step of normalization. This way, we get the variance of the data [17].

This resultant variance is the score of the feature with a set of all data records. This represents how the data is scattered in the maximum to minimum scale for the feature, which will determine the features that should be suitable for classification.

3.3.2.3 Position Update Logic

The penguins will usually go towards the location that has a better food quantity than the others. Here, the penguins would initially go to a feature respective to that fitness value acquired $f(x^{(i)})$, which is in a scale of 0 to 1. This fitness value is multiplied with m, where m is the total number of features, and we get the new position for our penguin for the next iteration.

$$\text{Position}, (n+1) = f(x(i)) * m \tag{3.3}$$

Based on this position update logic, the penguins will keep updating their position until they find the optimal amount of food.

3.3.2.4 Oxygen Update Logic

For the penguins, the quantity of oxygen is a limiting factor that restricts them to continue the whole searching procedure for an infinite amount of time. The oxygen is a physical quantity that is basically limited to all beings. Whenever their oxygen reserve is depleted, they return back and end the search for food. In our approach of PeSOA, we have used the idea of oxygen to limit the number of iterations the penguins will go, until they stop to find out the food that they have. We have a predefined value of oxygen that is same for all the penguins in the beginning and end of each iteration, and the amount of oxygen gets reduced to a fixed value so that after a definite amount of time is complete, it brings an end to the series of iterations. This can also be considered as the number of generations our iteration will go on until an optimal solution is reached.

3.4 RESULT ANALYSIS

The implementation of the algorithm has been done on Python programming language using Anaconda as the software and on Jupiter notebook, and the experiments are tested on a computer with system specifications of 4 GB RAM, Intel i3 core Processor, and 500 GB hard disk memory. Seven real-world "UCI Machine Learning Repository"-approved datasets are used to assess the efficiency of our proposed algorithm [18]. Some of them have about 4–5 features and some have about 30–40 features, which make it appropriate for us to perform the feature subset selection. These datasets are also diverse in the context of number of classes and samples. These datasets have been represented in Table 3.1.

These datasets include Ion, Pima, Iris, Vehicle, Wisconsin, Glass, and Wine. Using our proposed algorithm, i.e., FS-PeSOA, we have performed the feature subset selection, and then used KNN, Random Forest, and SVM classifiers to perform the classification. We have also tested the datasets using other algorithms like PCA and LDA, and then we compared the results hoping to get better performance.

These are the UCI-approved datasets that have been used to check the efficiency of our algorithm. According to the UCI repository of machine learning [18], Iris is the flower dataset that is perhaps the best-known database to be found in the pattern recognition field. It has three classes and four attributes, and the attributes include sepal length, petal length, sepal width, and petal width. The classes are Iris Setosa, Iris Versicolour, and Iris Virginica.

Ion dataset or ionosphere dataset is a collection of radar data collected in Goose Bay, Labrador. The system has a phased array of 16 high-frequency antennas with a transmitting power on the order of 6.4 kW, and the targets are free electrons that are present in the ionosphere. This dataset consists of 33 attributes and two classes, namely good radar and bad radar.

TABLE 3.1
List of Datasets from UCI [17] which Are Tested Using PESOA

Datasets	No. of Observations	No. of Classes
Iris	150	3
Glass	214	6
Ion	351	2
Pima	768	2
Vehicle	846	4
Wine	178	3
Wisconsin	569	2

Good radar is that which shows some structure in the ionosphere, and the bad radar is that whose signals cannot pass through the ionosphere. Pima dataset is collected by a survey done on 768 people by the National Institute of Diabetes and Digestive and Kidney Diseases. The motive of the dataset or the machine learning part of the dataset is to find out whether a particular patient has diabetes or not on the basis of various diagnostic measurements. In the dataset, the survey is done on Indian females who are above 21 years and are of Pima Indian Heritage. This dataset has two classes: either the patient is suffering or not. According to the "UCI repository of Machine Learning," the wine dataset contains the results of a chemical analysis of wines. It has three classes and multiple attributes like fixed acidity, volatile acidity, citric acid, residual sugar, chlorides, free sculpture dioxide, total sculpture dioxide, density, pH, sulfates, alcohol, etc. The number of features before subset selection was 13 and that after feature subset selection was reduced to 4.

According to the UCI Repository, the vehicle dataset classifies a given silhouette as one of four types of vehicles with 18 features, some of which are Compactness, Circularity, Radius Ratio, Elongatedness, etc. This data was originally gathered in 1986–1987 by J.P. Siebert. According to the UCI Repository of Machine Learning, the study of classification of types of glass is motivated by a criminological investigation. At the scene of the crime, the glass left can be used as evidence. It has six classes and multiple attributes such as Id number: 1–214, refractive index, Sodium (unit measurement: weight percent in corresponding oxide, as are attributes 4–10), Magnesium, Aluminum, Silicon, Potassium, Calcium, Barium, Iron, Type of glass: (class attribute): building_windows_float_processed,building_windows_non_float_processed,vehicle_windows_float_processed, vehicle_windows_non_float_processed (none in this database), containers, tableware, and headlamps. The number of features before subset selection was 10 and that after feature subset selection was reduced to 4. According to the UCI Repository of Machine Learning, the features are computed from a digitized image of a fine needle aspirate (FNA) of a breast mass. They describe the characteristics of the cell nuclei present in the image. The Wisconsin dataset has two classes and multiple attributes like ID number, Diagnosis (M = malignant, B = benign), and ten real-valued features that are computed for each cell nucleus like radius (mean of distances from center to points on the perimeter), texture (standard deviation of grayscale values), perimeter, area, smoothness (local variation in radius lengths), compactness (perimeter^2/area – 1.0), concavity (severity of concave portions of the contour), concave points (number of concave portions of the contour), symmetry, and fractal dimension ("coastline approximation" – 1). The number of features before subset selection was 30 and that after feature subset selection was reduced to 6. Table 3.2 introduces us to the results, i.e., Accuracy, Precision, Recall, and F1 score of our algorithm when tested upon the datasets in Table 3.1. The table also explains the total number of features that were present before and after the feature subset selection. The table gives a detailed information or results when different classifiers have been used.

Figure 3.2 shows the performance of accuracy, precision, recall, and the F1 score of a Glass dataset under KNN, Random Forest, and SVM, when previously operated by PCA, LDA, and our proposed FS-PeSOA algorithm. The following graph is of Ion dataset and its results.

TABLE 3.2
Proposed Algorithm Result Analysis

Dataset	Total Feature Before	Total Feature After	Classification	Accuracy	Precision	Recall	F1 Score
Iris	4	1	KNN	90	0.94	0.90	0.90
	4	1	Random Forest	90	0.94	0.90	0.90
	4	1	SVM	90	0.94	0.90	0.90
Glass	10	4	KNN	81.39	0.82	0.81	0.80
	10	4	Random Forest	79.06	0.81	0.79	0.79
	10	4	SVM	81.39	0.80	0.81	0.79
Ion	33	8	KNN	95.91	0.88	0.86	0.85
	33	8	Random Forest	91.54	0.92	0.92	0.91
	33	8	SVM	91.54	0.93	0.92	0.91
Pima	8	1	KNN	59.74	0.58	0.60	0.59
	8	1	Random Forest	64.28	0.64	0.64	0.64
	8	1	SVM	65.58	0.62	0.66	0.60
Vehicle	18	5	KNN	57.05	0.58	0.57	0.57
	18	5	Random Forest	52.35	0.52	0.52	0.52
	18	5	SVM	58.82	0.59	0.59	0.58
Wine	13	4	KNN	69.44	0.69	0.69	0.69
	13	4	Random Forest	55.55	0.64	0.56	0.57
	13	4	SVM	38.88	0.22	0.39	0.28
Wisconsin	30	6	KNN	79.04	0.83	0.82	0.84
	30	6	Random Forest	84.61	0.79	0.81	0.82
	30	6	SVM	95.61	0.96	0.96	0.96

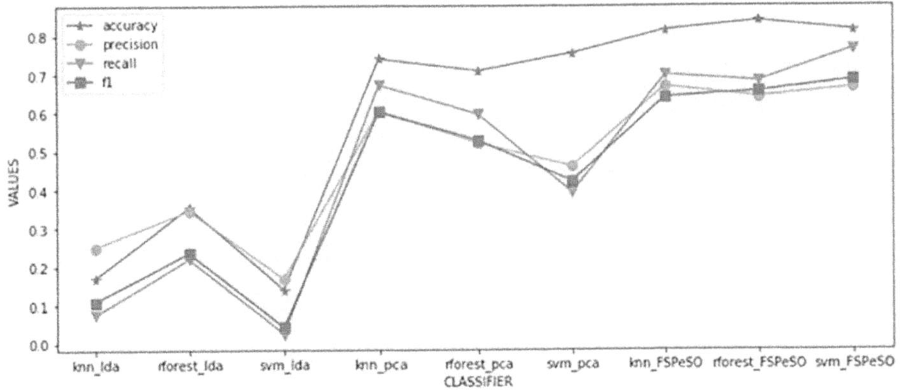

FIGURE 3.2 Performance analysis of the proposed FS-PeSOA with other algorithms on a Glass dataset.

Figure 3.3 shows the performance of accuracy, precision, recall, and the F1 score of an Ion dataset under KNN, Random Forest, and SVM when previously operated by PCA, LDA, and our proposed FS-PeSOA algorithm. The following graph is of Pima dataset and its results.

Figure 3.4 shows the performance of accuracy, precision, recall, and the F1 score of a Pima dataset under KNN, Random Forest, and SVM when previously operated by PCA, LDA, and our proposed FS-PeSOA algorithm. The following graph is of Wine dataset and its results.

Figure 3.5 shows the performance of accuracy, precision, recall, and the F1 score of a Wine dataset under KNN, Random Forest, and SVM when previously operated by PCA, LDA, and our proposed FS-PeSOA algorithm. The following graph is of Wisconsin dataset and its results.

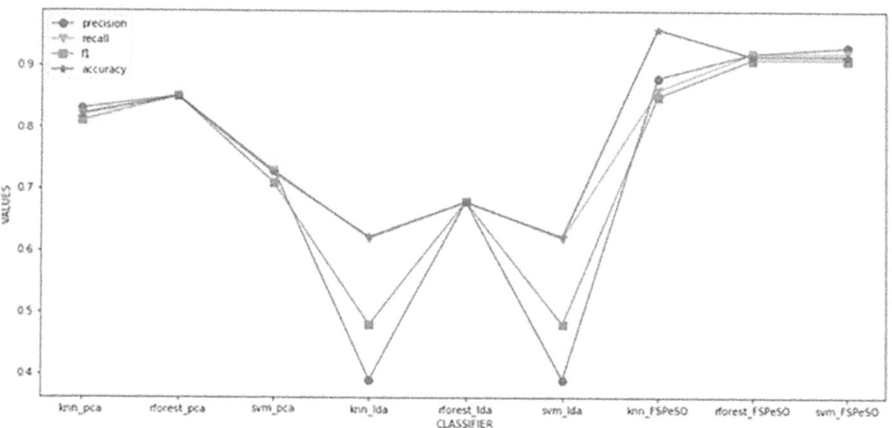

FIGURE 3.3 Performance analysis of the proposed FS-PeSOA with other algorithms on an Ion dataset.

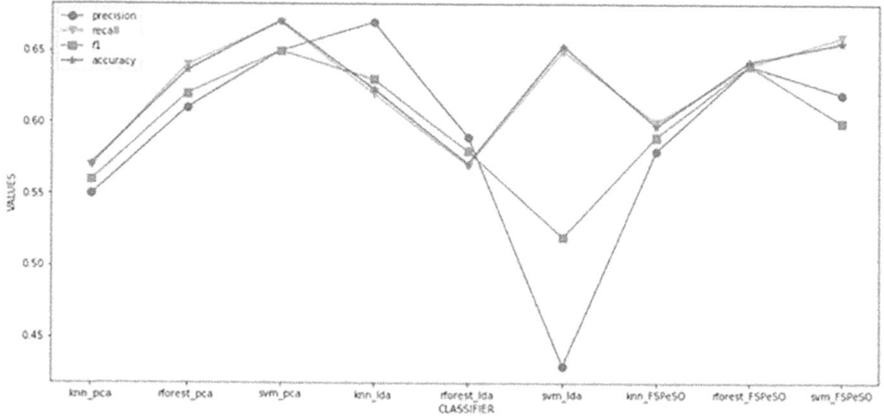

FIGURE 3.4 Performance analysis of the proposed FS-PeSOA with other algorithms on a Pima dataset.

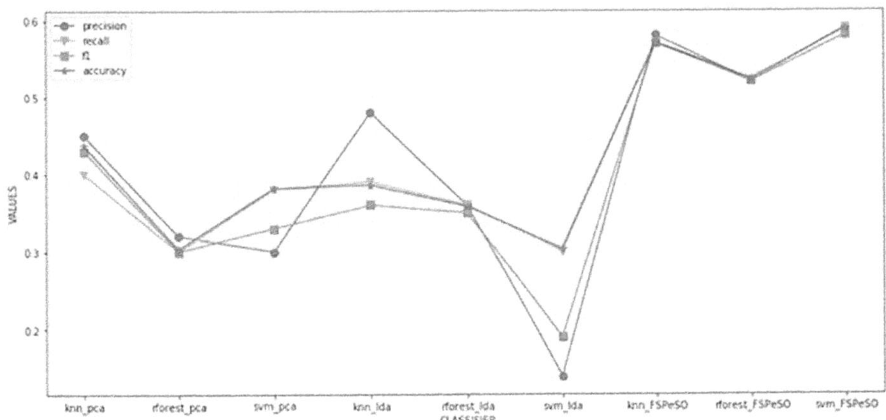

FIGURE 3.5 Performance analysis of the proposed FS-PeSOA with other algorithms on a Wine dataset.

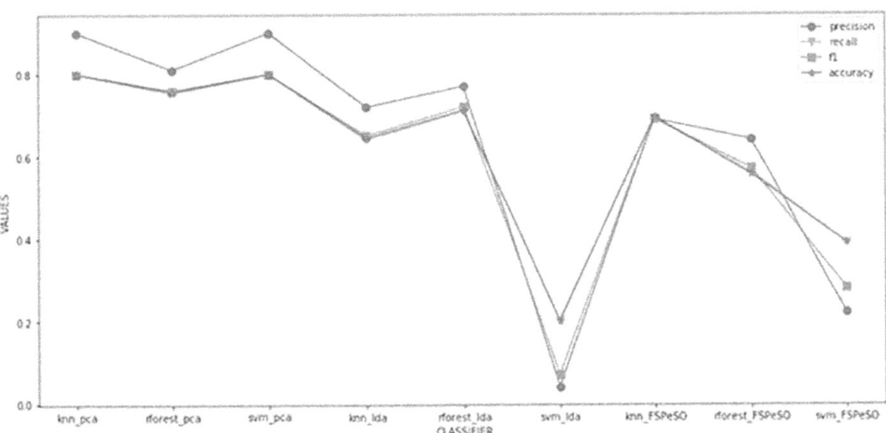

FIGURE 3.6 Performance analysis of the proposed FS-PeSOA with other algorithms on a Wisconsin dataset.

Figure 3.6 shows the performance of accuracy, precision, recall, and the F1 score of a Wisconsin dataset under KNN, Random Forest, and SVM when previously operated by PCA, LDA, and our proposed FS-PeSOA algorithm. The following graph is of Vehicle dataset and its results.

The Figure 3.7 shows the performance of accuracy, precision, recall, and the F1 score of a Vehicle dataset under KNN, Random Forest, and SVM when previously operated by PCA, LDA, and our proposed FS-PeSOA algorithm. These are the comparison graphs of different algorithms and our FS-PeSOA algorithm. We have gathered a comparison result in Table 3.3, which gives a data representation for these graphs.

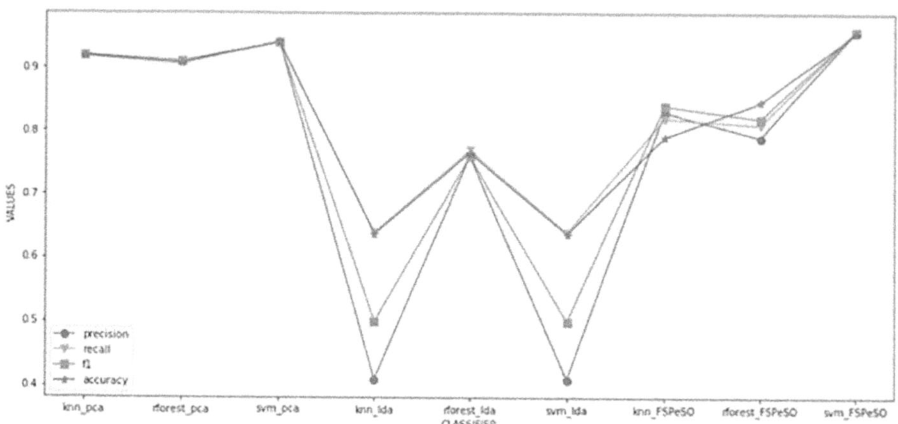

FIGURE 3.7 Performance analysis of the proposed FS-PeSOA with other algorithms on a Vehicle dataset.

Tables 3.3 and 3.4 show that our proposed FS-PeSOA algorithm generates the most number of "Win"s for supervised feature selection. To summarize, in the context of the seven UCI datasets used in the experiments,

- FS-PeSOA gives better average accuracy than LDA and PCA algorithms.
- FS-PeSOA has more number of "Win"s than any other algorithm (refer Table 3.4).

TABLE 3.3
Comparison Based on Accuracy, Precision, Recall, and F1 Score Parameters

Dataset	Algorithm	Accuracy	Precision	Recall	F1 Score
		A			
Iris	KNN with PCA	80	0.90	0.80	0.80
	KNN with LDA	60	0.47	0.60	0.50
	KNN with FSPeSOA	**90**	0.94	0.90	0.90
	Random Forest with PCA	77.77	0.89	0.78	0.77
	Random Forest with LDA	60	0.47	0.60	0.50
	Random Forest with FSPeSOA	**91.22**	0.94	0.90	0.90
	SVM with PCA	80	0.90	0.80	0.80
	SVM with LDA	60	0.47	0.60	0.50
	SVM with FSPeSOA	**90**	0.94	0.90	0.90
Glass	KNN with PCA	64.61	0.62	0.65	0.62
	KNN with LDA	53.84	0.67	0.54	0.59
	KNN with FSPeSOA	**81.39**	0.82	0.81	0.80
	Random Forest with PCA	66.15	0.66	0.66	0.65

(Continued)

TABLE 3.3 (*Continued*)
Comparison Based on Accuracy, Precision, Recall, and F1 Score Parameters

Dataset	Algorithm	Accuracy	Precision	Recall	F1 Score
	Random Forest with LDA	55.38	0.60	0.55	0.56
	Random Forest with FSPeSOA	**79.06**	0.81	0.79	0.79
	SVM with PCA	69.23	0.62	0.69	0.64
	SVM with LDA	44.61	0.22	0.45	0.29
	SVM with FSPeSOA	**81.39**	0.80	0.81	0.79
Ion	KNN with PCA	82.07	0.83	0.82	0.81
	KNN with LDA	62.26	0.39	0.62	0.48
	KNN with FSPeSOA	**95.91**	0.88	0.86	0.85
	Random Forest with PCA	84.9	0.85	0.85	0.85
	Random Forest with LDA	67.92	0.68	0.68	0.68
	Random Forest with FSPeSOA	**91.54**	0.92	0.92	0.91
	SVM with PCA	72.64	0.73	0.73	0.71
	SVM with LDA	62.26	0.39	0.62	0.48
	SVM with FSPeSOA	**91.54**	0.93	0.92	0.91
Pima	KNN with PCA	57.14	0.55	0.57	0.56
	KNN with LDA	62.33	0.67	0.62	0.63
	KNN with FSPeSOA	59.74	0.58	0.60	0.59
	Random Forest with PCA	63.63	0.61	0.64	0.62
	Random Forest with LDA	57.14	0.59	0.57	0.58
	Random Forest with FSPeSOA	64.28	0.64	0.64	0.64
	SVM with PCA	67.09	0.65	0.67	0.65
	SVM with LDA	65.36	0.43	0.65	0.52
	SVM with FSPeSOA	65.58	0.62	0.66	0.60
		B			
Vehicle	KNN with PCA	43.70	0.45	0.40	0.43
	KNN with LDA	38.58	0.48	0.39	0.36
	KNN with FSPeSOA	**57.05**	0.58	0.57	0.57
	Random Forest with PCA	30.31	0.32	0.30	0.30
	Random Forest with LDA	35.82	0.36	0.36	0.35
	Random Forest with FSPeSOA	**52.35**	0.52	0.52	0.52
	SVM with PCA	38.18	0.30	0.38	0.33
	SVM with LDA	30.31	0.14	0.30	0.19
	SVM with FSPeSOA	**58.82**	0.59	0.59	0.58
Wine	KNN with PCA	80	0.90	0.80	0.80
	KNN with LDA	64.4	0.72	0.64	0.65
	KNN with FSPeSOA	69.44	0.69	0.69	0.69
	Random Forest with PCA	75.5	0.81	0.76	0.76
	Random Forest with LDA	71.1	0.77	0.71	0.72
	Random Forest with FSPeSOA	55.55	0.64	0.56	0.57
	SVM with PCA	80	0.90	0.80	0.80
	SVM with LDA	20	0.04	0.20	0.07
	SVM with FSPeSOA	38.88	0.22	0.39	0.28

(*Continued*)

TABLE 3.3 (*Continued*)
Comparison Based on Accuracy, Precision, Recall, and F1 Score Parameters

Dataset	Algorithm	Accuracy	Precision	Recall	F1 Score
Wisconsin	KNN with PCA	91.81	0.92	0.92	0.92
	KNN with LDA	63.74	0.41	0.64	0.50
	KNN with FSPeSOA	79.04	0.83	0.82	0.84
	Random Forest with PCA	90.64	0.91	0.91	0.91
	Random Forest with LDA	76.6	0.76	0.77	0.76
	Random Forest with FSPeSOA	84.61	0.79	0.81	0.82
	SVM with PCA	94.15	0.94	0.94	0.94
	SVM with LDA	63.74	0.41	0.64	0.50
	SVM with FSPeSOA	95.61	0.96	0.96	0.96

TABLE 3.4
Win–Loss Ratio of Algorithms for Different Datasets

Dataset	Classifier	Accuracy	Precision	Recall	F1 Score
Iris	KNN	Win	Win	Win	Win
	Random Forest	Win	Win	Win	Win
	SVM	Win	Win	Win	Win
Glass	KNN	Win	Win	Win	Win
	Random Forest	Win	Win	Win	Win
	SVM	Win	Win	Win	Win
Ion	KNN	Win	Win	Win	Win
	Random Forest	Win	Win	Win	Win
	SVM	Win	Win	Win	Win
Pima	KNN				
	Random Forest	Win		Win	Win
	SVM				
Vehicle	KNN	Win	Win	Win	Win
	Random Forest	Win	Win	Win	Win
	SVM	Win	Win	Win	Win
Wine	KNN				
	Random Forest				
	SVM				
Wisconsin	KNN				
	Random Forest				
	SVM	Win		Win	Win
Win/Loss		14 Win/7 Loss	12 Win/9 Loss	14 Win/7 Loss	14 Win/7 Loss

3.5 CONCLUSIONS

In this chapter, a new supervised feature selection approach based on penguin search optimization has been presented. In this chapter, we have summarized the workflow of penguins to an algorithm and tried to use them in a manner that has been beneficial to the task of feature selection for different datasets. In terms of performance of the proposed FS-PeSOA algorithm, experiments using seven publicly available datasets have shown that it has given better results than the full feature set and the benchmark algorithms of feature selection. It has been compared against PCA and LDA for supervised feature selection. As a future work, this proposed algorithm can be used for science and development purposes and also in biomedical research areas where data analysis is required for the identification of various patterns of different diseases.

REFERENCES

1. Stewart, S., & Thomas, M. (2007). Eigenvalues and eigenvectors: Formal, symbolic, and embodied thinking. In *The 10th Conference of the Special Interest Group of the Mathematical Association of America on Research in Undergraduate Mathematics Education* (pp. 275–296), San Diego, California
2. Tibrewal, B., Chaudhury, G. S., Chakraborty, S., & Kairi, A. (2019). Rough set-based feature subset selection technique using Jaccard's similarity Index. In *Proceedings of International Ethical Hacking Conference 2018* (pp. 477–487). Springer, Singapore.
3. Goswami, S., Das, A.K., Guha, P. et al. (2017). An approach of feature selection using graph-theoretic heuristic and hill climbing. *Pattern Analysis and Applications*, Springer. doi:10.1007/s10044-017-0668-x.
4. Goswami, S., Das, A.K., Guha, P. et al. (2017). A new hybrid feature selection approach using feature association map for supervised and unsupervised classification. *Expert Systems with Applications*, Elsevier, 88, 81–94. doi:10.1016/j.eswa.2017.06.032.
5. Al-Kassim, Z., Memon, Q. (2017). Designing a low-cost eyeball tracking keyboard for paralyzed people. *Computers & Electrical Engineering*, 58, 20–29.
6. Ng, A. (2000). CS229 Lecture notes. *CS229 Lecture Notes*, 1(1), 1–3.
7. Goswami, S., Chakraborty, S., Guha, P., Tarafdar, A., & Kedia, A. (2019). Filter-based feature selection methods using hill climbing approach. In *Natural Computing for Unsupervised Learning* (pp. 213–234). Springer, Cham.
8. Guyon, I., & Elisseeff, A. (2003). An introduction to variable and feature selection. *Journal of Machine Learning Research*, 3(Mar), 1157–1182.
9. Kohavi, R., & John, G. H. (1997). Wrappers for feature subset selection. *Artificial Intelligence*, 97(1–2), 273–324.
10. Gheraibia, Y., & Moussaoui, A. (2013, June). Penguins search optimization algorithm (PeSOA). In *International Conference on Industrial, Engineering and Other Applications of Applied Intelligent Systems* (pp. 222–231). Springer, Berlin, Heidelberg.
11. Chandrasekhar, G., & Sahin, F. (2014). A survey on feature selection methods. *Computers & Electrical Engineering*, 40(1), 16–28.
12. Al-Ani, A. (2005). Feature subset selection using ant colony optimization. *International Journal of Computational Intelligence*, 2(1), 53–58.
13. Sahu, B., & Mishra, D. (2012). A novel feature selection algorithm using particle swarm optimization for cancer microarray data. *Procedia Engineering*, 38, 27–31.

14. Mafarja, M., & Mirjalili, S. (2018). Whale optimization approaches for wrapper feature selection. *Applied Soft Computing*, 62, 441–453.
15. Rashedi, E., & Nezamabadi-pour, H. (2014). Feature subset selection using improved binary gravitational search algorithm. *Journal of Intelligent & Fuzzy Systems*, 26(3), 1211–1221.
16. Parsopoulos, K. E., & Vrahatis, M. N. (2002). Particle swarm optimization method for constrained optimization problems. *Intelligent Technologies–Theory and Application: New Trends in Intelligent Technologies*, 76(1), 214–220.
17. Kotsiantis, S. B., Zaharakis, I., & Pintelas, P. (2007). Supervised machine learning: A review of classification techniques. *Emerging Artificial Intelligence Applications in Computer Engineering*, 160, 3–24.
18. Lichman, M., & Bache, K. (2013). UCI Machine Learning Repository. University of California, Irvine, School of Information and Computer Sciences. In [Online]. Available: http://archive.ics.uci.edu/ml.

4 A Physical Design Strategy on a NoSQL DBMS

Marcos Jota, Marlene Goncalves, and Ritces Parra
Universidad Simón Bolívar

CONTENTS

4.1 INTRODUCTION

Massive amounts of data are being generated by users and electronic devices every day. Sensors connected to electronic devices provide an enormous amount of accurate information in real time, while the data generated by users on social networks produces valuable information about them. In an era of big data, organizations need to gather and efficiently manage such amount of data for their decision-making processes.

Currently, the graph technology is become increasingly important, and the graphs are used to model dynamic and complex relationships of the data in order to generate knowledge. For example, Google has incorporated a Knowledge Graph to its search algorithm to significantly improve semantic searches; the searches are disambiguated and contextualized [1].

Although the data have been stored in relational databases traditionally, they have serious problems in terms of scalability in information management [2]. Thus, a new paradigm has emerged, and it is called NoSQL (not only SQL) databases. NoSQL databases provide better scalability because it is easier to add, delete, or perform operations in the system without affecting performance. Particularly, Neo4j is a NOSQL database management system (DBMS) that is graph-oriented. In this chapter, our main objective is to introduce a physical design strategy that improves the query execution for a specific workload in Neo4j. A physical design consists of proposing several data structures or guidelines in order to improve the data access in a database. It is worth noting that the structures used for a physical design will depend on the DBMS. Therefore, the first step of a physical design is to have a good knowledge of the DBMS. We selected Neo4j because it is a DBMS pioneered in the use of graphs as a physical structure for storing data.

This chapter is structured as follows: Section 4.2 presents a motivation example for our problem of physical design in a graph-based DBMS, Neo4j. Section 4.3 describes Neo4j as a graph-based DBMS and its execution plans and physical operators. Section 4.4 proposes a set of guidelines for a physical design on databases stored in Neo4j. Section 4.5 explains the application of our physical design guidelines for some example queries. The performance of the proposed strategy will be empirically evaluated in Section 4.6. Related work is presented in Section 4.7. Some limitations of our guidelines are discussed in Section 4.8. Finally, the future research directions and conclusion of this chapter will be discussed in Sections 4.9 and 4.10, respectively.

4.2 MOTIVATION EXAMPLE

Suppose a movie database contains 171 nodes. Each node can be a movie or a person. There are 38 movie nodes and 133 person ones. Additionally, this database has six relationships between these nodes. Figure 4.1 introduces a schema of our example database that includes one bucle between two Person nodes and five relationships between Person and Movie. Not all nodes of a given type have to be in any of the relationships shown in the schema of Figure 4.1.

In addition, consider a query to find information about the matrix movie. We can specify this query in Cypher language, as is shown in Figure 4.2.

The query in Figure 4.2 was executed and its execution plan retrieved by Neo4j is presented in Figure 4.3. We can observe in Figure 4.3 that the first operator,

AllNodeScan, performs a scan on each of the nodes in the graph to check whether it fulfills the query condition, the Filter operator evaluates the query condition, selecting those results that meet this condition, and the ProduceResults operator finally generates the results for the user. It is noteworthy that if the database comprises thousands or millions of nodes instead of 171 nodes, the query in Figure 4.2 will be way more costly.

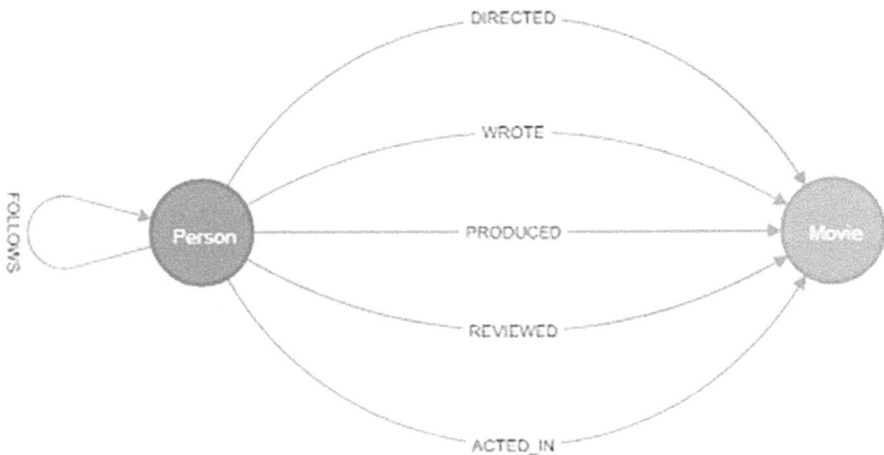

FIGURE 4.1 Database schema example.

```
MATCH (movie {title: "The Matrix"})
RETURN movie;
```

FIGURE 4.2 An example query for finding the matrix movie.

FIGURE 4.3 Basic plan generated for executing a query.

We can also help the Neo4j planner incorporating additional information in our query. Since we know that the node is a Movie node instead of a Person node, we can pose this query in such a way that it is only executed for searches on the nodes of the Movie type. Figure 4.4 illustrates an example query where the type of node is specified into the match clause.

When the type of node is incorporated into the query, a better plan is obtained. In Figure 4.5, it can be noted that the *NodeByLabelScan* operator is applied rather than the *AllNodeScan* node. Since the *NodeByLabelScan* operator is only executed on Movie nodes, it produces 38 estimated nodes, i.e., it approximately discards 78% nodes in its search.

Intuitively, a small sample of physical design strategies has been introduced in this section as an example in Neo4j [3]. A physical design produces an implementation-level description of a database: the storage structures and methods used for efficient access to the data [4]. An adequate physical design is the key for quickly accessing the information. It depends on specific aspects of the DBMS on which it needs to be implemented and important clues must be followed to design an optimal physical database. Thus, we must know the internal workings of our DBMS well in order to understand what is happening and how we can optimize a query. In our motivating example, a rewritten query improved its performance; however, not all classic physical design strategies are successful in Neo4j. Particularly, Neo4j offers a native index version whose implementation is not clearly specified by Neo4j and is limited for certain data types such as spatial, temporal, and numeric. For the other data types, Neo4j uses Lucene which supports inverted indexing. Neo4j indexing differs from traditional systems such as PostgreSQL and Oracle, which include

```
MATCH (movie:Movie {title: "The Matrix"})
RETURN movie
```

FIGURE 4.4 Query for finding the movie matrix using the node label.

FIGURE 4.5 Plan generated for executing a query using node labels.

several types of indexes including B-Trees. In consequence, Neo4j does not have a simple way of improving range comparisons, because the B-Tree use is usually a recommended strategy for this purpose; Neo4j supports a limited use of indices on range comparisons [5]. In addition, Neo4j lacks collection operators, and it only has UNION and UNION ALL operators that reduce the possibilities of rewriting a query to avoid a certain operator.

Despite some limitations of Neo4j, it also presents a different way to manipulate the data, e.g., it introduces a list data structure in order to manipulate data in several ways. In this sense, a new challenge is created for database administrators that have to adapt themselves to new technologies including NoSQL DBMSs. Additionally, the physical database design is a fundamental phase to ensure a good performance for database applications. When a database application grows in scale, the need for improving query performance also increases. If the queries begin to worse their performance, it is necessary to adequately handle significant amounts of information and efficiently execute different operations on the data.

Our goal in this chapter is to propose a physical database design strategy that improves the query performance for Linked Data Benchmark Council (LDBC) Social Network Benchmark (SNB) [6] using a graph NoSQL DBMS, Neo4j. Neo4j is one of the main NoSQL DBMSs on graph databases that are being widely used because they are the most natural way to understand any system that surrounds us and to represent its connections through graphs. For example, Facebook and Google support their business by means of graphs. Facebook is built on the value of relationships between people by capturing all these connections in a social graph [7] while Google improves its searches using a Knowledge Graph [1].

4.3 NEO4J

Neo4j is an open-source, native graph DBMS that provides an ACID (Atomicity, Consistency, Isolation, Durability)-compliant transactional backend [8]; its source code is written in Java and Scala. Particularly, Neo4j efficiently implements a property graph model. A property graph is built using nodes and edges, but it has two additional definitions: properties and labels. A property is an attribute of a node or a relationship. A label is used to tag a group of nodes or relationships in order to represent its role on the model. An example of a property graph can be observed in [8].

Additionally, Neo4j is able to generate a physical or execution plan for a query. An execution plan is a tree-like structure composed of operators [9]. These operators can be classified into the following five categories:

- Expanding operators: They are those used to traverse the graph. Given a set nodes, it will traverse the relationships, outgoing or ingoing, and retrieve the end nodes.
- Node-retrieving operators: They retrieve the starting point of a search in a graph. If an index is not specified, a scan over all nodes will be performed, else an index will be used. These operators are leaf nodes in an execution plan.

- Filter operators: They select a subset of data. For example, removing all nodes that don't fulfill a certain condition or removing all nodes except the first X nodes. These operators are usually found between operators.
- Join operators: They integrate the result of two branches with already expanded paths. These operators are performed when the execution plan has to merge two branches to produce one result.
- Aggregation operators: These operators correspond to usual aggregation operations such as count, sum, average, etc.

It can be noted that there are other types of operators for writing, such as create, update, or delete, but they are not considered in this chapter; additional details of Neo4j operators can be found in [9].

To generate an execution plan, Neo4j requires knowing certain statistics to determine which is the best plan according to the planner. These statistics are the number of nodes having a certain label; the number of relationships by type; selectivity per index; the number of relationships by type, and the number of relationships grouped by the label of the ending or starting node [9]. Then, these statistics are applied to estimate the number of records that is expected to be produced by each operator. Once the best plan is built by the planner, then it is executed. In addition, the execution plan is stored in cache to prevent any unnecessary recalculation. However, the execution plan is generated again if the computed statistics change.

Once a plan is executed, Neo4j shows actual records for each operator and an abstract unit called DB (Database) Hits. A DB Hit is an abstract unit of storage engine work and could correspond to looking up a node, property, or relationship. Since the amount of DB Hits is somewhat proportional to the amount of time it takes to execute the query, it is obvious why one of the main objectives should minimize the DB Hits.

Finally, it is important to note that some considerations must be taken into account to improve the execution time of a query. The operators of an execution plan, the amount of records each operator process and its outputs, and the amount of DB Hits generated while executing a plan are clues for a database administrator on how he/she can make physical design and reduce the processing time of a query.

4.4 DESIGN GUIDELINES

A good physical design can be achieved if the designer understands how the DBSM works, especially what are the query processing techniques offered by it. The following guidelines for a physical design in Neo4j summarize techniques that can be applied by a database designer:

- **Guideline 1 (Query Rewriting)**: A designer may improve the query performance if he appropriately modifies the current query text. During this work, we found out some patterns in a query text that usually lead to improvements on the query performance. These patterns are as follows:
  - Path redundancy: A same path or part of it can be repeated in a query text, and therefore, there is a path redundancy in a query text. If the

designer detects a path redundancy in the query text, he can simplify the query through other constructions that the query language provides. For example, a designer can group repeated paths using a case expression to avoid unnecessary calculations or relationship expansions. Additionally, if the path has at least three nodes, the designer can consider materialize it by means of relationship creation, and then, he/she can rewrite the query replacing the repeated path with the new materialized relationship.
  - Minimal query: A path is not minimal if the designer identifies unnecessary paths. If the designer eliminates a part of a path in the query text and the resulting query is still correct, then the query is not minimal. This part of a path is not really required since it does not contribute in the query result.
- **Guideline 2 (Path Materialization)**: A designer can also decide to create a materialized path. The first step is to identify a path with the highest amount of DB Hits. The second step is to materialize a relationship to create a shorter alternative path. Since Neo4j does not manage materialized views, the third step is to rewrite the query replacing the path identified in the first step by an alternative path. For example, consider the nodes A, B, and C, and the relationships r and q, if a path s with the highest amount of DB Hits in a query is (A)–[r]–(B)–[q]–(C), then a new relationship p can be created as (A)–[p]–(C), and the query can be rewritten replacing the path composed of r and q by p. In consequence, the DB Hits are reduced because the path was precomputed and materialized, and the cost of computing the path is avoided. Finally, when we create a materialized path, additional space is required to be stored. Moreover, if the memory is not enough, then the path cannot be created.
- **Guideline 3 (Index Creation)**: The designer must analyze each query and decide the creation of an index on those nodes where a high selectivity condition has been defined. An index can reduce the number of nodes to be explored, since a scan on all nodes is avoided, and in consequence, a smaller number of paths are expanded. Moreover, if an index needs to be created, the designer must select the type of index to be defined. The Neo4j index is simple or composite. A simple index corresponds to an index on a single property, and a composite index defines an index on several properties. The composite indexes have certain limitations. For example, a composite index cannot be used for range conditions or substring searches using the CONTAINS operator. Additionally, Neo4j implements two types of indexes. The first type is the native index that uses an unknown implementation and is limited to certain types of values (spatial, temporal, numeric, etc.). The second type of index is based on Lucene, and it uses inverted indexing for the remaining data types that are not supported by the native index [10]. Finally, although indices can improve the performance of queries, they require additional space to be stored, and if the database is highly updatable, the index update will impact into the database performance since the DBMS must update the index entries removing the old ones and inserting the new ones.

4.5 PHYSICAL DESIGN

Before starting a physical design of a database, its data must be analyzed in conjunction with the queries that are frequently executed, in order to propose strategies and structures that can improve the processing time of its queries. Thus, the dataset selected for our physical design was generated from the SNB [11] of the LDBC. An SNB is a data generator that produces synthetic social network datasets. In this experimental study, we work with the Interactive Workload and Business Intelligence Workload (BIW), which is the only workload of this benchmark, developed for Cypher and consists of analytical queries. The specification for each of the queries of this workload can be found in [12,13].

Based on the choke points, we selected 13 from 25 queries of the Interactive Workload and BIW. A choke point is an aspect observed during the evaluation of queries and corresponds to an opportunity of physical design to improve the query processing time [13]. Moreover, the choke points may be classified in terms of aggregation performance, join performance, data access locality, expression calculation, correlated subqueries, parallelism and concurrency, and RDF (Resource Description Framework) and Graph specifics.

Each of the queries of the Interactive Workload and BIW is associated with at least one choke point. In this work, we will only consider 22 choke points, although the benchmark subsequently changed, adding six additional check points.

The process for selecting 13 queries from the Interactive Workload and BIW consisted of randomly choosing a reduced number of 13 combinations from a set of 25 queries. It is noteworthy that $\frac{25}{13} = 5,200,300$ is a very large number. In this sense, we applied two criteria to reduce the total number of combinations. These criteria are the following: (i) For each choke point, there must be at least one query associated with it; (ii) There cannot be a choke point that is associated with all of the queries in each combination.

Since we want to test the design guidelines against all the choke points available, we defined the first criterion. Thus, we decided that the final combination must associate each choke point with at least one query. Figure 4.6 illustrates a table that contains the choke points for each query in combination with the queries: BI 3, BI 4, BI 6, BI 7, BI 9, BI 12, BI 14, BI 16, BI 18, BI 19, BI 21, BI 22, and BI 23, where the BI acronym means business intelligence that is assigned by SNB. In Figure 4.6, each row represents a query, and each column corresponds to a choke point. It can be noted that each of the columns has at least one dot, i.e., each choke point is in this combination.

The second criterion was defined to avoid having one choke point shared between all of the queries in a combination. To prove the physical design guidelines in several situations, we avoided having a common characteristic between all of the queries in a combination. We can observe in Figure 4.6 that no column is completely filled with dots.

With these two criteria, the total amount of combinations was reduced to 276,194. At last, for our experimental study, we randomly chose one of these combinations

	1.1	1.2	1.3	1.4	2.1	2.2	2.3	2.4	3.1	3.2	3.3	4.1	4.2	4.3	5.1	5.2	5.3	6.1	7.1	7.2	7.3	7.4
BI 3								●	●	●		●		●			●	●				
BI 4	●	●	●		●	●		●			●											
BI 6		●					●															
BI 7		●					●			●	●							●				
BI 9		●	●		●		●	●														
BI 12		●				●			●									●				
BI 14		●				●	●			●										●	●	●
BI 16		●	●				●	●			●								●	●	●	
BI 18	●	●		●						●			●	●								
BI 19	●		●		●		●	●			●				●						●	●
BI 21		●			●		●	●		●	●				●		●					
BI 22			●	●	●				●		●				●	●	●					
BI 23				●			●	●			●			●								

FIGURE 4.6 Choke points associated with each query for a combination. (Source: Ref. 12.)

was a combination with the following queries: BI 3, BI 4, BI 6, BI 7, BI 9, BI 12, BI 14, BI 16, BI 18, BI 19, BI 21, BI 22, and BI 23.

After understanding the dataset and determining the 13 queries on which we are going to apply our guidelines, we analyze the execution plan of each selected query. In the following sections, we will specify one query in natural language for each guideline, and we will briefly describe the execution plan for each query, and finally, we will propose the application of one of the guidelines that we consider to be better for the case. For reasons of space, only one execution plan is presented graphically.

4.5.1 Query Rewriting Using Path Redundancy Pattern

The query rewriting guideline was employed for the query BI 3 of the SNB benchmark. This query consists of counting those messages that were created for a given month and year and tagged during the next month of the selected month and year. Before deciding which guideline will be used, we analyze the execution plan shown in Figure 4.7.

We can observe in Figure 4.7 that the first operator of the execution plan is a scan on the Tag nodes, then there is an expand operator on the relationship between Message and Tag filtering the messages that have been created for a given month and year. Subsequently, an aggregation operator is identified to count the number of messages. Afterwards, there is another expand operator on the relationship between Message and Tag to retrieve a message that has been created in the month following the selected month and year. Also, the plan has an additional aggregation operator to count the number of messages. Finally, the Tag name and the number of messages are projected, and the first 100 results are returned because of a Top operation.

Based on the execution plan of Figure 4.7, we decide to apply the first guideline in terms of the path redundancy pattern. In the red box of Figure 4.7, the relationship between Message and Tag is expanded twice. Particularly, the query BI 3 includes an optional match clause to retrieve the messages created for a given month and year, and another optional match clause to get the messages created in the month

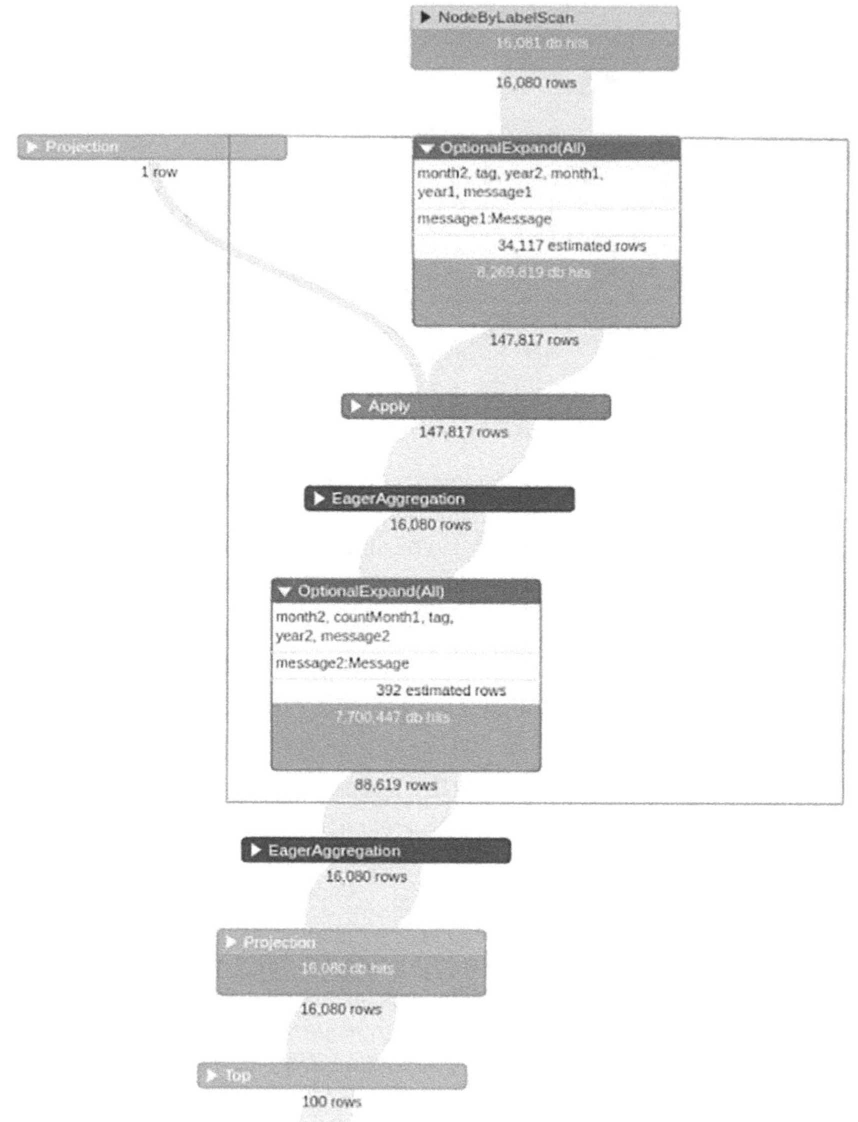

FIGURE 4.7 Execution plan of query BI 3.

following the selected month and year. Thus, we rewrite the query BI 3 to expand the relationship between Message and Tag only once by means of a case expression. The lines of the query BI 3 associated to the resulting rewriting are shown in Figure 4.8, where the common path in the two optional match clauses is highlighted in bold.

Since the version of Neo4j that we are using did not support date formats, the date was stored as string. Therefore, the expression message. creation-Date/100,000,000,000 = (year1 * 100) + month1 corresponds to an operation to

```
OPTIONAL MATCH (tag)<-[:HAS_TAG]-(message:Message) WITH tag,
        SUM( CASE WHEN message.creationDate/10000000000 = (year1*100) + month1
                THEN 1
                ELSE 0
                END) AS countMsg1 ,
        SUM( CASE WHEN message.creationDate/10000000000 = (year2*100) + month2
                THEN 1
                ELSE 0
                END) AS countMsg2
```

FIGURE 4.8 Fragment of the BI 3 query.

extract the month and year according to the position in the string in order to compare them.

Finally, we discard the third guideline. Particularly, the query BI 3 has optional match clauses. Since this kind of clause does not support indexes, the third guideline cannot be applied.

4.5.2 Query Rewriting Using Minimal Query Pattern

The query BI 7 was rewritten by means of identifying the minimal query pattern in it. This query finds all people who have created a message with a given tag and counts the number of likes that have been received from the people who like any of the message created with the given tag. Its execution plan is characterized by (i) scanning on the Tag nodes; (ii) filtering on the nodes Tag, Person, and Message; (iii) expanding on the relationship between Tag and Message twice; (iv) expanding the relationship between Message and Person four times; (v) a hash join; (vi) expanding the relationship between Person and Message; (vii) a nested loop to identify the people who like any of the messages from the people who created a message with a certain tag; (viii) an aggregation operation; (ix) a Top operator to produce the first 100 results.

In this scenery, the minimal query pattern for the first guideline was employed because the path Tag, Message, and Person are unnecessarily expanded twice in the query BI 7. The lines of the query BI 7 corresponding to these two expansions are shown in Figure 4.9, where the common path in the two optional match clauses is highlighted in bold.

If we removed the second line, the query is still correct, and therefore, it is unnecessary for the query BI 7. Though, the first guideline improved the query time when the unnecessary match clause was removed, the query still requires a lot of execution time. In consequence, we decide to apply the second guideline. The performance of query BI 7 becomes much better when we materialized a new relationship with the path between Person and Message, which is the most expensive path in terms of DB Hits.

```
MATCH (tag)<-[:HAS_TAG]-(message1:Message)-[:HAS_CREATOR]->(person1:Person)
MATCH (tag)<-[:HAS_TAG]-(message2:Message)-[:HAS_CREATOR]->(person1)
```

FIGURE 4.9 Fragment of the BI 7 query.

4.5.3 Path Materialization

We materialized one of the paths of the query BI 4, considering the second guideline for this case. The query BI 4 returns all the forums created in a given country, which contains at least one post with a given tag (Tagclass); the forum location is the same as the location of the forum moderator. Its execution plan is characterized by (i) two scan operators on the nodes Country and TagClass; (ii) six expand operators on the relationship between Country and City, City and Person, Person and Forum, Forum and Post, Post and Tag, and Tag and TagClass; (iii) three filter operators on the nodes Country, Tag, and TagClass; (iv) a hash join between the Tag obtained from the path between Post and Tag and the Tag retrieved from the path between Tag and TagClass; (v) an aggregation operator to count the number of posts; (vi) a projection and a sort operator by two properties; (vii) a Top operator to return the first 20 results.

It is important to note that the execution plan of the query BI 4 has a filter operator on a unique key, the country name. The country name is a good candidate for an index creation following the first guideline, because one country node will be retrieved instead of scanning 112 country nodes. Nevertheless, an index on the country name does not significantly improve the query performance because the scan on all country nodes is cheap. Another index that can be defined for this query is an index on the name property of Tagclass node. However, the execution time of the query BI 4 considerably worsened when we created an index on Tagclass nodes, since the engine starts to expand from the Tagclass nodes rather than the country ones, even though the number of Tagclass nodes is much higher than the number of countries, despite having created an index on the country nodes. One of the aspects to be highlighted is that the index on Tagclass can be created to improve the queries BI 9, BI 16, and BI 19 but it would have a negative impact on the query BI 4.

Finally, since the index creation could improve a little or worsen the processing time of the query BI 4, we decided to follow the second guideline. In particular, the most expensive path considering DB Hits is the one composed of the nodes Person, Forum, and Post. Nevertheless, this path was not created because the Forum node is required for the projection operator, because the query BI 4 returns the id, the title, and the creation date of the forum. Thus, if we materialize a path comprised by Person, Forum, and Post, this path would not be used by the planner. In consequence, the path with the second highest DB Hits was materialized, i.e., the path composed of Post, Tag, and TagClass. Lastly, this same guideline was applied for the following queries: BI 9, BI 16, BI 18, BI 19, BI 21, B1 22, and BI 23. It is worth noting that some relationships created for some queries were also used by other queries. Thus, the new relationship created for query BI 4 was employed in the plan of the query BI 16, and the materialized path for query BI 7 was used in the plans of queries BI 22 and BI 21, and the relationship defined for the query BI 18 was utilized in the plan of query BI 19.

4.5.4 Index Creation

The third guideline was applied for the query BI 6, which identifies people who have created at least one message with a given tag. Also, it counts the number of likes and replies that the message has received. The execution plan built by Neo4j for this query has the following operators: (i) a scan on the Tag node; (ii) three filters on the nodes Tag, Message, and Person; (iii) four expands on Tag and Message, Message and Person, Person and Message, Message and Comment; (iv) a sort operator on Tag name; (v) an aggregation to count the number of likes that the message has received and the number of replies that the message has had; (vi) a projection; (vii) a Top for the first 100 results.

On this case, an index on the Tag name was created by applying the third guideline, considering that the Tag name is a unique key. Contrary to the query BI 4, the Tag node has a notably large number of instances, and therefore, the query performance improvement is significant because a scan on Tag nodes will be highly costly. Lastly, this same guideline was applied for the query BI 14.

4.6 EXPERIMENTAL STUDY

In this section, the experimental study is reported. First, it begins by describing the metrics and the configuration that were considered in this experimental study. Second, it continues by comparing the physical design strategy following the guidelines proposed in this work against the default strategy of Neo4j.

4.6.1 Experimental Design

Datasets: This study was conducted on synthetic datasets. Synthetic datasets consist of data produced by a data generator of SNB called DATAGEN [14]. DATAGEN generates data based on a scheme of a social network, and it extracts some data collected from DBpedia [2], which allows the data to be realistic and correlated; the posts constitute the largest amount of data in the generated dataset and contain textual data extracted from DBpedia. The database schema contains people who live in a city, which is part of a country or continent. These people know others and write messages that are tagged according to a tag class. Also, people are members of forums that contain messages, they can work in a company, or they can study in a university. Additionally, the nodes in the schema have a degree of distribution similar to a power law. Thus, to generate the data, DATAGEN creates all relationships from an amount of people to ensure that data and link distribution are generated as in a real social network like Facebook [2]. According to the documentation [12], the DATAGEN is able to generate datasets at scales of 1 GB, 3 GB, 10 GB, 30 GB, etc. Even though DATAGEN allows any scale factor, its documentation also specifies that the maximum recommended factor for a dataset should be 1 TB. In our experimental, the dataset size was set to 1, 10, and 100 GB. We limited the dataset size because of space requirements. First, the DATAGEN runs on Hadoop, which needs a lot of disk space to execute MapReduce jobs. Second, the DATAGEN generates CSV (comma separated values) files with considerable sizes. For example, 1.2 GB of space is required for a 1 GB dataset. Third, our physical design involves materialized paths increasing the database size to 2.7 GB for a 1 GB dataset.

Queries: Besides DATAGEN, benchmark provides a BIW. This workload was designed on choke points that are considered to test the DBMS performance on different scenarios. The workload consists of 25 queries in terms of choke points specified to create a unique scenario of optimization for the planner. All the queries require certain input parameters to be executed. These parameters are provided by the DATAGEN such that any of the parameters should make the query having similar runtime and same optimal logical plan. In this experimental study, we applied our guidelines on 13 BIW queries. Although not all the queries were compared in our experiments, we plan to study the characteristics of the remaining 12 queries as a future work.

Metrics: To study the query performance, DB Hits and the total execution time were reported. Since the DB Hits are only bounded to operations on the graph, they are considered invariant because they will only change if the original database also changes. Contrary to DB Hits, the time results can vary for each query execution.

Experimental Environment: The experimental study was conducted on an AWS dedicated instance of i3.4xlarge. This instance has 16 virtual CPUs, 122 GB of RAM, two 1.9 TB NVMe SSD, and 8 GB EBS storage using Ubuntu 18.04 Server 64-bit as an operating system. We selected NVMe SSD because it is the latest technology on storage devices. Also, we chose one of the two solid-state disks to isolate the database from the datasets, i.e., one solid-state disk stores the datasets and the other contains the Neo4j metadata. Additionally, we have used the Neo4j community edition version 3.3.5. The memory was configured according to the recommendation of Neo4j: a Java initial heap size of 30.9 GB and a maximum heap size of 100 GB.

4.6.2 Impact of the Proposed Physical Design on Query Performance for a 1 GB Database

This experiment was performed to empirically determine that the runtime and DB Hits with a physical design strategy are better with respect to the traditional query evaluation by Neo4j when the database size is 1 GB.

Figure 4.10 shows the execution time on a 1 GB database for 13 BIW queries of SNB. The *X* axis has queries BI *i,* where *i* represents the query number and the *Y* axis corresponds to the normalized average time to the range (0,1). Also, Table 4.1 includes the percentage of improvements in runtime for each query.

We can observe in Figure 4.10 and Table 4.1 that runtimes improve when our physical design strategy is applied, except for the queries BI 3 and BI 21 in which the execution time slightly worsened. The percentages of improvements in runtime vary from 9.6% to 98.14%, and the execution time decreases at most 1.5%.

The highest performance improvements were for the queries BI 7 and BI 22 because we materialized the most expensive paths. The other operators belonging to the execution plans for these two queries were cheap. Therefore, the second guideline was really useful to decrease the execution time from 97.53% to 98.14%. Nevertheless, the performance was little improved for the queries BI 18, BI 19, and BI 23 despite the materialized path. For these queries, the percentage of improvements in runtime is between 9.6% and 11.6%. For one hand, there are still costly scan operators in the execution plan of the query BI 18. On the other hand, there are expand operators in the plans of the queries BI 19 and BI 23 that

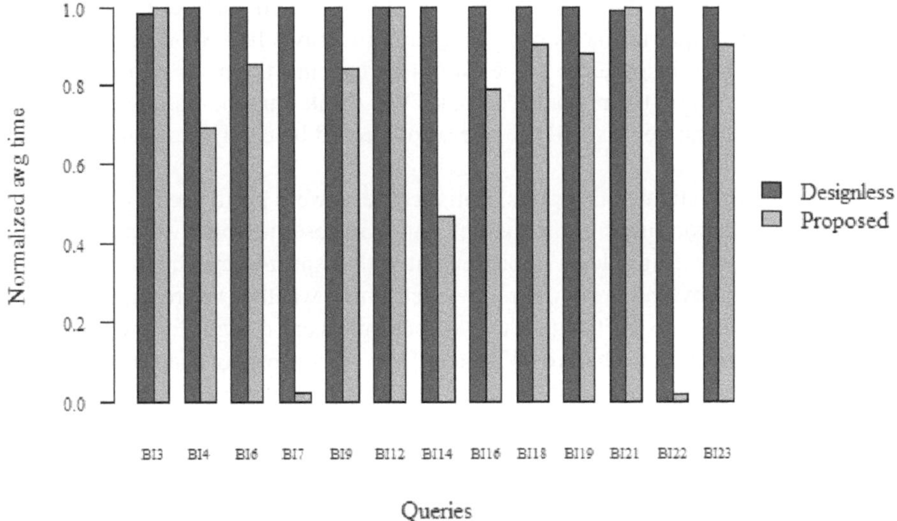

FIGURE 4.10 Time executed by each query on a 1 GB database using physical design and the default Neo4j evaluation.

TABLE 4.1
Percentage of Improvements in Runtime of Our Physical Design Strategy w.r.t. the Default Neo4j Evaluation for Each Query on a 1 GB Database

Query	BI 3	BI 4	BI 6	BI 7	BI 9	BI 14	BI 16	BI 18	BI 19	BI 21	BI 22	BI 23
Time improv. (%)	−1.48	31.05	14.53	97.53	15.67	53.19	20.92	9.77	11.83	−0.94	98.14	9.6
DB Hits improv. (%)	10.55	13.91	17.54	96.02	18.04	53.32	62.63	5.66	2.15	0.3	95.91	0.36

are unavoidable, and they still impact on the final execution plan. Additionally, path materialization was more effective for queries BI 4, BI 9, and BI 16 if we compared them against queries BI 18, BI 19, and BI 23. Their improvement percentages vary from 15.67% to 31.05% because the materialized paths have a higher impact on the query performance.

The third highest performance improvement was obtained for the query BI 14 in half the time of the strategy without physical design. Its improvement is 53.19%, because an index structure was defined for a high selectivity condition. Also, the initial execution plan expanded two different paths, and then they were joined with the remaining nodes. However, the planner improves its execution plan filtering intermediate results instead of completely expanded paths. In consequence, unnecessary traversals are avoided significantly reducing the cost of retrieving nodes. In addition, the performance of query BI 6 was a little better when the index was created, and its improvement percentage was 14.53%.

Finally, the queries BI 3 and BI 21 slightly worsened with execution time. Rewriting of these queries included aggregation functions like SUM or additional structures like lists in order to force Neo4j to execute them on main memory instead of processing them on hard disk. We think that our strategy cannot decrease the runtime because all the operations could be easily performed on the main memory.

To statistically validate our results, statistical tests were performed to determine whether the execution times of the queries on a database in which we apply physical design represent a significant improvement with respect to execution times on a database without physical design. First, we determine whether the resulting execution times for both possibilities (with or without physical design) follow a normal distribution performing a Shapiro–Wilk test [15]. According to Shapiro–Wilk test [15], the execution times obtained in our experiments (samples) can be considered to follow a normal distribution when their p-value is greater than $\alpha = 0.05$. The p-value is a measure of statistical significance that represents the probability of obtaining, by likelihood or chance, a difference as large or greater than that observed, meeting that there is no real difference in the population from which the samples come. It is usually established that if this probability value is less than 5% (0.05), it is sufficiently unlikely to be due to the chance to reject with reasonable certainty our initial hypothesis, and thus, it affirms that the difference is real. If it is greater than 5%, we will not have the necessary confidence to deny that the difference observed is the work of chance.

According to the results obtained from the Shapiro–Wilk test, there is at least one of the execution times whose p-value is lower than $\alpha = 0.05$, and therefore, there is a statistical significant evidence that the execution times (samples) do not follow a normal distribution.

As a consequence of not having data that follow a normal distribution, we proceed to apply a test that does not require this condition. The Wilcoxon signed-rank paired test [16] is a nonparametric test that allows to compare the means of the two related samples, i.e., average of the query execution times on the initial database without physical design and the average of the query execution times on the database with physical design. After performing this test, the resulting p-value is 0.006836, and thus, we can statistically conclude that, with a significance level of 0.05, the mean of the query execution time on a 1 GB database without physical design is significantly higher than the mean of the query execution time on a 1 GB database with physical design.

In addition to runtime, we also report the DB Hits in Figure 4.11. The X axis contains the query identifiers, and the Y axis corresponds to the DB Hits normalized in the range (0, 1). Also, Table 4.2 includes the percentage of improvements in DB Hits for each query.

We do not apply statistical tests because the DB Hits are invariant. It can be noted in Figure 4.11 and Table 4.2 that the results of DB Hits are similar to the execution time. However, we can observe that the DB Hits improve for the queries BI 3 and BI 21. Since a DB Hit is an abstract unit, it is possible that the execution time has been impacted by the computer configuration instead of the actual design strategy applied to improve the query performance.

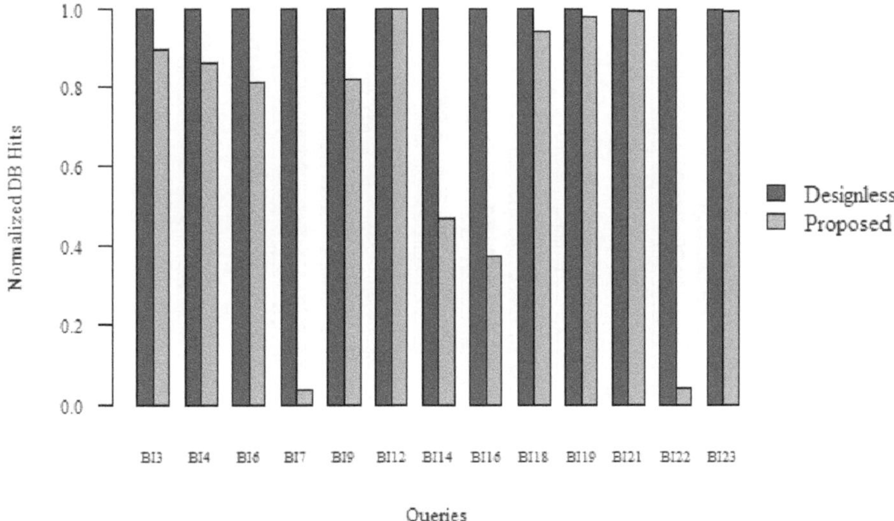

FIGURE 4.11 DB Hits by each query on a 1 GB database using physical design and the default Neo4j evaluation.

TABLE 4.2
Percentage of Improvements in DB Hits of Our Physical Design Strategy w.r.t. the Default Neo4j Evaluation for Each Query on a 1 GB Database

Query	BI 3	BI 4	BI 6	BI 7	BI 9	BI 14	BI 16	BI 18	BI 19	BI 21	BI 22	BI 23
Time improv. (%)	−1.48	31.05	14.53	97.53	15.67	53.19	20.92	9.77	11.83	−0.94	98.14	9.6
DB Hits improv. (%)	10.55	13.91	17.54	96.02	18.04	53.32	62.63	5.66	2.15	0.3	95.91	0.36

4.6.3 Impact of the Proposed Physical Design on Query Performance for a 10 GB Database

Figure 4.12 shows the execution time on a 10 GB database for the 13 BIW queries of SNB. The *X* axis has queries BI *i*, where *i* represents the query number, and the *Y* axis corresponds to the normalized average time to the range (0, 1). Also, Table 4.3 includes the percentage of improvements in runtime for each query.

We can observe in Figure 4.12 and Table 4.2 that runtimes improve when our physical design strategy is applied. The percentages of improvements in runtime vary from 7.76% to 97.75%. It can be noted that the queries BI 3 and BI 21 do not worsen. Since the database size is bigger, the operations could be easily performed in main memory for the 1 GB database, but they require more time to be executed for the 10 GB database.

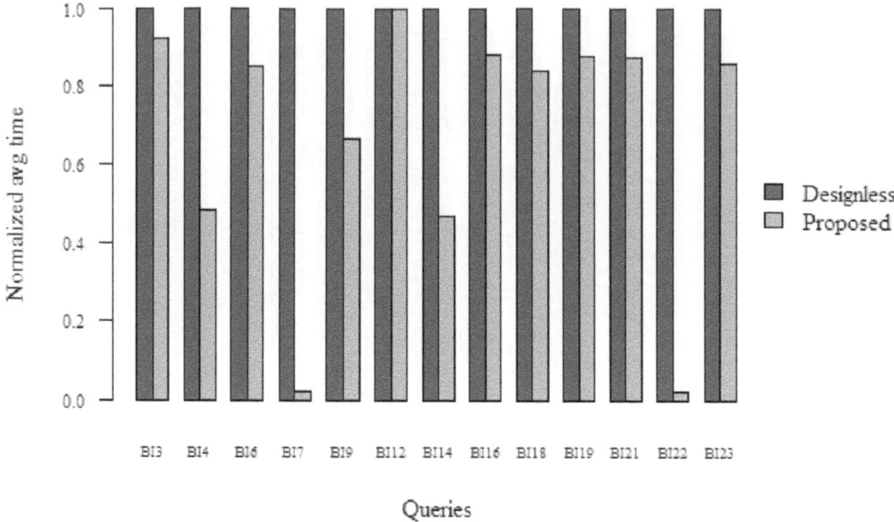

FIGURE 4.12 Time executed by each query on a 10 GB database using physical design and the default Neo4j evaluation.

TABLE 4.3
Percentage of Improvements in Runtime of Our Physical Design Strategy w.r.t. the Default Neo4j Evaluation for Each Query on a 10 GB Database

Query	BI 3	BI 4	BI 6	BI 7	BI 9	BI 14	BI 16	BI 18	BI 19	BI 21	BI 22	BI 23
Time improv. (%)	7.76	51.83	14.86	97.56	33.48	53.28	11.91	16.16	12.29	12.67	97.75	14.11
DB Hits improv. (%)	9.14	13.95	2.84	97.71	32.49	38.18	24.63	24.57	18.79	0.13	95.6	0.14

To validate the results, we applied a Wilcoxon signed-rank paired test. With a *p*-value of 0.0002441, we can statistically conclude that, with a significance level of 0.05, the mean of the query execution times on a 10 GB database without physical design is significantly higher than the mean of the query execution times on a 10 GB database with physical design.

Additionally, Figure 4.13 shows the DB Hits for each query and strategy. The *X* axis contains the query identifiers, and the *Y* axis corresponds to the DB Hits normalized in the range (0, 1). Also, Table 4.4 includes the percentage of improvements in DB Hits for each query.

It can be noted in Figure 4.13 and Table 4.4 that the results of DB Hits are similar to the execution time. However, we can observe that the performance of the queries BI 6 were lower w.r.t. the execution of this query for a 1 GB database. Particularly, the runtime worsens with index as the database size augments because of the number of relationships increasing significantly versus the increase of the number of nodes. In contrast to the query BI 14, the nodes of its index did not increase exponentially.

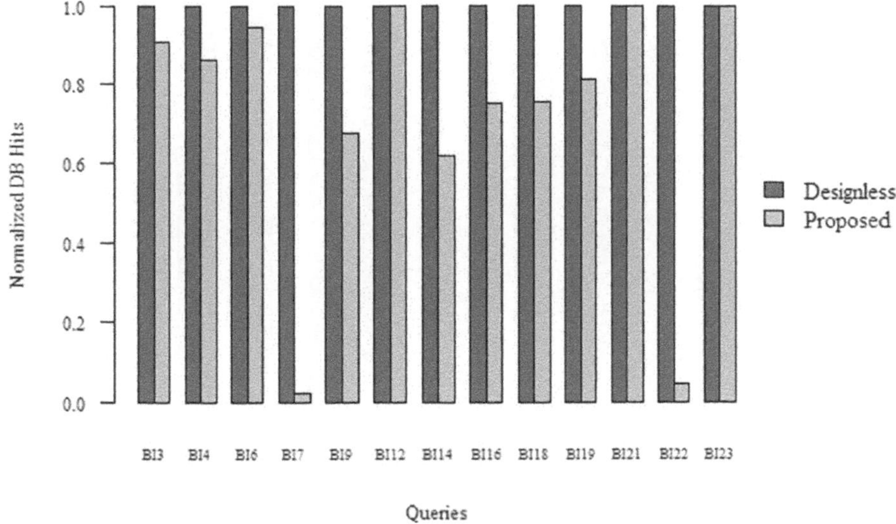

FIGURE 4.13 DB Hits by each query on a 10 GB database using physical design and the default Neo4j evaluation.

TABLE 4.4
Percentage of Improvements in DB Hits of Our Physical Design Strategy w.r.t. the Default Neo4j Evaluation for Each Query on a 10 GB Database

Query	BI 3	BI 4	BI 6	BI 7	BI 9	BI 14	BI 16	BI 18	BI 19	BI 21	BI 22	BI 23
Time improv. (%)	7.76	51.83	14.86	97.56	33.48	53.28	11.91	16.16	12.29	12.67	97.75	14.11
DB Hits improv. (%)	9.14	13.95	2.84	97.71	32.49	38.18	24.63	24.57	18.79	0.13	95.6	0.14

4.6.4 Impact of the Proposed Physical Design on Query Performance for a 100 GB Database

Figure 4.14 shows the execution time on a 100 GB database for the 13 BIW queries of SNB. The *X* axis has the queries BI *i*, where *i* represents the query number, and the *Y* axis corresponds to the normalized average time to the range (0, 1). Also, Table 4.5 includes the percentage of improvements in runtime for each query. In particular, the queries BI 7 and BI 22 are not shown in Figure 4.14 and Table 4.5 because they could not be executed due to a lack of memory.

We can observe in Figure 4.14 and Table 4.5 that runtimes improve when our physical design strategy is applied. The percentages of improvements in runtime vary from 36.74% to 98.65%. Also, the queries BI 3 and BI 21 do not worsen.

To validate our results, we performed a Wilcoxon signed-rank paired test. With a *p*-value of 0.0009766, we can statistically conclude that, with a significance level of

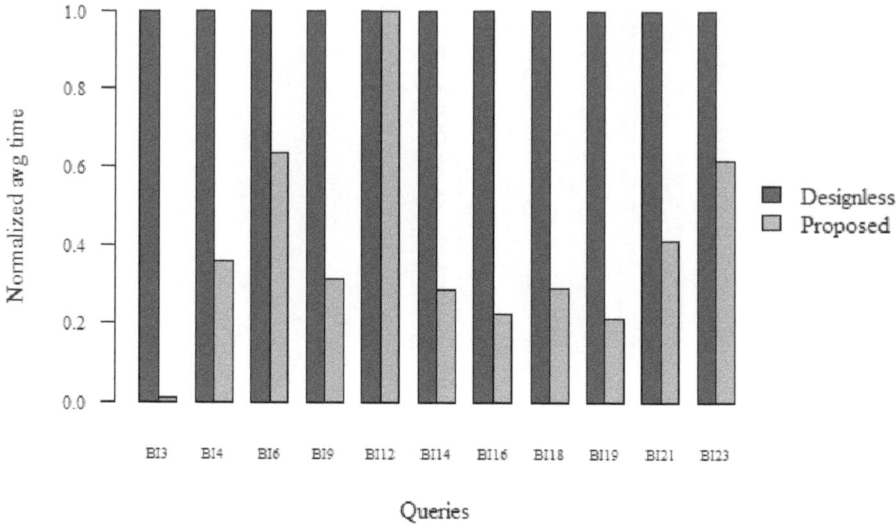

FIGURE 4.14 Time executed by each query on a 100 GB database using physical design and the default Neo4j evaluation.

TABLE 4.5
Percentage of Improvements in Runtime of Our Physical Design Strategy w.r.t. the Default Neo4j Evaluation for Each Query on a 100 GB Database

Query	BI 3	BI 4	BI 6	BI 9	BI 14	BI 16	BI 18	BI 19	BI 21	BI 23
Time improv. (%)	98.65	64.42	36.74	68.82	71.48	77.82	70.78	78.71	58.89	38.39
DB Hits improv. (%)	8.34	16.65	0.32	39.78	42.6	60.55	2.64	1.06	0.08	0.1

0.05, the mean of the query execution times on a 100 GB database without physical design is significantly higher than the mean of the query execution times on a 100 GB database with physical design.

In addition, Figure 4.15 shows the DB Hits for each query and strategy. The *X* axis contains the query identifiers, and the *Y* axis corresponds to the DB Hits normalized in the range (0, 1). Also, Table 4.6 includes the percentage of improvements in DB Hits for each query.

It can be noted in Figure 4.15 and Table 4.6 that the DB Hits results are similar to the execution time. However, we can observe that the performance of queries BI 6 were lower w.r.t. the execution of this query for a 1 GB database and a 10 GB database.

4.7 RELATED WORK

Physical database design has attracted much interest in the database community during decades [17,18,19,20,21]. These works propose guidelines to follow in the field

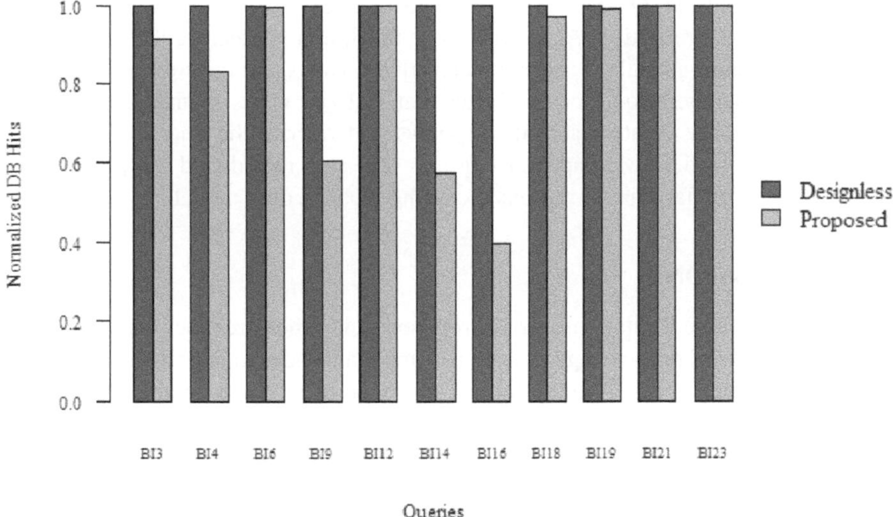

FIGURE 4.15 DB Hits by each query on a 100 GB database using physical design and the default Neo4j evaluation.

TABLE 4.6
Percentage of Improvements in DB Hits of Our Physical Design Strategy w.r.t. the Default Neo4j Evaluation for Each Query on a 100 GB Database

Query	BI 3	BI 4	BI 6	BI 9	BI 14	BI 16	BI 18	BI 19	BI 21	BI 23
Time improv. (%)	98.65	64.42	36.74	68.82	71.48	77.82	70.78	78.71	58.89	38.39
DB Hits improv. (%)	8.34	16.65	0.32	39.78	42.6	60.55	2.64	1.06	0.08	0.1

of relational databases that allow to improve the performance of queries executed on relational databases. Recently, some works have been introduced for the physical design of NoSQL databases because of the boom that has awakened this type of databases nowadays. In [22], the authors adapted star schemas of data warehouses to HBase (Hadoop database) by means of three physical designs; HBase is a column-oriented NoSQL DBMS. Also, Goncalves and Mendoza [23] used denormalization and attributes, ordering as physical design strategies on another column-oriented NoSQL DBMS, MonetDB. Imam et al. [24] defined physical design guidelines for NoSQL document-store databases. NoSQL document-store databases store data as semistructured documents such as XML (Extensible Markup Language), JSON (JavaScript Object Notation), among others. Asaad et al. [25] presented a uniform database methodology on the face of enormous varieties of NoSQL databases. This methodology does not include physical design. Mior et al. [26] developed a recommendation system to map from a NoSQL application conceptual data model to a NOSQL database schema, and this system was tested on the key-value DBMS called Cassandra. Unlike these works, we define physical design guidelines for graph-oriented NoSQL DBMSs. Our guidelines are based on materialization, indexes, and query rewriting. Agarawal, Chaudhuri, and Narasayya [27] addressed the problem of automatically selecting materialized views and indexes in SQL databases. Unfortunately, Neo4j does not offer materialized views management. Thus, we had to manually create materialized views and rewrite the queries to use the defined materialized views. Papakonstantinou [28] also introduced the problem of query rewriting on database variations including NoSQL data model.

4.8 DISCUSSION

In this section, some limitations of our proposed guidelines are discussed. The first limitation is that the proposed guidelines assume that all designers own basic physical database design skills. With no basic knowledge of physical database design, it will be a challenge to start designing with the proposed guidelines. In this sense, path materialization can be difficult to apply because Neo4j does not provide support for materialized views. It is the designer's responsibility to create and update materialized paths, in addition to rewriting the query by replacing the portion of the path to be materialized with the materialized relationship. However, a designer can effectively apply the guidelines through experimentation and determine which one is the best guideline according to the query characteristics. Also, our proposed guideline may be the basis for automating the physical design process on graph-oriented NoSQL databases from scratch, which may not require further technical training. The optimizer can automatically select the best materialized path for a query, and the database engine can keep updated all the materialized paths.

The second limitation relates to the few structures provided by the DBMS. The indices are not used by Neo4j if they are defined on a property in the condition of an optional match clause. Sometimes, a hash index can be advantageous when a condition involves an exact value instead of a range of values. Thus, the number of structures limits the number of guidelines that can be defined for the database physical design.

The third limitation is the amount of NoSQL DBMS. The guidelines described herein do not necessarily apply to other DBMSs. For example, the path materialization no longer makes sense for other types of DBMSs. Also, there is no standard for the query language as with relational DBMSs. The physical design guidelines associated with the query rewriting is tied to the Cypher language of Neo4j. When the DBMS changes, the query language becomes different, and therefore, query rewriting guidelines must be rethought.

4.9 FUTURE RESEARCH DIRECTIONS

In this chapter, we proposed a physical design for social network databases, considering 13 queries of the SNB. Although this benchmark has 25 queries, we could only analyze the execution plan for 13 queries. Thus, we plan to study the other 12 SNB queries as future work. Additionally, new versions of Neo4j emerged while we are performing our experimental study. The newest version of Neo4j supports B-Tree and indexes on arrays and list. Therefore, we can perform a physical design on the newest version of Neo4j as future work to define new guidelines to help the designer to improve performance queries.

In addition, the proposed physical design guidelines are based on query rewriting, path materialization, and indexing. These guidelines help to improve those queries characterized by optional match clauses and large search spaces due to a considerable number of relationships and nodes. Some queries worsened or slightly improved mainly because indexing or materialization could not be applied or they just did not contribute enough to reduce the processing time. In this sense, the strategy of how Neo4j evaluates such queries must be analyzed in depth to propose better physical design guidelines that achieve a better query performance.

Lastly, multimodel NOSQL systems are being more popular. To propose a physical design on a multimodel NOSQL system can be highly challenging because they support several different data models. Also, this type of DBMS can establish a standard of query language, and thus, guidelines such as query rewriting can be sustained over time when not having to be rethought for another query language.

4.10 CONCLUSION

This chapter has unveiled a physical design proposal for a database of SNB and has also presented an experimental study indicating that there is a substantial improvement for the query performance in Neo4j when the physical design guidelines proposed in this work are applied; performance is measured in terms of execution time and DB Hits.

The proposed guidelines defined in this work are based on query rewriting, materialization, and indexes. There is a cost associated with the creation of an index or path materialization, because the index requires additional space, and the path must be precomputed and stored. However, the SNB queries are analytical and are frequently executed, and therefore, paying this price seems worthwhile.

The proposed guidelines were applied for queries on databases of different sizes to empirically study the query performance. In addition, statistical tests were performed to validate our results. Nonparametric tests were chosen because they statistically demonstrate that sampled data were not normally distributed. Based on statistical tests, it was proved that the resulting runtimes for 13 queries using the proposed physical design guidelines were better with a statistically significant difference. Additionally, our experimental study shows that the execution time improves as the database size increases.

REFERENCES

1. A. Singhal, "Introducing the Knowledge Graph: Things, Not Strings," Official Google Blog. May, 2012. [Online]. Available at: www.blog.google/products/search/introducing-knowledge-graph-things-not/. [Accessed Jan. 30, 2019].
2. A. Castro, J. González, and M. Callejas, "Utilidad y funcionamiento de las bases de datos NoSQL," *Revista Facultad De Ingeniería*, vol. 21, pp. 21–32, Jan. 2013.
3. P. Selmer and M. Needham, "Tuning Your Cypher: Tips & Tricks for More Effective Queries. Neo4j Graph Database Platform," 2016, [Online]. Available at: https://neo4j.com/blog/tuning-cypher-queries/. [Accessed Oct. 30, 2018].

4. R. Elmasri, and S. Navathe, "Physical database design and tuning," in *Fundamentals of Database Systems*. Reading, MA: Addison-Wesley, 2017, pp. 601–654.
5. "Neo4j Indexes," [Online]. Available at: www.neo4j.com/docs/developer-manual/3.4/cypher/schema/index/. [Accessed Oct. 30, 2018].
6. O. Erling, A. Averbuch, J. Larriba-Pey, H. Chafi, A. Gubichev, A. Prat, M.-D. Pham, and P. Boncz, "The LDBC Social Network Benchmark: Interactive Workload," in Proceedings of the ACM SIGMOD International Conference on Management of Data, 2015, pp. 619–630, Melbourne, Australia
7. C. McCarthy, "Facebook: One Social Graph to Rule Them All?," April, 2010, [Online]. Available at: www.cbsnews.com/news/facebook-one-social-graph-to-rule-them-all. [Accessed Oct. 31, 2018].
8. "What Is a Graph Database," [Online]. Available at: https://neo4j.com/developer/graph-database/#_what_is_neo4j/. [Accessed Jan. 28, 2019].
9. "Neo4j Execution Plans," [Online]. Available at: https://neo4j.com/docs/cypher-manual/current/execution-plans/. [Accessed Jan. 28, 2019].
10. "11.2.2. Schema indexes - 11.2. Index configuration," Neo4j.com. [Online]. Available at: https://neo4j.com/docs/operations-manual/3.5/performance/index-configuration/schema-indexes/#index-configuration-index-providers. [Accessed Jan. 30, 2019].
11. "Social Network Benchmark," [Online]. Available at: http://ldbcouncil.org/developer/snb. [Accessed Jan. 28, 2019].
12. "LDBC Council. The LDBC Social Network Benchmark (Version 0.3.2)," 2018, pp. 12–28, 26–28, 87–90.
13. G. Szárnyas, A. Prat-Pérez, A. Averbuch, J. Marton, M. Paradies, M. Kaufmann, and J. Antal, "An early look at the LDBC social network benchmark's business intelligence workload," in *Proceedings of the* 1st *ACM SIGMOD Joint International Workshop on Graph Data Management Experiences & Systems (GRADES) and Network Data Analytics (NDA)*, 2018, Houston, Texas
14. "LDBC Social Network Data Generator," [Online]. Available at: https://github.com/ldbc/ldbc_snb_datagen. [Accessed Jan. 25, 2019].
15. S. Shapiro and M. Wilk, "An Analysis of Variance Test for Normality (Complete Samples)," [Online]. vol. 52. Available at: www.bios.unc.edu/~mhudgens/bios/662/2008fall/Backup/wilkshapiro1965.pdf. [Accessed Jan. 26, 2019].
16. F. Wilcoxon, "Individual comparisons by ranking methods," *Biometrics Bulletin*, vol. 1, no. 6, pp. 80–83, 1945. Available at: www.jstor.org/stable/3001968. [Accessed Apr. 04, 2019].
17. M. Schkolnick, "Physical database design techniques," in *NYU Symposium on Data Base Design Techniques*, pp. 229–252, 1979, New York
18. M. Schkolnick and P. Sorenson, "The effects of denormalization on database performance," *Australian Computer Journal*, vol. 14, no. 1, pp. 12–18, 1982.
19. S. Finkelstein, M. Schkolnick, and P. Tiberio, "Physical database design for relational databases," *ACM Transactions on Database Systems (TODS)*, vol. 13, no. 1, pp. 91–128, 1988.
20. S. Chaudhuri, and V. R. Narasayya, "An efficient, cost-driven index selection tool for Microsoft SQL server," in *VLDB*, pp. 146–155, 1997, San Francisco
21. C. C. Fleming, and B. V. von Halle, *Handbook of Relational Database Design*. Boston, MA: Addison-Wesley, 1989.
22. L. C. Scabora, J. J. Brito, R. R. Ciferri, and C. D. D. A. Ciferri, "Physical data warehouse design on NoSQL databases," in *Proceedings of the* 18th *International Conference on Enterprise Information Systems*. SCITEPRESS-Science and Technology Publications, Lda, pp. 111–118, 2016, Rome Italy

23. M. Goncalves and J. N. Mendoza, "A physical design strategy for datasets with multiple dimensions," in *Intelligent Multidimensional Data Clustering and Analysis*, Editors: Siddhartha Bhattacharyya , Sourav De , Indrajit Pan, Paramartha Dutta, IGI Global, pp. 1–27, 2017.
24. A. A. Imam, S. Basri, R. Ahmad, J. Watada, M. T. Gonzalez-Aparicio, and M. A. Almomani, "Data modeling guidelines for NoSQL document-store databases," *International Journal of Advanced Computer Science and Applications*, vol. 9, no. 10, pp. 544–555, 2018.
25. C. Asaad and K. Baïna, "NoSQL databases–seek for a design methodology," in *International Conference on Model and Data Engineering*, Springer, Cham, pp. 25–40, 2018.
26. M. J. Mior, K. Salem, A. Aboulnaga, and R. Liu, "NoSE: Schema design for NoSQL applications," *IEEE Transactions on Knowledge and Data Engineering*, vol. 29, no. 10, pp. 2275–2289, 2017.
27. S. Agarawal, S. Chaudhuri, and V. Narasayya, "Automated selection of materialized views and indexes for SQL databases," in *Proceedings of 26th International Conference on Very Large Databases*, Cairo, Egypt, pp. 191–207, 2000.
28. Y. Papakonstantinou, "Polystore query rewriting: The challenges of variety," in *EDBT/ICDT Workshops*, 2016, Bor-deaux, France

5 Large-Scale Distributed Stream Data Collection Schemes

Tomoya Kawakami
Nara Institute of Science and Technology

Tomoki Yoshihisa
Osaka University

Yuuichi Teranishi
Osaka University
National Institute of Information and Communications Technology

CONTENTS

5.1 INTRODUCTION

The Internet of Things (IoT) [1] has attracted greater interest and attention with the spread of network-connected small devices such as sensors, smartphones, and wearable devices. In the data science field, stream data generated from IoT devices are analyzed to get various information. A larger amount of data can lead to high-quality information as faster stream data collection is one of the main techniques in the data science field, and various schemes have been proposed. To enable IoT applications for data collection, pub/sub messaging [2] is considered to be a promising event delivery method that can achieve asynchronous dissemination and collection of information in real time in a loosely coupled environment. For example, sensor devices correspond to publishers, and IoT application corresponds to a subscriber. Topic-Based Pub/Sub (TBPS) protocols such as MQTT (Message Queuing Telemetry Transport) [3] and AMQP (Advanced Message Queuing Protocol) [4] are widely used by many IoT applications. These systems have a broker server for managing topics. The broker gathers all the published messages and forwards them to the corresponding subscribers.

In IoT and Big Data applications, collecting all of the raw (unfiltered) sensor data is important for conducting various forms of analysis [5]. In this case, the larger the number of sensors treated on the application for analysis, the larger the number of messages that need to be received per time unit on the broker and subscribers in TBPS. For example, when there are publishers corresponding to a certain kind of sensor that publishes sensor data every 10 s, the number of target sensors in an application is 10,000, and the broker must receive 1,000 messages per second on an average. Thus, the number of messages received by the broker and subscribers tends to explode in IoT and Big Data applications. In general, the number of messages sent and received per unit time affect the network process load, because tasks such as adding/removing headers and serializing/deserializing payloads are required for each message. Therefore, even though the size of each sensor data is small, the increase in the number of publishers can cause network process overloads on the broker and subscribers. This leads to a loss of data or unusual increases in delivery latency problems that have an adverse effect on IoT and Big Data applications.

Many existing studies tackle the problem of scalability in TBPS systems. The approach of these studies is based on distributed brokers in which brokers are run as peers in a peer-to-peer system. The brokers construct an overlay network among themselves. For example, there are Distributed Hash Table (DHT)-based approaches [6,7], hybrid overlay approaches [8], and Skip Graph (SG)-based [9] approaches [10,11]. These approaches can keep the number of connections that each broker needs to accept small by multihop message forwarding on overlays. However, the aim of these existing methods is to deliver messages from one publisher to multiple subscribers in a scalable manner. Thus, they are unable to avoid network process overloads caused by collection, such as when messages are received from a large number of publishers. In addition, the existing techniques do not assume different intervals at the same time to periodically collect data from publishers.

Therefore, we define continuous sensor data with different intervals (cycles) as a sensor data stream and have proposed collection methods for distributed sensor data streams as a TBPS system [12,13]. Especially in [12], we have proposed a message forwarding scheme on overlays called "Collective Store and Forwarding" (CSF). The scheme can reduce a load of network processes dramatically when there are a large number of publishers (nodes), maintaining the delivery time constraints given for the messages. In addition, we have also proposed a flexible tree construction method called "Adaptive Data Collection Tree" (ADCT) on Chord# [14] that can adjust the maximum load of network processes on distributed brokers to avoid overloads caused by CSF. Moreover, we have proposed an expanded method assigning phase differences to balance the collection time among the same or specific collection cycle nodes [15]. We call this novel approach "phase shifting (PS)." The PS approach enables the SG method to decrease the probability of load concentration to the specific time or node. Assigning phase differences at random to the nodes, the collection times are distributed even if there are same collection cycle nodes. We have evaluated our proposed methods in simulation. Our experimental results show that our proposed method can reduce the loads of nodes and realize highly scalable systems to periodically collect distributed sensor data. The scalability of the data collection systems is significantly important to accommodate a huge number of objects and encourage the growth of the data science field.

In the following, the data collection scheme for distributed TBPS is described in Section 5.2. The data collection scheme considering phase differences is described in Section 5.3. We describe the discussion and related work in Sections 5.4 and 5.5, respectively. Finally, the conclusion of the chapter is presented in Section 5.6.

5.2 DATA COLLECTION SCHEME FOR DISTRIBUTED TBPS

In this section, we describe a data collection scheme for distributed TBPS messaging that can prevent overloads in network processes in large-scale IoT applications.

5.2.1 Assumed Environment

In this section, we present an overview of the distributed TBPS environment that we assume.

5.2.1.1 Assumed TBPS Architecture

Figure 5.1 shows the distributed broker architecture for the TBPS we assume in this study, which is same as the existing distributed TBPS methods [10,11] that achieve

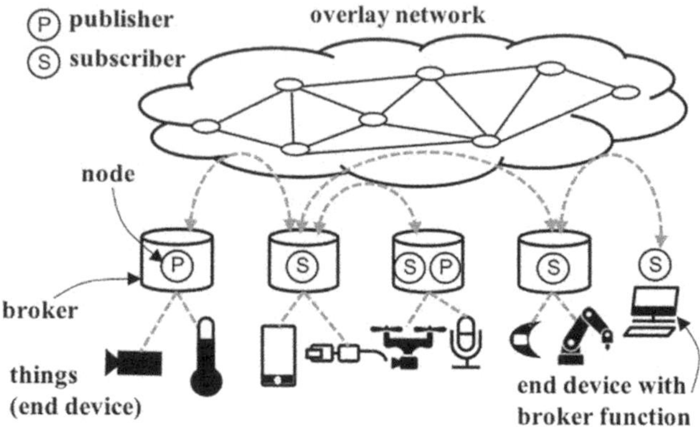

FIGURE 5.1 Distributed TBPS system by overlay network.

scalability. Each broker corresponds to the network entity located physically close to each end device. The end devices correspond to IoT entities such as sensors, smartphones, and appliances.

The broker can run on any networked computer that is located close to the end device. The computer can be a server at the distributed data center, a Wi-Fi access point with computing power, and a personal desktop computer. The model is compatible with the hierarchical cloud architectures recently proposed for IoT (the so-called "edge computing" architecture).

A broker contains a subscriber or a publisher of a topic when one of the devices accommodated on the broker attempts to subscribe/publish to the topic. The brokers then construct an overlay network for message delivery on each topic. Each broker joins the overlay network to handle distributed message deliveries. The end device can be a broker if it is able to handle an overlay function. Hereafter, an entity that joins the overlay is referred to as a "node." The node can be a publisher, a subscriber, or both. Brokers can contain multiple publishers or subscribers across different topics.

5.2.1.2 Assumed Overlay for Distributed TBPS

The article [10] proposed an SG-based TBPS (hereafter ST) method. ST constructs a "strong relay-free" topology on a key order-preserving structured overlay SG [9] to deliver pub/sub messages.

The basic overlay structure for the distributed TBPS that we use in this study is the same as the existing ST method. The ST method can be applied to any key order-preserving structured overlay. As a key order-preserving structured overlay, we assume Chord# [14] is used. Figure 5.2 shows an example structure of Chord# and TBPS on it created by the ST method. In Chord#, each node is connected by bidirectional pointers at "level 0" in the order of the keys owned by nodes. The node with the minimum/maximum key has pointers to the node with the maximum/minimum key. Thus, the level 0 becomes a ring structure. Each node has multiple levels of pointers. The pointer on level 1 becomes a unidirectional pointer to the

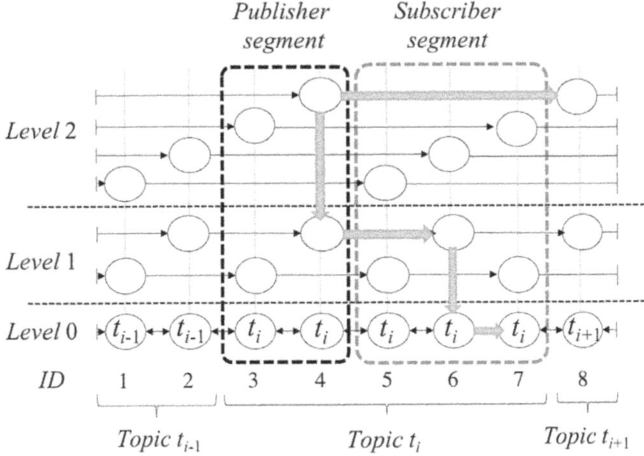

FIGURE 5.2 TBPS overlay structure on Chord# and its forwarding path.

2^l-th skipped node on the right side. These pointers are maintained as a routing table of the node. For example, the node with ID 1 has a pointer to 3 at levels 1 and 5 at level 2, as shown in the figure. This pointer set is called a "finger table." The structure of the finger table is constructed by using the periodic message exchange between the neighbor nodes on each level.

A message toward a key is forwarded using upper-level pointers. By using this structure, any key can reach the corresponding node at most $\log_2 N$ hops, where N is the total number of nodes. Moreover, Chord# can maintain the same level of performance even when there are a small number of nodes, as long as the routing tables are updated sufficiently. Refer article [14] for more details about Chord#.

In ST, each subscriber or publisher joins a structured overlay network based on its key, which includes a string for the topic name. One broker can have multiple keys to contain multiple publishers or subscribers of different topics. To distinguish subscribers with the same topic name, a unique suffix (ID) is added to the topic name on each key. There is a subscriber segment and a publisher segment on each topic. The subscriber segment must exist on the right side of the publisher segment on Chord#, because it only has right-side pointers on upper levels. This structure enables publishers to detect the existence of subscribers using the neighbor link at level 0, thereby reducing the amount of redundant traffic from publishers when there are no subscribers [10].

In this TBPS overlay structure, the following theorem holds, as it uses Chord# as the base structure.

Theorem 1

The maximum number of hops required to deliver a message for a topic and the number of connections each node need to accept are $\log_2 M$, where M is the number of publishers and subscribers for a topic.

By this theorem, the message delivery for a large number of subscribers is executed in a scalable manner. However, the broker that accommodates a subscriber can cause network process overloads when an application needs to accept a large number of publishers. The number of messages the node needs to receive in a short period becomes at most as many as the number of publishers, because each publisher runs asynchronously. In addition, a few publishers need to forward a larger number of messages than others due to the path concentration of data forwarding on structured overlays. As a result, the load of the network process easily exceeds the capacity of the node.

5.2.2 Proposed Method

In this section, we describe the details of our proposed TBPS scheme for solving the problems associated with large-scale data collection.

5.2.2.1 Methodology Principle

The basic idea of our proposal is to reduce the overhead of message forwarding for a large number of messages by using a mean known as "message merging."

In general, if the size of data that is needed to be transferred is the same, the performance is better when multiple items of data are sent as one large chunk segment than they are sent in multiple small segments [16]. Therefore, if there are a large number of small messages that are needed to be delivered, the message transfer time is reduced by merging the messages into one large message.

Figure 5.3 shows the results of a preliminary experiment that measured the transmission speed between two virtual machines on a cloud test bed [17] using a Java-based messaging framework [18]. The specifications of the experiment environment are listed in Table 5.1.

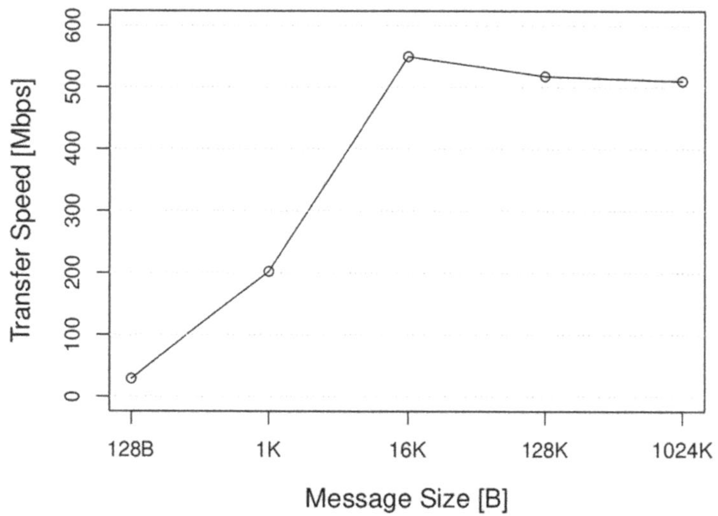

FIGURE 5.3 Message size vs. transmission speed.

TABLE 5.1
Preliminary Experiment Setup

Parameters	Value
Virtual CPU	1 core/2.1 GHz
No. of virtual machines	1/Host
Memory	2 Gbyte
Network	1000 BASE-T
OS	Ubuntu 12.04
Protocol	TCP/IP
Hypervisor	KVM 3.5

The average of ten times per experiment for each data size is plotted. As shown in the figure, when the message size is small, the network speed is limited. As message size grows, the network speed rises. In this experimental environment, if the message size is larger than 16 kbyte, the network speed is saturated. The cause of the speed limitation is the overhead needed to handle each message, such as adding/removing headers and serializing/deserializing payloads.

In TBPS systems, publishers act asynchronously, and the messages are published at the timing set by the publisher. To merge messages in this situation, we propose a message forwarding scheme called CSF on Chord#-based distributed TBPS. In CSF, we assume that each message has its own "delivery deadline." In IoT applications, the sensor data has its own expiration time, which refers to the duration for which the data is considered to be valid. For example, if an image sensor observes the density of people in a certain area every 1 min, the observed value may be valid for 1 min. In this case, the deadline of the message is 1 min. The delivery deadline is defined by publishers based on the time constraints of the message content. The relay nodes on the message delivery tree on CSF store the received messages and forward merged messages, observing the delivery deadline.

We also propose an adaptive tree construction method called ADCT. ADCT adaptively adjusts the maximum overhead for the nodes to forward messages as the CSF process. By sharing the message process loads among publisher nodes that are close to the subscriber segment, the ADCT can flexibly reduce the maximum overhead and prevent the concentration of network process loads.

5.2.2.2 Collective Store and Forwarding

In CSF, each publisher has the *time to publish*, and the published message has a *delivery deadline*. The time to publish is the future time when the next message will be published. Therefore, the publisher needs to know when the next message will be generated. This is a realistic assumption in IoT applications because sensors, which generate periodic sensor data, know when the next message will be published.

Figure 5.4 shows the pseudocode of a CSF algorithm. It describes the behavior when a published message arrives at a node and the behavior of the publish function. When a message arrives, if the deadline of the received message is after the next time to publish, the message is stored on *storage*, which in this case is a database on

```
Algorithm 1: Collective Store & Forwarding on u
upon receiving ⟨publish, message⟩
begin
    if (timeToPublish + T < message.deadline) then
        store(storage, message);
    else
        v ← nextForwardingNode(message.topic);
        send ⟨publish, message⟩ to v;

function publish ⟨message⟩
begin
    foreach m in storage do
        if (m.topic = message.topic) then
            message.payload ←
            append(message.payload, m.payload);
            if (m.deadline < message.deadline) then
                message.deadline ← m.deadline;

    v ← nextForwardingNode(message.topic);
    send ⟨publish, message⟩ to v;
    clear(storage, topic);
    update next timeToPublish;
```

FIGURE 5.4 Pseudocode of a CSF algorithm.

memory or disk. In other cases, the message is forwarded to the next node. *T* is the margin time for message transfers, which should be specified if the message must strictly meet the deadline. The value should be larger than the node-to-node latency multiplied by the maximum number of hops. The maximum number of hops can be estimated by the maximum finger table entry level that points to a node within the same topic.

In the publish function, the append function creates a payload of the messages merging the stored messages. At this time, the deadline is set to the earliest possible deadline (a process which we refer to as "deadline reduction"). The message is then forwarded to the next node, and the *storage* is cleared to store messages that will be received until the next instance of publishing on the topic. As described earlier, CSF requires an extra process to store and merge messages compared with the normal overlay implementations. However, it is a simple algorithm to implement. If the physical storage size is limited, the algorithm can be easily modified to remain within the predefined storage size.

Figure 5.5 shows the cases of CSF. In these sequence diagrams, the small circle indicates the timing for invoking the publish function.

There are three publishers, and two of them (publishers 2 and 3) also act as relays. Figure 5.5a shows a typical case in which the messages can be merged. In this example, the deadline of message 1 is after the generation time of message 4 on publisher 2. Therefore, message 1 is merged with message 4. Likewise, message 4 is merged with message 8 on publisher 3. As a result, the subscriber receives only one

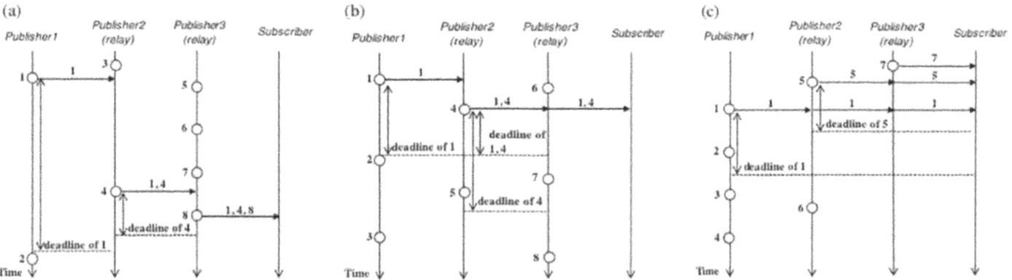

FIGURE 5.5 Message merging. (a) Merge without deadline reduction, (b) Merge with deadline reduction, (c) Pass through (no merge).

message when receiving the contents of 1, 4, and 8. Figure 5.5b shows another typical case in which the deadline of the published message on the relay node is reduced (deadline reduction). Because message 1 is merged with 4, the deadline is reduced to that of 1 since it has an earlier deadline. Message 4 could not be merged on publisher 3 because the deadline was reduced.

If the deadline of the received message is earlier than the next time to publish, the message must be forwarded immediately. Figure 5.5c provides an example of this case.

According to Theorem 1, the messages published for a topic t have chances to be merged $\log_2 |N_t|$ times at the maximum and $\log_2 |N_t|/2$ times on average, in which N_t is the number of publishers for t when there is only one subscriber for t.

5.2.2.3 Adaptive Data Collection Tree

In addition to CSF, we propose a flexible collection tree construction method called ADCT.

ADCT defines a *responsive subscriber* and one or more *boundary nodes*. The responsive subscriber receives the merged message of publishers from boundary nodes and becomes the root of the delivery tree for all subscribers.

In ADCT, the messages published for topic t are forwarded by searching for a node that has the largest key on the publisher segment. To search for the largest key, the maximum value of the publisher segment (the maximum ID value is set as a suffix) is specified as a key for a search on Chord#. The search ends when the neighbor node is larger than the maximum value of the publisher segment. We refer to the node that has the largest key on the publisher segment as the boundary node. The boundary node forwards messages directly to the leftmost subscriber, which corresponds to the responsive subscriber. The network process load to receive and forward messages on the boundary node becomes the highest among the publishers.

If the network process load exceeds a certain threshold, the publisher node next to the boundary node is invited as part of the boundary nodes.

To adjust the number of boundary nodes in a heterogeneous environment, the published message from the boundary node to the responsive subscriber includes overload/underload information. The responsive subscriber invites or withdraws

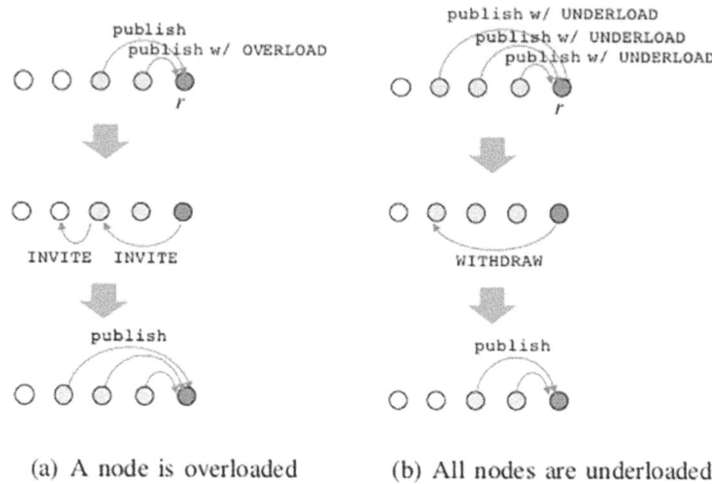

FIGURE 5.6 A boundary node adjustment protocol.

boundary nodes based on this information. Figure 5.6 shows a boundary node adjustment protocol on ADCT. The responsive subscriber is hereafter written as *r*. If one of the boundary nodes becomes overloaded, an INVITE message is sent to the node that has the minimum key among boundary nodes with the pointer to *r*. After that, the message is forwarded to an adjacent node. The node that receives the INVITE message acts as a boundary node. If all boundary nodes are underloaded, then a WITHDRAW message is sent to the leftmost boundary nodes, and the number of boundary nodes is reduced.

On the boundary nodes, the following theorem holds.

Theorem 2

The number of messages needed to be treated on a boundary node becomes $1/|B|$ of the total number of messages received on *r*, where *B* is the set of boundary nodes.

Proof: When $|B| = 1$, the boundary node receives messages via all levels of links. The probability that the message for a topic *t* is received via link level l ($l = 0, 1, \ldots, \log_2 |P_t|$) becomes $\Sigma 1/2^{i+1} = 1/2^{l+1}$ ($i = l + 1, \ldots, \log_2 |P_t|$). That is, the probability that the 2^k-th node from the boundary node forwards message becomes $1/2^k$. The 2^i-th node in *B* receives $1/2^{i+1}$ of the total number of messages. Therefore, when $|B|$ increases to $2|B|$, the number of forwarded messages on the nodes in old *B* becomes 1/2. Thus, on each node in *B*, the number of forwarded messages becomes $1/|B|$.

According to this theorem, a load of network processes, such as a load to receive and send messages, is balanced among the boundary nodes, if the CSF performs uniformly among nodes. Figure 5.7 shows examples of ADCT when $|B|$ is 1, 2, and 4. As shown in the figure, the number of nodes under the boundary nodes is nearly balanced when $|B|$ is 2 and 4.

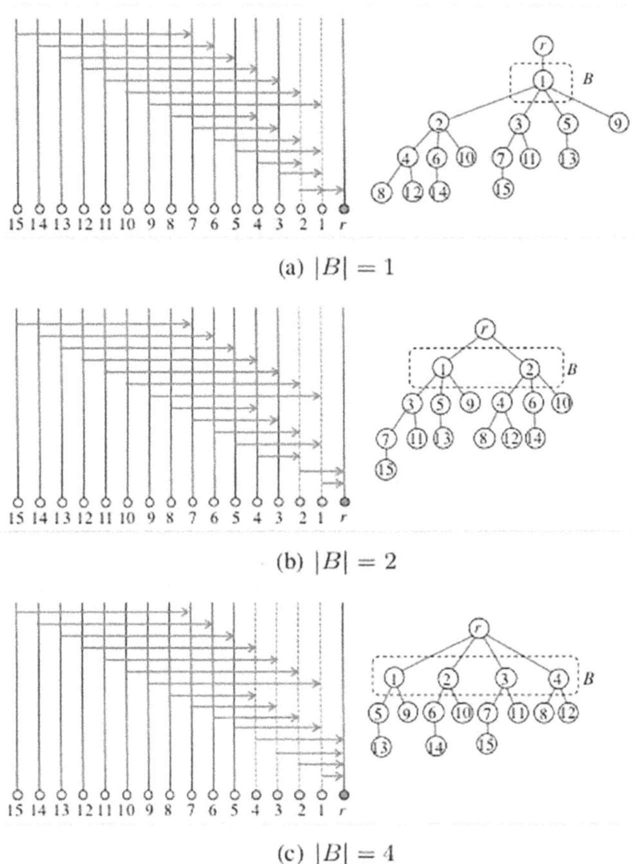

(a) $|B| = 1$

(b) $|B| = 2$

(c) $|B| = 4$

FIGURE 5.7 Examples of ADCT.

5.2.3 Evaluation

In order to show the effectiveness of our TBPS method, we conducted simulations to evaluate its performance. We developed an original simulator program to evaluate the performance on a 10,000 node-scale peer-to-peer system, because existing network simulators do not have the ability to evaluate this scale. Instead of simulating underlay physical networks, we conducted a peer-to-peer simulation in which we calculated the network performance using real data obtained from experiments.

We evaluated the received number of messages on a publisher/subscriber, the received message size on a subscriber, and the estimated network occupation time for message handling on a publisher/subscriber.

5.2.3.1 Simulation Parameters

The simulation parameters are shown in Table 5.2. To evaluate the basic performance of our proposal, the messages were published periodically with the same cycle (1 time unit), the deadline was set to be the same as the length of the cycle,

TABLE 5.2
Simulation Setup

Parameters	Value
Publish cycles	1 (time unit)
Deadline	1 (time unit)
Message data size	128 (bytes)
Transmission speed	Follows Figure 5.3
Number of nodes	1,000–10,000
Elapsed time	128
T	0
Tree structures	DAT, RQ, ADCT

and T is set as zero (ignore the transmission latency). The message size is set as 128, which is a relatively small message but enough for sensor data. To estimate the network occupation time, the result of the preliminary experiment on Figure 5.3 was used. The simulations were executed ten times for each parameter, and the average value was plotted.

As a comparison tree structure for ADCT, the aggregation tree structure of Distributed Aggregation Tree (DAT) and range query by Split Forward Broadcasting (SFB) (denoted as RQ) were used. Though DAT is proposed as an aggregation method, it is a basic tree construction method and can be applied to CSF on Chord#. RQ is the reverse tree structure of a range query from *r*. The query path was constructed for all publishers from *r* by SFB. RQ requires periodic range queries to be executed, but DAT and ADCT do not require this process. Example structures of DAT and RQ are shown in Figure 5.8.

5.2.3.2 Simulation Results

Figures 5.9–5.11 show the results when the number of publishers was changed from 1,000 to 10,000. In these cases, |*B*| in ADCT was set as 2. Figures 5.12–5.14 show the results when |*B*| was changed in ADCT. DAT and RQ become a constant value in these cases. The number of publishers was set as 10,000.

Figures 5.9 and 5.12 show the number of messages on *r* and the publisher node that received the maximum number of messages (hereafter, *p*). As shown in Figure 5.9, CSF can reduce the load dramatically. When there are 10,000 nodes,

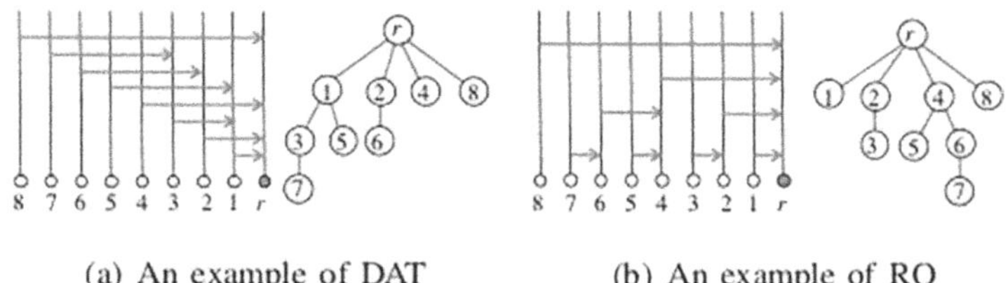

FIGURE 5.8 The tree structures for comparison.

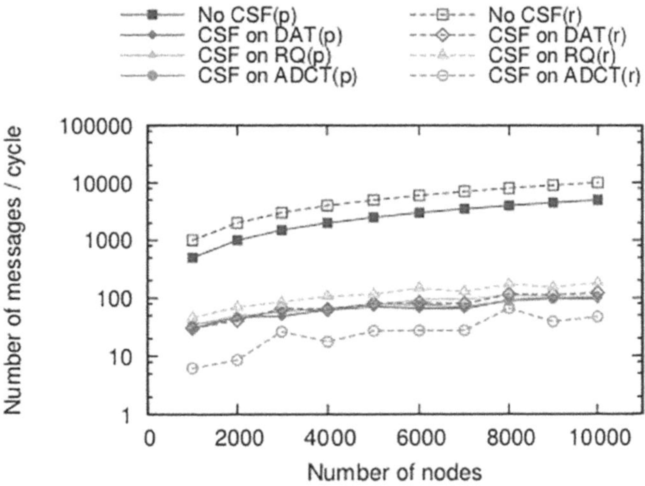

FIGURE 5.9 Number of messages ($|B| = 2$).

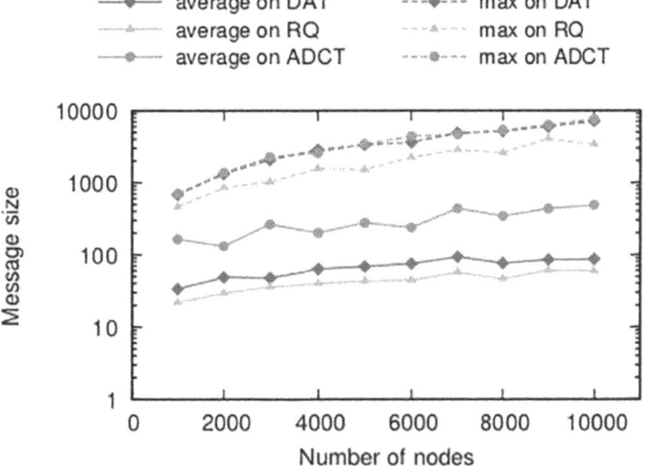

FIGURE 5.10 Message size on r ($|B| = 2$).

nearly 98%–99% of the number of messages was reduced on both r and p. When $|B|$ increased, the load on p was reduced as expected.

Figures 5.10 and 5.13 show the maximum and average size of the messages received on r. Though the maximum size of the messages exhibits a similar tendency for all tree structures, the average size of the message on ADCT becomes larger than other structures, especially when $|B|$ is small.

Figures 5.11 and 5.14 show the network occupation time to receive messages on r and p. The performance using CSF seems similar in each tree structure. Compared with the structure without CSF, about 90% of the network occupation time was reduced on both r and p. When CSF is not used, about 350 ms was consumed on r.

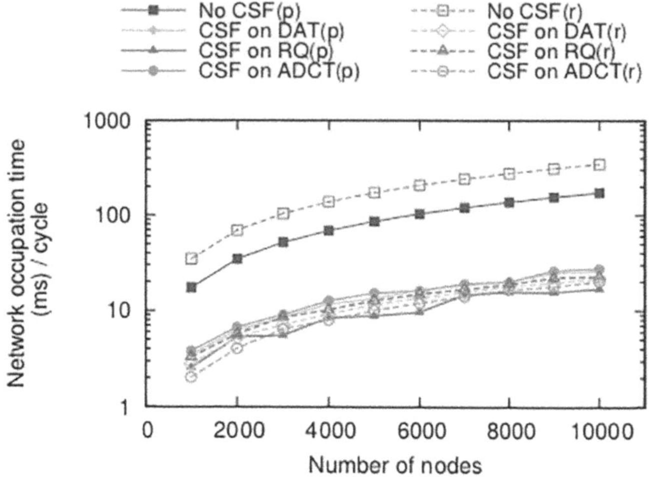

FIGURE 5.11 Network occupation time ($|B| = 2$).

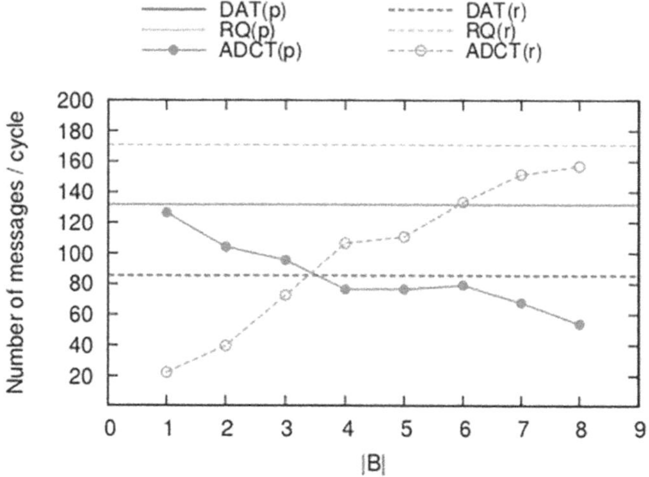

FIGURE 5.12 Number of messages (10,000 publishers).

It is too long as a network process since the result was only for one topic, but as a TBPS system, multiple topics need to be handled.

The network occupation time on p became lower than other methods when $|B|$ is larger than 4. The results show that the boundary node adjustment protocol can properly control the network load. On the other hand, though the number of messages on r was increased, the network occupation time increased slightly. That is because the average size of ADCT is kept as large as other methods, even when $|B|$ is increased as shown in Figure 5.13.

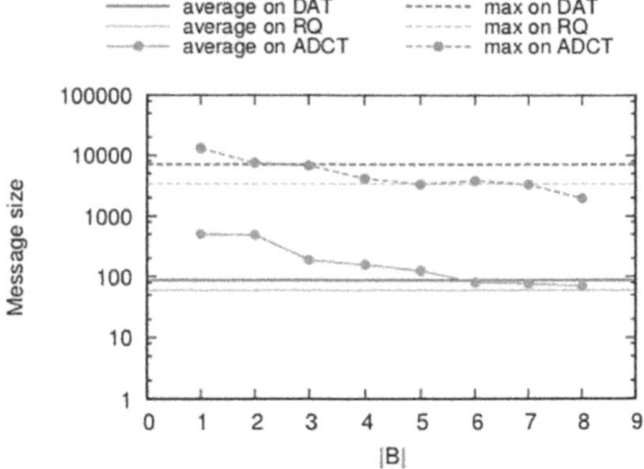

FIGURE 5.13 Message size on *r* (10,000 publishers).

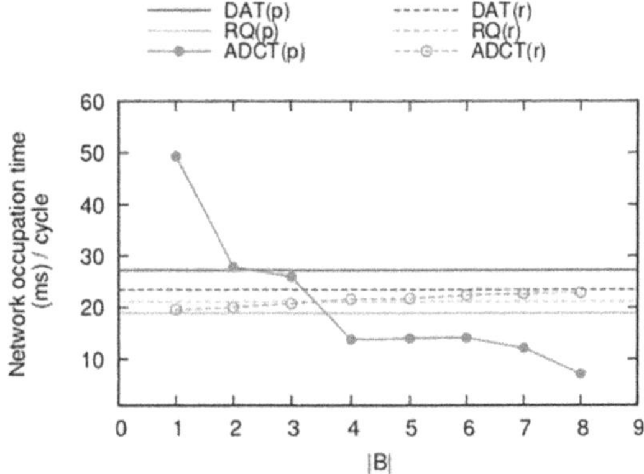

FIGURE 5.14 Network occupation time (10,000 publishers).

5.3 DATA COLLECTION SCHEME CONSIDERING PHASE DIFFERENCES

In this section, we describe a data collection scheme considering phase differences.

5.3.1 Problems Addressed

In this section, we present the addressed problems for our data collection scheme considering phase differences.

5.3.1.1 Assumed Environment

The purpose of this study is to disperse the communication load in the sensor stream collections that have different collection cycles. The source nodes have sensors so as to gain sensor data periodically. The source nodes and collection node (sink node) of those sensor data construct P2P (Peer to Peer) networks. The sink node searches source nodes and requires a sensor data stream with those collection cycles in the P2P network. Upon reception of the query from the sink node, the source node starts to deliver the sensor data stream via other nodes in the P2P network. The intermediate nodes relay the sensor data stream to the sink node based on the routing tables.

5.3.1.2 Input Setting

The source nodes are denoted as N_i ($i = 1, \ldots, n$), and the sink node of sensor data is denoted as S. In addition, the collection cycle of N_i is denoted as C_i.

In Figure 5.15, each node indicates a source node or sink node, and the branches indicate collection paths for the sensor data streams. Concretely, they indicate communication links in an application layer. The branches are indicated by dotted lines because there is a possibility that the branches may not collect a sensor data stream depending on the collection method. The sink node S is at the top, and the four source nodes $N_1, \ldots, N_4$ ($n = 4$) are at the bottom. The figure in the vicinity of each source node indicates the collection cycle, and $C_1 = 1$, $C_2 = 2$, $C_3 = 2$, and $C_4 = 3$. This corresponds to the case where a live camera acquires an image once for every second, and N_1 records the image once every second, N_2 and N_3 record the image once every 2 s, and N_4 records the image once every 3 s, for example. Table 5.3 shows the collection cycle of each source node and the sensor data to be received in the example in Figure 5.15. The purpose of this study is to disperse the communication load in the sensor stream collections that have different collection cycles. The source nodes have sensors so as to gain sensor data periodically. The source nodes and the collection node (sink node) of those sensor data construct P2P networks. The sink node searches source nodes and requires a sensor data stream with those collection cycles in the P2P network. Upon reception of the query from the sink node, the source node starts to deliver the sensor data stream via other nodes in the P2P network. The intermediate nodes relay the sensor data stream to the sink node based on their routing tables.

5.3.1.3 Definition of a Load

The communication load of the source nodes and sink node is given as the total of the load due to the reception of the sensor data stream and the load due to the transmission. The communication load due to reception is referred to as the reception load, the reception load of N_i is I_i, and the reception load of S is I_0. The communication load due to the transmission is referred to as the transmission load, the transmission load of N_i is O_i, and the transmission load of S is O_0.

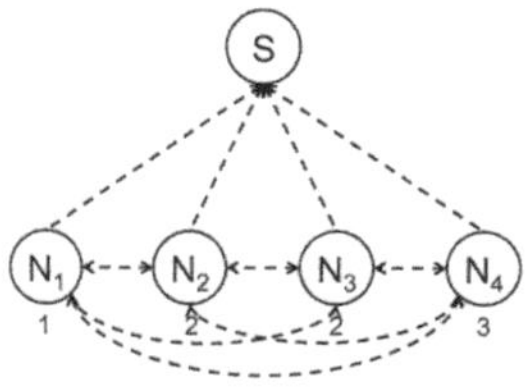

FIGURE 5.15 An example of input setting.

TABLE 5.3
An Example of the Sensor Data Collection

Time	N_1 (Cycle = 1)	N_2 (Cycle = 2)	N_3 (Cycle = 2)	N_4 (Cycle = 3)
0	*	*	*	*
1	*			
2	*	*	*	
3	*			*
4	*	*	*	
5	*			
6	*	*	*	*
7	*			
…	…	…	…	…

In many cases, the reception load and the transmission load are proportional to the number of sensor data pieces per unit hour of the sensor data stream to be sent and received. The number of pieces of sensor data per unit hour of the sensor data stream that is to be delivered by N_p to N_q ($q \neq p$; p, $q = 1, \ldots, n$) is $R(p, q)$, and the number delivered by S to N_q is $R(0, q)$.

5.3.2 Proposed Method

In this section, we present our SG-based method considering phase differences.

5.3.2.1 Skip Graphs

In this study, we assume an overlay network for the SG-based TBPS, such as Banno et al. [10].

SGs are overlay networks where the skip lists are applied in the P2P model [9]. Figure 5.16 shows the structure of an SG. In Figure 5.16, squares show entries of routing tables on peers (nodes), and the number inside each square shows a key of the peer. The peers are sorted in ascending order by those keys, and bidirectional links are created among the peers. The numbers below the entries are called

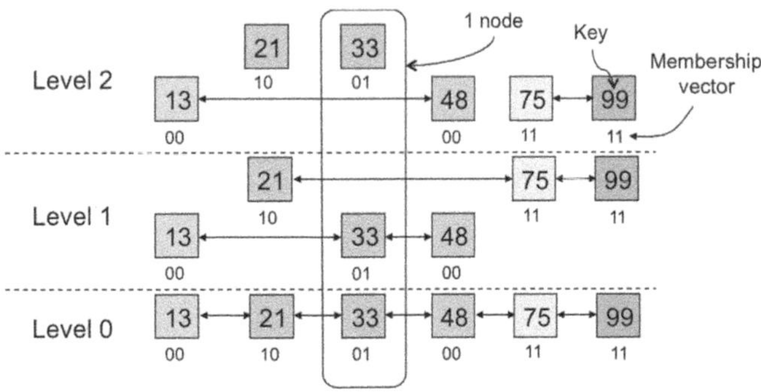

FIGURE 5.16 A structure of an SG.

"membership vector." The membership vector is an integral value assigned to each peer when the peer joins. Each peer creates links to other peers on multiple levels based on the membership vector. In SGs, queries are forwarded by higher level links to other peers when a single key and its assigned peer are searched. This is because higher level links can efficiently reach the searched key with less hops than the lower level links. In case of range queries that specify the beginning and end of keys to be searched, the queries are forwarded to the peer whose key is within the range or less than the end of the range. The number of hops to key search is represented to O(log n) when n is denoted as the number of peers. In addition, the average number of links on each peer is represented as log n.

5.3.2.2 Phase Differences

Currently, we have proposed a large-scale data collection schema for distributed TPBS [12]. In [12], we employ "CSF" that stores and merges multiple small size messages into a large message along a multihop tree structure on a structured overlay for TBPS, taking into account the delivery time constraints. This makes it possible to reduce the overhead of network process even when a large number of sensor data is published asynchronously. In addition, we have proposed a collection system considering phase differences [13]. In the proposed method, the phase difference of the source node N_i is denoted as d_i ($0 \leqq d_i < C_i$). In this case, the collection time is represented as $C_i p + d_i$ ($p = 0, 1, 2, \ldots$). Table 5.4 shows the time to collect data in the case of Figure 5.15, where the collection cycle of each source node is 1, 2, or 3. By considering phase differences like Table 5.4, the collection time is balanced within each collection cycle, and the probability of load concentration to the specific time or node is decreased. Each node sends sensor data at the time base on his collection cycle and phase difference, and other nodes relay the sensor data to the sink node. In this chapter, we call phase differences as "PS." Figures 5.17 and 5.18 show an example of the data forwarding paths on SGs without and with PS, respectively.

5.3.3 Evaluation

In this section, we describe the evaluation of the proposed SG-based method with PS by simulation.

TABLE 5.4
An Example of the Collection Time Considering Phase Differences

Cycle	Phase Difference	Collect Time
1	0	0, 1, 2, 3, 4, …
2	0	0, 2, 4, 6, 8, …
	1	1, 3, 5, 7, 9, …
3	0	0, 3, 6, 9, 12, …
	1	1, 4, 7, 10, 13, …
	2	2, 5, 8, 11, 14, …

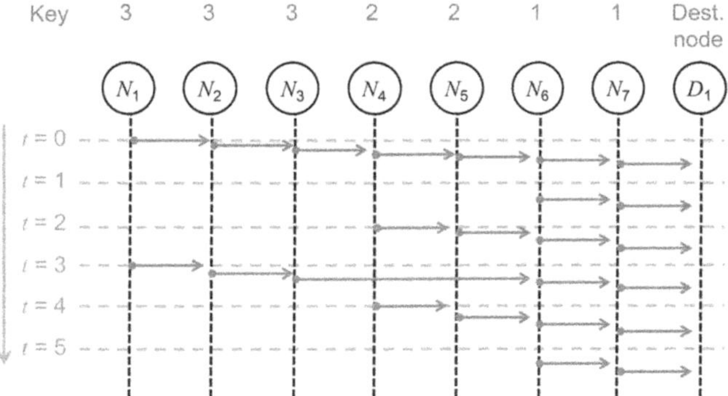

FIGURE 5.17 An example of the SG-based method without PS.

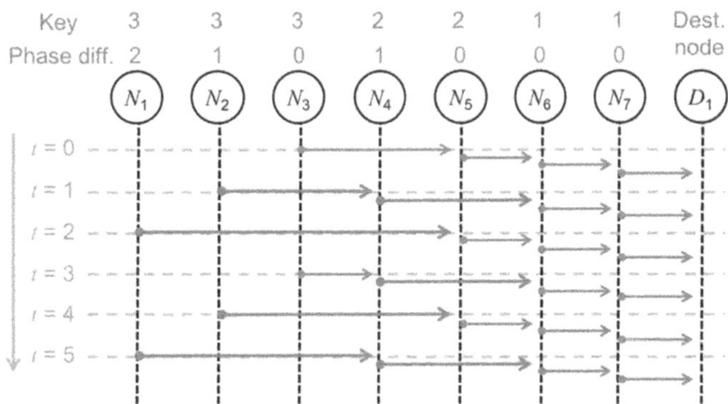

FIGURE 5.18 An example of the SG-based method with PS.

5.3.3.1 Collection Target Nodes

Table 5.5 shows the simulation environments. We evaluate our proposed system in two environments by a combination of collection cycles. The number of the sink node is one, and the collection cycle C_i and phase difference d_i of each node is determined at random. In the simulations, we measure the number of nodes targeted to collect data from time 0 to 99 and compare the results with the case of not considering phase differences.

Figures 5.19 and 5.20 show the results of the number of nodes targeted to collect data from time 0 to 99. The horizontal axis shows the time, and vertical axis shows

TABLE 5.5
Simulation Environments

Environment	No. of Nodes	Cycles
1	1,000	1, 2, 3
2	1,000	1, 2, …, 10

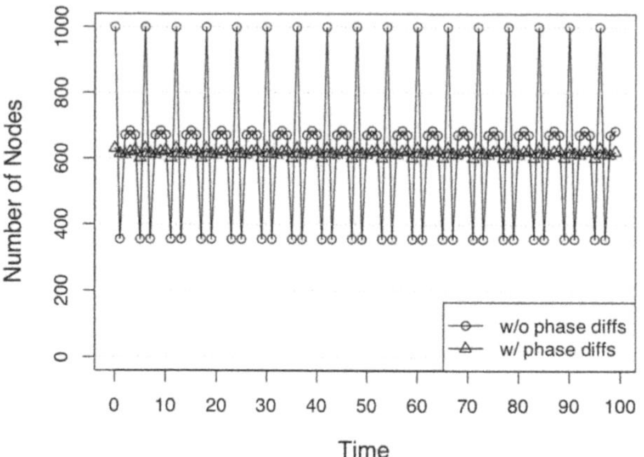

FIGURE 5.19 The number of related nodes in environment 1.

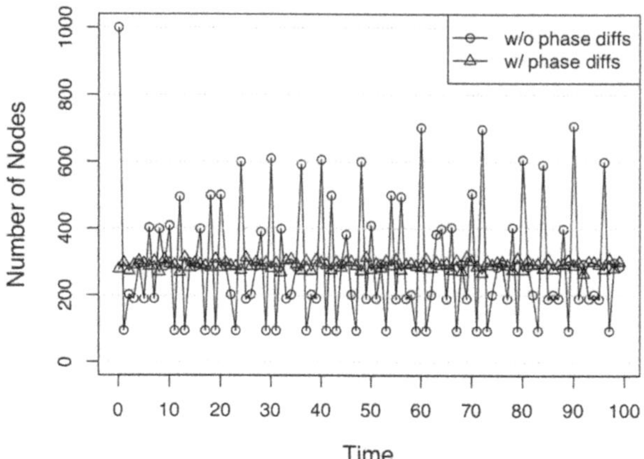

FIGURE 5.20 The number of related nodes in environment 2.

the number of targeted nodes at each time. In the simulation environment 1, shown by Figure 5.19, the case of not considering phase differences collects data from all 1,000 nodes at time 0, 6, 12, …, 96. This is because the collection cycle is 1, 2, or 3, and the lowest common multiple is 6. At other times in the case of not considering phase differences, the number of nodes extremely and constantly increases/decreases. On the other hand, the collection time is shifted by the phase difference in our proposed system, and the number of nodes is probabilistically equalized each time if the phase difference of each node is determined at random. Therefore, the probability of load concentration is decreased. Also in the simulation environment 2 shown by Figure 5.20, our proposed system can cause high balancing similar to the results in the simulation environment 1, while the case of not considering phase differences changes the number of nodes complexly by a combination of cycles from 1 to 10.

5.3.3.2 Communication Loads and Hops

In this simulation environment, the collection cycle of each source node denoted as C_i is determined at random between 1 and 10. The simulation time denoted as t is related to the combination of collection cycles and is between 0 and 2,519. In addition, this simulation has no communication delays among nodes although there are various communication delays in the real world. As comparison methods, we compare the proposed method with SG-based method without PS shown in Figure 5.15, the method in which all source nodes send data to the destination node directly (Source Direct), and the method in which all source nodes send data to the next node for the destination node (Daisy Chain, DC). Figures 5.21 and 5.22 show an example of SD and DC with PS, respectively.

Figures 5.23 and 5.24 show the results for the maximum instantaneous load and total load of nodes by the number of nodes, respectively. The number of node is the value on the lateral axis, and the allowable number of stream aggregation is under 11. Figure 5.23, the proposed method, SGs with PS, has a lower instantaneous load compared with SD-based methods, where the destination node receives data directly from the source node. Although the larger the allowable number of stream

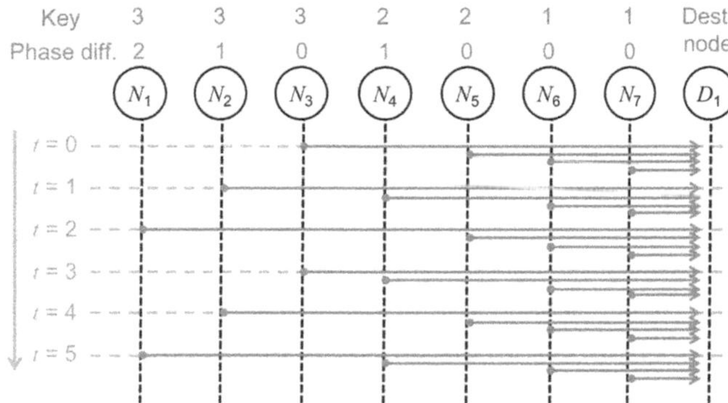

FIGURE 5.21 An example of Server Direct (SD) method.

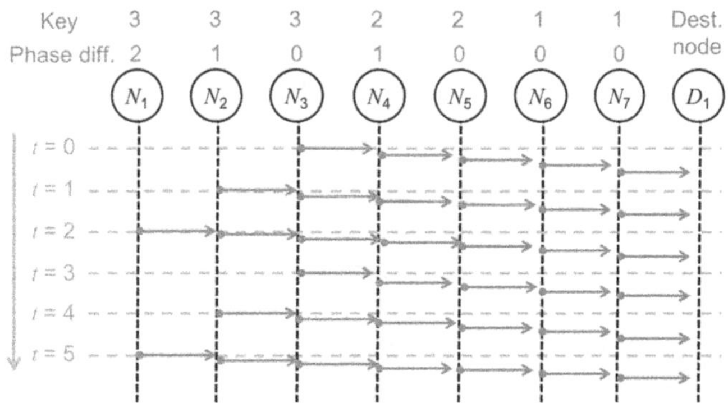

FIGURE 5.22 An example of DC method.

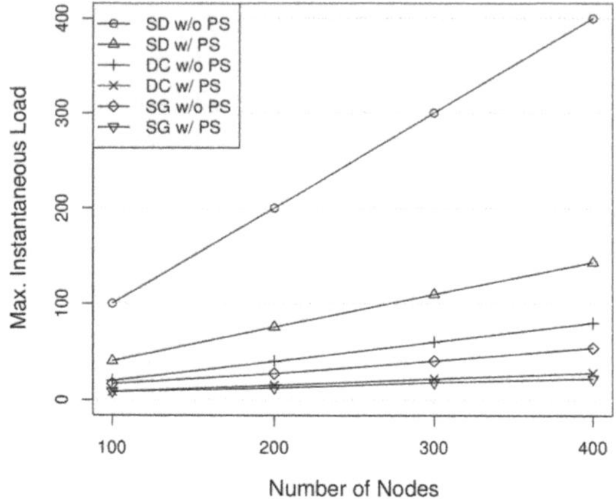

FIGURE 5.23 The maximum instantaneous load by the number of nodes.

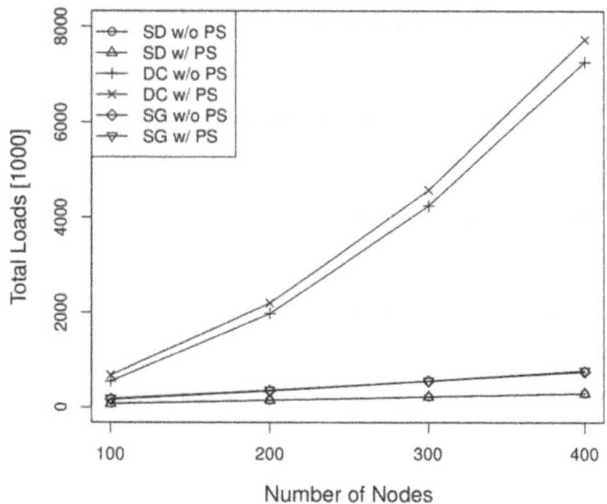

FIGURE 5.24 The total loads by the number of nodes.

aggregation in DC-based methods, the smaller the number of transmission and reception. In this simulation environment, however, the proposed method has a lower instantaneous load than the results of DC-based methods. In addition, the proposed method has a lower instantaneous load compared with SG without PS, because each node has different transmission and reception timing by its phase difference even if another node is configured in the same collection cycle. In Figure 5.24, on the other hand, SD-based methods have the lowest total loads. However, the proposed method has lower total loads compared with DC-based methods in this simulation environment.

Similar to the results for the maximum instantaneous load and total loads of nodes, Figures 5.25 and 5.26 show the results for the average number and maximum number of hops by the number of nodes under 11 stream aggregation, respectively. In Figures 5.25 and 5.26, SD-based methods have only one hop as the average

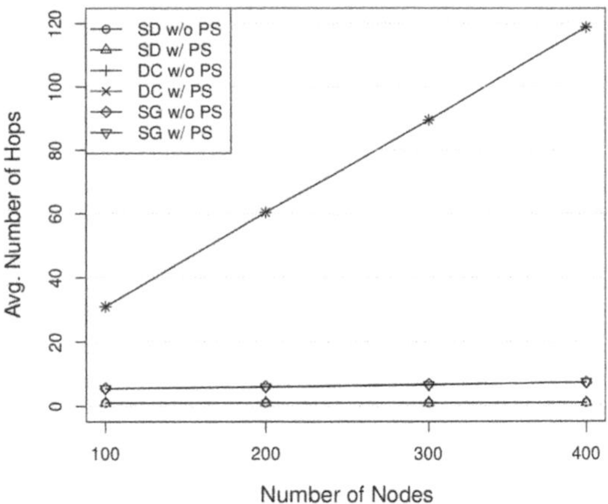

FIGURE 5.25 The average hops by the number of nodes.

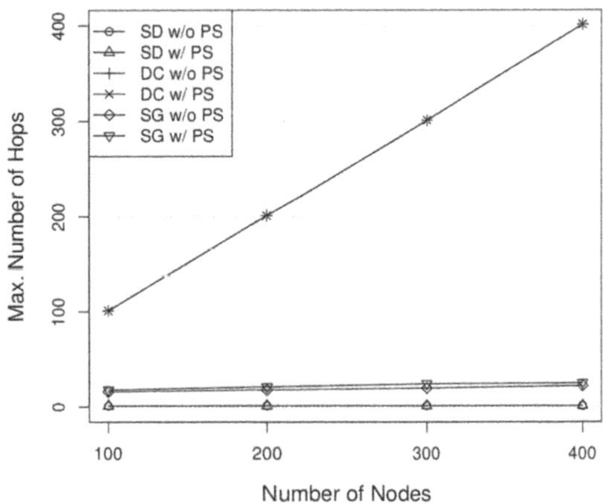

FIGURE 5.26 The maximum hops by the number of nodes.

number and maximum number, although those instantaneous loads described in Figure 5.23 are high. The proposed method has log n as the average number of hops, while n is denoted as the number of nodes and DC-based methods are affected linearly by n.

Figures 5.27 and 5.28 show the results for the maximum instantaneous load and total loads of nodes by the allowable number of stream aggregation, respectively. The allowable number of stream aggregation is the value on the lateral axis, and the number of node is 200. SD-based methods have a constant value as the maximum instantaneous load not affected by the allowable number of stream aggregation, because the source nodes send data to the destination node directly. In Figures 5.27 and 5.28, most of the results decrease by an increase of the allowable number of stream aggregation. The proposed method, SG with PS, has lower results for both of the maximum instantaneous load and total load even in the

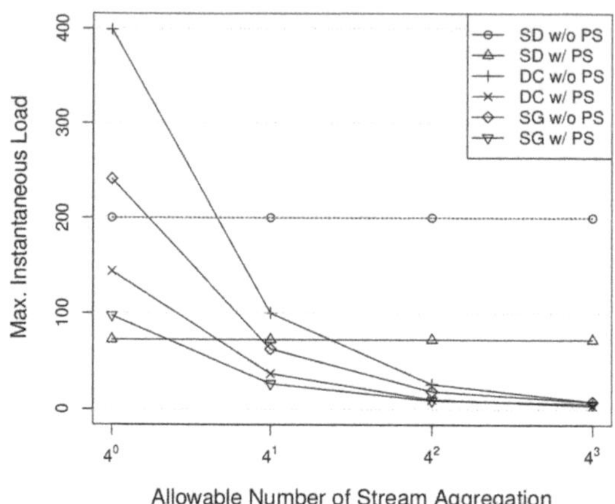

FIGURE 5.27 The maximum instantaneous load by the allowable number of stream aggregation.

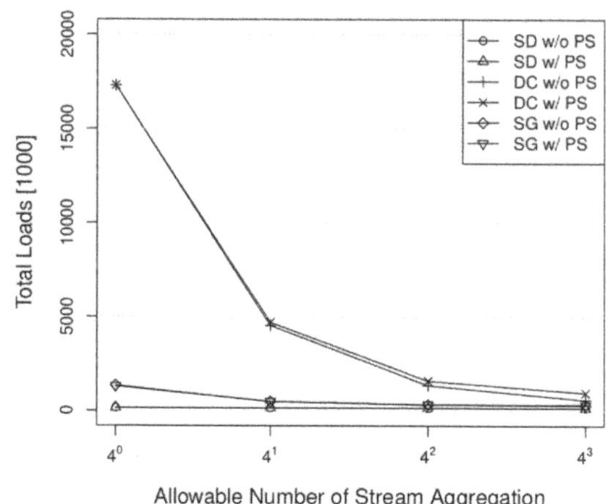

FIGURE 5.28 The total loads by the allowable number of stream aggregation.

realistic situation, 4^1 stream aggregation, compared with DC-based methods that require many stream aggregation to reduce those loads. In addition, the average number and maximum number of hops are the same to the results of 200 nodes in Figure 5.25 and 5.26, because they are unaffected by the allowable number of stream aggregation.

5.4 DISCUSSION

We described the data collection scheme for distributed TBPS and PS approach in Sections 2 and 3, respectively. Our experimental results show that our proposed method can reduce the loads of nodes and realize highly scalable systems to periodically collect distributed sensor data.

As the limitations of our current study, we assume that the pieces of data are not so different from each other. In the real world, however, various types of data are published at the same time, such as texts, images, and audio. Those pieces of data have different sizes and loads to be processed. We can clear this limitation by considering not only the number of data pieces (or transmission/reception) but also the types of them. Similar to the inconsideration of data types, the inconsideration of nodes' performances is another limitation of this study. We can clear this limitation by considering the nodes' performance, such as processing power, memory size, and network environment. In addition, this study has a limitation in a viewpoint of security or privacy. For example, private data are preferred to be sent to the subscriber via fewer nodes. Encryption of the data or communication is one of the common approaches, and arrangement of the data forwarding paths considering security/privacy is another solution to clear this limitation, e.g., the forwarding paths are directly connected to those valid subscribers for private data.

5.5 RELATED WORK

Related to the distributed stream data collection, various techniques have been proposed to disperse the communication loads for stream delivery [19].

P2P stream delivery techniques have been proposed to use a P2P architecture and disperse the communication loads among the processing computers (nodes) [20–24]. The P2P stream delivery techniques are divided into pull type and push type. In the pull-type technique such as PPLive, DONet [20], and SopCast, the reception nodes request data from other nodes and receive them. The reception nodes find the nodes that have not yet been receiving the requested data, and hence redundant communications do not occur. In the push-type techniques such as AnySee, data are sent from the transmission node to other nodes [21]. The transmission nodes find the nodes that have not yet received the requested data, and redundant communications do not occur. Also, the techniques combining pull type and push type have been proposed, such as PRIME [22].

Data delivery path construction techniques been proposed, as a multicast tree has to prevent the concentration of communication loads to the specific node [25–29]. In the ZIGZAG method, nodes construct clusters, and the multicast tree is constructed by the clusters [25]. The number of clusters included in each depth of a multicast tree is made the same, and thus, the loads are dispersed. Multicast trees are constructed only from information gained in the application layer, and it is not necessary to understand the physical network structure.

In the MSMT (Mean Scheduled Maintenance Time)/MBST (Main Board Sleep Time) method, the concentrations of the communication loads are more prevented than the ZIGZAG method by considering the physical network structure [27]. However, the implementability of the MSMT/MBST method was poor because it is necessary to understand all the network structures between the nodes. In locality aware clustering (LAC), the loads are more dispersed than the ZIGZAG method by considering part of the nodes, though the physical network structure cannot be understood [28].

In the aforedescribed P2P stream delivery technique, the same data streams are assumed to be sent to a number of reception nodes. In the delivery of sensor data streams, however, the same sensor data stream is assumed to have different delivery cycles to be delivered. In this case, those sensor data streams are delivered as different data streams for each delivery cycle. Thus, the communication loads cannot be efficiently dispersed. On the other hand, our proposed methods consider the different frequencies or cycles of each data stream and construct delivery paths to efficiently collect them.

As distributed stream data collection systems, an existing method to reduce the number of messages to receive data from large-scale nodes is to execute a range query on key order-preserving overlays. For example, " SFB" [30] is an efficient way to construct tree structures for range queries. The data collection from publishers on a subscriber in TBPS corresponds to the execution of range query for nodes that have keys with a topic. It can reduce the number of messages by merging responses from nodes along the reverse path of the query delivery tree structure. However, this method loses the asynchronous real-time feature of TBPS. The latest sensor data is not delivered until the subscriber executes a range query. Once the tree structure is constructed, it can be reused, but the periodic execution of range queries is needed to catch up with the joins/leaves of publishers and subscribers.

Some existing works address the "aggregation problem" on structured overlays. DAT [31] constructs a tree to aggregate data from distributed nodes using the Chord [32] overlay structure. DAT computes the aggregated value of all the local values, applying a given aggregation function on the distributed nodes. DAT can be used for data collections if the nodes execute the message merging function as an aggregation function. However, to merge and collect published messages, the publishers need to publish messages at exactly the same time, which is not a realistic assumption. Moreover, the aforementioned methods cause path concentration of data being forwarded on the nodes that are located close to the subscriber nodes on the overlay. In addition, once the tree structure is decided, it cannot be changed dynamically. As a result, a network process overload tends to occur on these nodes. On the other hand, our proposed ADCT method can construct a flexible collection tree and adaptively adjust the maximum overhead for the nodes to merge and forward messages.

5.6 CONCLUSION

The chapter gives information about the latest techniques of large-scale data collection schemes to the readers. We define continuous sensor data with different intervals (cycles) as a sensor data stream and described the techniques for TBPS systems called "CSF," "ADCT," and "PS." Our experiment results show that our proposed method can reduce the loads of nodes and realize highly scalable systems to periodically collect distributed sensor data. The scalability of the data collection systems is significantly important to accommodate a huge number of objects and encourage the growth of the data science field.

In future, we will try to clear the current limitations described in Section 5.4. More specifically, we will consider other information to determine the data forwarding paths such as data types, node performances, and security/privacy.

ACKNOWLEDGEMENTS

This work was supported by JSPS KAKENHI Grant Number 16K16059, 17K00146, 18K11316, Hoso Bunka Foundation, and I-O DATA Foundation.

REFERENCES

1. Steve Hodges, Stuart Taylor, Nicolas Villar, James Scott, Dominik Bial, and Patrick Tobias Fischer, Prototyping Connected Devices for the Internet of Things, *IEEE Computer*, Vol. 46, No. 2, pp. 26–34, Feb. 2013.
2. Patrick Th. Eugster, Pascal A. Felber, Rachid Guerraoui, and Anne-Marie Kermarrec, The Many Faces of Publish/Subscribe, *ACM Computing Surveys*, Vol. 35, No. 2, pp. 114–131, June 2003.
3. MQTT Version 3.1.1. http://docs.oasis-open.org/mqtt/mqtt/v3.1.1/mqtt-v3.1.1.pdf (accessed Jan. 31, 2019).
4. Advanced Message Queuing Protocol. www.amqp.org/ (accessed Jan. 31, 2019).
5. Nik Bessis and Ciprian Dobre, *Big Data and Internet of Things: A Roadmap for Smart Environments*, Studies in Computational Intelligence, Springer, Vol. 546, 2014, Berlin
6. Miguel Castro, Peter Druschel, Anne-Marie Kermarrec, and Antony Rowstron, SCRIBE: A Large-Scale and Decentralized Application-Level Multicast Infrastructure, *IEEE Journal on Selected Areas in Communications (JSAC)*, Vol. 20, No. 8, 2002, pp. 1489 - 1499
7. Sylvia Ratnasamy, Mark Handley, Richard M. Karp, and Scott Shenker, Application-Level Multicast Using Content-Addressable Networks, in *Proceedings of the* 3rd *International COST264 Workshop on Networked Group Communication (NGC* 2001*)*, pp. 14–29, Nov. 2001, Seoul, Korea
8. Fatemeh Rahimian, Sarunas Girdzijauskas, Amir H. Payberah, and Seif Haridi, Vitis: A Gossip-Based Hybrid Overlay for Internet-Scale Publish/Subscribe Enabling Rendezvous Routing in Unstructured Overlay Networks, in *Proceedings of the* 25th *IEEE International Parallel and Distributed Processing Symposium (IPDPS* 2011*)*, pp. 746–757, May 2011.
9. James Aspnes and Gauri Shah, Skip Graphs, *ACM Transactions on Algorithms (TALG)*, Vol. 3, No. 4 (37), pp. 1–25, Nov. 2007.
10. Ryohei Banno, Susumu Takeuchi, Michiharu Takemoto, Tetsuo Kawano, Takashi Kambayashi, and Masato Matsuo, Designing Overlay Networks for Handling Exhaust Data in a Distributed Topic-based Pub/Sub Architecture, *Journal of Information Processing (JIP)*, Vol. 23, No. 2, pp. 105–116, Mar. 2015.
11. Yuuichi Teranishi, Ryohei Banno, and Toyokazu Akiyama, Scalable and Locality-Aware Distributed Topic-Based Pub/Sub Messaging for IoT, in *Proceedings of the* 2015 *IEEE Global Communications Conference (GLOBECOM* 2015*)*, pp. 1–7, Dec. 2011, San Diego, CA, USA
12. Yuuichi Teranishi, Tomoya Kawakami, Yoshimasa Ishi, and Tomoki Yoshihisa, A Large-Scale Data Collection Scheme for Distributed Topic-Based Pub/Sub, in *Proceedings of the* 2017 *International Conference on Computing, Networking and Communications (ICNC* 2017*)*, 6 pages, Jan. 2017.
13. Tomoya Kawakami, Yoshimasa Ishi, Tomoki Yoshihisa, and Yuuichi Teranishi, A Skip Graph-Based Collection System for Sensor Data Streams Considering Phase Differences, in *Proceedings of the* 8th *International Workshop on Streaming Media Delivery and Management Systems (SMDMS* 2017*) in Conjunction with the* 12th *International Conference on P2P, Parallel, Grid, Cloud and Internet Computing (3PGCIC* 2017*)*, pp. 506–513, Nov. 2017.

14. Thorsten Schütt, Florian Schintke, and Alexander Reinefeld, Range Queries on Structured Overlay Networks, *Computer Communications*, Vol. 31, No. 2, pp. 280–291, Feb. 2008.
15. Tomoya Kawakami, Tomoki Yoshihisa, and Yuuichi Teranishi, A Load Distribution Method for Sensor Data Stream Collection Considering Phase Differences, in *Proceedings of the 9th International Workshop on Streaming Media Delivery and Management Systems (SMDMS 2018) in Conjunction with the 13th International Conference on P2P, Parallel, Grid, Cloud and Internet Computing (3PGCIC 2018)*, pp. 357–367, Oct. 2018, Taichung, Taiwan
16. Kazuyuki Shudo, Collective Forwarding on Structured Overlays, *IPSJ Transactions on Advanced Computing Systems*, Vol. 2, No. 3, pp. 39–46, Mar. 2009 (In Japanese).
17. Y. Teranishi, Y. Saito, S. Murono, and N. Nishinaga, JOSE: An Open Testbed for Field Trials of Large-Scale IoT Services, *NICT Journal*, Vol. 6, No. 2, pp. 151–159, Mar. 2016.
18. Y. Teranishi, PIAX: Toward a Framework for Sensor Overlay Network, in *Proceedings of CCNC 2009*, pp. 1–5, 2009, Las Vegas, US
19. Zhijie Shen, Jun Luo, Roger Zimmermann, and Athanasios V. Vasilakos, Peer-to-Peer Media Streaming: Insights and New Developments, *Proceedings of the IEEE*, Vol. 99, No. 12, pp. 2089–2109, Oct. 2011.
20. Xinyan Zhang and Jiangchuan Liu and Bo Li and Tak-Shing Peter Yum, CoolStreaming/DONet: A Data-Driven Overlay Network for Peer-to-Peer Live Media Streaming, in *Proceedings of the 24th Annual Joint Conference of the IEEE Computer and Communications Societies (INFOCOM 2005)*, pp. 2102–2111, Mar. 2005, Miami, US
21. Xiaofei Liao, Hai Jin, Yunhao Liu, Lionel M. Ni, and Dafu Deng, AnySee: Peer-to-Peer Live Streaming, in *Proceedings of the 25th IEEE International Conference on Computer Communications (INFOCOM 2006)*, pp. 1–10, Apr. 2006.
22. Nazanin Magharei and Reza Rejaie, PRIME: Peer-to-Peer Receiver-Driven Mesh-Based Streaming, in *Proceedings of the 26th IEEE International Conference on Computer Communications (INFOCOM 2007)*, pp. 1415–1423, May 2007.
23. Linchen Yu, Xiaofei Liao, Hai Jin, and Wenbin Jiang, Integrated Buffering Schemes for P2P VoD Services, *Peer-to-Peer Networking and Applications*, Vol. 4, No. 1, pp. 63–74, 2011.
24. Suguru Sakashita, Tomoki Yoshihisa, Takahiro Hara, and Shojiro Nishio, A Data Reception Method to Reduce Interruption Time in P2P Streaming Environments, in *Proceedings of the 13th International Conference on Network-Based Information Systems (NBiS 2010)*, pp. 166–172, Sept. 2010, Takayama, Gifu Japan
25. Ashwin R. Bharambe, Mukesh Agrawal, and Srinivasan Seshan, Mercury: Supporting Scalable Multi-Attribute Range Queries, in *Proceedings of the ACM Conference on Applications, Technologies, Architectures, and Protocols for Computer Communications (SIGCOMM 2004)*, pp. 353–366, Aug. 2004.
26. Duc A. Tran, Kien A. Hua, and Tai Do, ZIGZAG: An Efficient Peer-to-Peer Scheme for Media Streaming, in *Proceedings of the 22nd Annual Joint Conference of the IEEE Computer and Communications Societies (INFOCOM 2003)*, pp. 1283–1292, Mar. 2003.
27. Xing Jin, W.-P. Ken Yiu, S.-H. Gary Chan, and Yajun Wang, On Maximizing Tree Bandwidth for Topology-Aware Peer-to-Peer Streaming, *IEEE Transactions on Multimedia*, Vol. 9, No. 8, pp. 1580–1592, Dec. 2007.
28. Kanchana Silawarawet and Natawut Nupairoj, Locality-Aware Clustering Application Level Multicast for Live Streaming Services on the Internet, *Journal of Information Science and Engineering*, Vol. 27, No. 1, pp. 319–336, 2011.

29. Tien Anh Le and Hang Nguyen, Application-Aware Cost Function and Its Performance Evaluation over Scalable Video Conferencing Services on Heterogeneous Networks, in *Proceedings of the IEEE Wireless Communications and Networking Conference: Mobile and Wireless Networks (WCNC 2012 Track 3 Mobile and Wireless)*, pp. 2185–2190, Apr. 2012, Paris, France
30. Ryohei Banno, Tomoyuki Fujino, Susumu Takeuchi, and Michiharu Takemoto, SFB: A Scalable Method for Handling Range Queries on Skip Graphs, *IEICE Communications Express*, Vol. 4 (2015), No. 1, pp. 14–19, Feb. 2015.
31. Min Cai and Kai Hwang, Distributed Aggregation Algorithms with Load-Balancing for Scalable Grid Resource Monitoring, in *Proceedings of the 21st IEEE International Parallel and Distributed Processing Symposium (IPDPS 2007)*, pp. 1–10, Mar. 2007, Long Beach, California
32. Ion Stoica, Robert Morris, David Liben-Nowell, David R. Karger, M. Frans Kaashoek, Frank Dabek, and Hari Balakrishnan, Chord: A Scalable Peer-to-Peer Lookup Protocol for Internet Applications, *IEEE/ACM Transactions on Networking*, Vol. 11, No. 1, pp. 17–32, Feb. 2003.

Part II

Data Design and Analysis

6 Big Data Analysis and Management in Healthcare

R. Dhaya and M. Devi
King Khalid University

R. Kanthavel and Fahad Algarni
University of Bisha

CONTENTS

6.1 INTRODUCTION

The guarantee of big data has gotten incredible attention in healthcare as it inquires about after medication revelation, treatment development, customized prescription, and ideal patient care that can lessen cost and enhance quiet results. Billions of dollars have been contributed to capture data laid out in big activities that are frequently detached. The Continuously Learning Healthcare System is additionally being supported by the Institute of Medicine to close the hole between logical disclosure, patient and clinician commitment, and clinical practice. Be that as it may, the big data guarantee has not yet been acknowledged to its potential as the minor accessibility of data does not convert into information or clinical practice. In addition, because of the variety in data unpredictability and structures, inaccessibility of computational advancements, and worries of sharing private patient data, few ventures of extensive clinical data sets are made accessible to scientists as a rule [1].

The idea of "big data" isn't new; the manner in which it is characterized is continually evolving. Different endeavors at characterizing big data basically portray it as an accumulation of data components, whose measure, speed, type, as well as multifaceted nature expect one to look for, receive, and create new equipment and programming systems, with the end goal to effectively store, break down, and imagine data [2]. Healthcare is a prime case of how the three V's of data – velocity, variety, and volume – are an inborn part of the data it produces. This data is spread among numerous healthcare frameworks, health backup plans, specialists, government substances, etc. Moreover, every one of these data archives is soloed and innately unequipped for giving a stage to worldwide data straightforwardness. To add to the three V's, which is shown in Figure 6.1, the veracity of healthcare data is likewise basic for its important use towards creating inspirational research.

Due to the characteristic complexities of healthcare data, there is a potential advantage in creating and actualizing big data arrangements inside this domain. Recorded ways to deal with medicinal research have commonly centered around the examination of malady states dependent on the adjustments in physiology as a restricted perspective of a certain particular methodology of data. Although the way to deal with understanding infections is fundamental, inquiring about it at this dimension quietens the variety and disconnectedness that characterize the genuine basic therapeutic instruments [3]. Following traditional (mechanical) procedures, the field of medication has started to adjust to the present advanced data age. New innovations make it conceivable to catch tremendous measures of data about every individual patient over an extensive timescale. Be that as it may, regardless of the approach of restorative hardware, the data caught and accumulated from these patients has remained immeasurably underutilized and squandered along these lines. Figure 6.2 shows the transition of care in hospital systems, pharmacy, EMS (Emergency Medical Services), healthcare department, and community support.

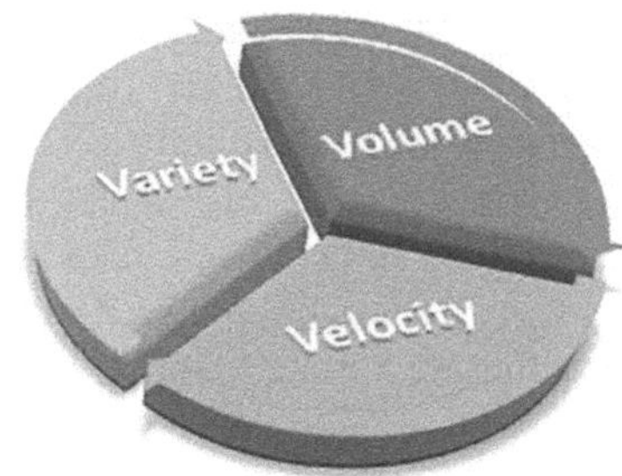

FIGURE 6.1 The three V's of data.

FIGURE 6.2 Transitions of care.

The loop behind an effective healthcare system is completely dependent on the close monitoring and aid of the hospital to identify diseases, extending support from the community, prescribing proper medications by the physician that can be had from the pharmacy, and providing emergency medical service during the golden hours. Thus, the transition of care has been bounded towards the analysis of healthcare data in every aspect.

Critical physiological and pathophysiological marvels are simultaneously showed as changes over different clinical streams. Along these lines, understanding and anticipating illnesses require an amassed methodology where organized and unstructured data arising from a heap of clinical and nonclinical modalities are used for a more complete point of view of the infection states [4]. A part of healthcare investigation that has as of late picked up footing is in tending to a portion of the developing agonies in presenting ideas of big data examination to medication. Specialists are contemplating the mind-boggling nature of healthcare data as far as the two attributes of the data itself and the scientific classification of examination that can be genuinely performed on them.

The structure of this book chapter has been organized into four main segments. First, the relevant research articles with regard to Big Data Analysis and Management in Healthcare have been put together as preliminary studies. Second, healthcare data and its classifications are analyzed, and in the third section, focus is given to the need of big data analysis in healthcare and its various challenges. The fourth section illustrates the collection and analysis of healthcare data in a detailed manner. The fifth part of this chapter discusses the management of healthcare data. Finally, the applications of big data in healthcare, conclusive remarks, and its future are elaborated in a lucid manner.

6.2 PRELIMINARY STUDIES

The relevant research papers on big data in healthcare applications have been surveyed in various streams that include basics of big data in healthcare, collection of data, and analysis of data and types.

Nada Elgendy and Ahmed Elragal (2014) analyzed various analytic methods and tools that can be embedded to big data, as well as the chances created by using big data analytics in several decision-making areas. M. D. Anto Praveena and B. Bharathi (2017) provided an overview of the big data analytics, issues, challenges, and different technologies that pertains to big data. E. A. Archenaaa and Mary Anitab (2015) gave an insight of how we can remove excess value from the information produced by healthcare and public sector offices. However, without genuine data analytic methods, these data will be of no use. Linnet Taylor, Ralph Schroeder, and Eric Meyer (2014) outlined the places, including the prediction and "nowcasting" of economic trends and mapping and predicting influence in the context of marketing or labor market data.

C. Lakshmi and V. Nagendra Kumar (2016) presented the meaning and explanation of big data and define the big data challenges. They also presented a proper framework to decompose big data systems into four sequential modules, namely data generation, data acquisition, data storage, and data analytics to form a big data value chain. Ticiana L. Coelho da Silva, Regis P. Magalhaes, et al. (2018) aimed at providing organizations in the choosing of technologies or platforms more related to their analytic processes by extending a short review as per some categories of Big Data problems by processing (streaming and batch), storage, data integration, analytics, data governance, and monitoring. Tasleem Nizam and Syed Imtiyaz Hassan (2017) presented the architecture of various technologies that can be applied for handling Big Data and can be defined in addition to the applications of Big Data system representation. McKinsey surveyed (2012) a lot of enterprise systems to understand that big data is something to handle, but the issue they face now is how to apply it effectively. Jafar Raza Alam, Asma Sajid, Ramzan Talib, and Muneeb Niaz (2014) studied the key issues why these organizations are not starting their planning stage to implement and execute the big data strategy, because they do not understand enough about the big data and its benefits.

Suriya Begum and Kavya Sulegaon (2016) discussed the analysis of Big Data Analytics concepts and the existing techniques and tools like Hadoop for data security. Martin Sarnovsky, Peter Bednar, and Miroslav Smatana (2018) designed a framework that enlarges the concept of Healthcare Industrial internet of things (IoT) for principles of edge and cognitive computing to improve the quality of healthcare, improve security, and minimize costs for cloud services and network traffic in future health IoT environments. Parupudi Aditya Vineet Kumar (2018) presented to classify the literature further in the most suitable phase it can apply to improve the use of big data analytics for the relevant phases of design science that have not been elaborated properly.

Quek Kia Fatt and Amutha Ramadas (2018) discussed the applications and challenges of Big Data in Healthcare in applying big data in healthcare, especially in relation to privacy, security, standards, governance, integration of data, data accommodation, data classification, incorporation of technology, etc. Gesundheit Österreich Forschungs and Und Planungs GmbH (2016) studied Big Data in Public Health, Telemedicine, and Healthcare, which helps to determine applicable examples of the use of Big Data in health and create recommendations for their implementation in the European Union. Ashwin Belle, Raghuram Thiagarajan, et al. (2015) discussed about the major challenges involved, with a focus on three upcoming and promising areas of medical research: image, signal, and genomic-based analytics. Potential areas of research within this domain that have the provision to a meaningful effect on healthcare delivery are also analyzed. Sanskruti Patel and Atul Patel (2016) fundamentally studied the effect of implementing big data solutions in the healthcare sector, the potential opportunities, challenges and available areas, and tools to implement big data analytics.

Revanth Sonnati (2017) focused on the analysis of data and the strengths and drawbacks compared with the conventional techniques available. Lidong Wang and Cheryl Ann Alexander (2015) introduced the Big Data concept and characteristics, healthcare data, and some vital issues of Big Data. Muhammad Umer Sarwar et al. (2017) surveyed the Big Data Analytics in healthcare and highlighted the limitations of existing machine learning algorithms that are outlined for big data analysis in healthcare. R. Hermon and P. A. Williams (2014) studied a systemic review methodology to provide a classification of big data use in healthcare, and their outcomes represented that the natural classification is not clinical application based, rather it comes in four broad categories: administration and delivery, clinical decision support (CDS), consumer behavior, and support services. W. Raghupathi and V. Raghupathi (2014) addressed the question of biomedical and health analyzers working in analytics that Big Data needs to know.

P. Groves, B. Kayyali, D. Knott, and S. Van Kuiken (2013) analyzed the effective tools used for visualization of big data and the implementation of new visualization tools to create the big data in healthcare industry to know the processes and use of big data in healthcare management.

6.3 HEALTHCARE DATA

Various cases in healthcare are appropriate for big data. Some scholastic or research-centered healthcare foundations are either trying different things with big data or utilizing it in cutting edge to inquire about ventures. Those organizations draw upon data researchers, analysts, graduates, so as to wrangle the complexities of big data. In the accompanying areas, we'll address a portion of those complexities and what's being done to streamline big data to make it more available.

In healthcare, we do have huge volumes of data coming in. EMRs (electronic medical record) alone gather immense measures of data. The greater part of that data is gathered for recreational purposes, as indicated by Brent James of Intermountain Healthcare. Be that as it may, neither the volume nor the speed of data in healthcare is really sufficiently high to require big data today. Our work with health frameworks demonstrates that just a little portion of the table in an EMR database is significant to the current routine with regard to medication and its related investigation in utilizing cases. Along these lines, by far, most of the data accumulation in healthcare today could be viewed as recreational. In spite of the fact that data may have an incentive not far off as the quantity of utilization cases grows, there aren't numerous genuine utilize cases for quite a bit of that data today.

There is absolute assortment in the data, yet most frameworks fundamentally gather the same data objects with an incidental change to the model. So, new utilize cases supporting genomics will absolutely require a big data approach. Figure 6.3 shows the types of healthcare data that shows electronic healthcare data (EHD), administrative reports, claims data, patient or disease registry data, health survey data, and clinical trial data.

A type of healthcare data is EHD, which needs to be saved as an electronic health record (EHR) to denote an advanced adaptation of a patient's paper diagram. EHRs are ongoing, quite focused records that make data accessible right away and safely to approved clients. While an EHR contains the medicinal and treatment chronicles of patients, an EHR framework is worked to go past standard clinical information gathered in a supplier's office and can be comprehensive of a more extensive perspective of a patient's consideration. EHRs can contain a patient's medicinal history, analyze meds, treatment designs, inoculation dates, hypersensitivities, radiology pictures, and lab and test outcomes; enable access to confirm-based apparatuses that suppliers can use to settle on choices about a patient's consideration; and robotize

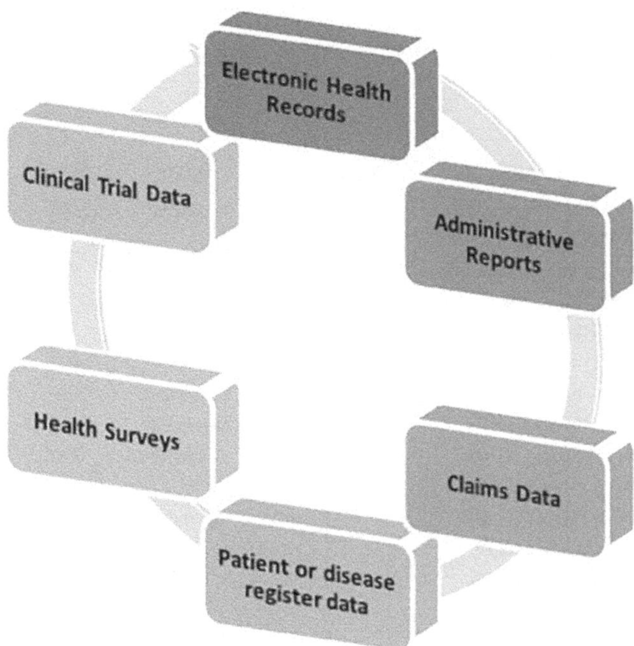

FIGURE 6.3 Data types of healthcare data.

and streamline supplier work process. One of the key highlights of an EHR is that health data can be made and overseen by approved suppliers in a computerized configuration fit for being imparted to different suppliers crosswise over more than one healthcare association. EHRs work to impart data to other healthcare suppliers and associations; for example, research facilities, pros, restorative imaging offices, drug stores, crisis offices, and school and working environment centers, and so they contain data from all clinicians engaged with a patient's consideration. The data set in Table 6.1 presents the age-adjusted death rates for the ten leading causes of death in the United States from 1999. (Courtesy: NCHS—Leading Causes of Death: United States, www.healthdata.gov [5].) The objective of presenting this data set is to categorize the disease name, reason for the disease, year of happenings, place in which the disease falls, number of deaths, and age-adjusted death rate to analyze sample healthcare data logically.

Data is based on information from all resident death certificates filed in the 50 states and in the District of Columbia using demographic and medical characteristics. Age-adjusted death rates (per 100,000 populations) are based on the 2000 U.S. standard population. Populations used for computing death rates after 2010 are postnasal estimates based on the 2010 census, estimated as of July 1, 2010. Rates for census years are based on populations enumerated in the corresponding censuses. Rates for noncensus years before 2010 are revised using updated intercostal population estimates and may differ from the rates previously published.

TABLE 6.1
Sample Healthcare Data Set

Year	Cause Name	Cause Name	State	Deaths	Age-Adjusted Death Rate
2016	Accidents (unintentional injuries) (V01-X59,Y85-Y86)	Unintentional injuries	Alabama	2755	55.5
2016	Accidents (unintentional injuries) (V01-X59,Y85-Y86)		Wyoming	371	61.9
2013	Alzheimer's disease (G30)	Alzheimer's disease	Alaska	72	18.9
2012	Alzheimer's disease (G30)			102	27.2
2002	Alzheimer's disease (G30)		Delaware	128	16.7
2016	Alzheimer's disease (G30)		District of Columbia	120	18.3
2012	Alzheimer's disease (G30)			129	20.5
2009	Alzheimer's disease (G30)		Rhode Island	321	21.7
2008	Alzheimer's disease (G30)			359	24.4
2007	Alzheimer's disease (G30)			328	22.2
2006	Alzheimer's disease (G30)			297	20.4
2005	Alzheimer's disease (G30)			298	20.9
2004	Alzheimer's disease (G30)			283	20.1
2003	Alzheimer's disease (G30)			303	22
2004	Malignant neoplasms (C00-C97)	Cancer	Vermont	1212	179.5
2003	Malignant neoplasms (C00-C97)			1210	183.6
2002	Malignant neoplasms (C00-C97)			1224	188.3
2001	Malignant neoplasms (C00-C97)			1249	196.4
2000	Malignant neoplasms (C00-C97)			1240	198
1999	Malignant neoplasms (C00-C97)			1255	203.7
2016	Malignant neoplasms (C00-C97)		Virginia	15027	156.1
2014	Malignant neoplasms (C00-C97)			14749	161.5
2013	Malignant neoplasms (C00-C97)			14414	162
2011	Malignant neoplasms (C00-C97)			14376	170.8
2016	Chronic lower respiratory diseases (J40-J47)	CLRD	Idaho	865	45.7
2015	Chronic lower respiratory diseases (J40-J47)			843	46.3
2014	Chronic lower respiratory diseases (J40-J47)			819	45.8
2013	Chronic lower respiratory diseases (J40-J47)			808	46.7
2012	Chronic lower respiratory diseases (J40-J47)			754	46
2011	Chronic lower respiratory diseases (J40-J47)			824	51.5
2010	Chronic lower respiratory diseases (J40-J47)			727	47

(*Continued*)

TABLE 6.1 (*Continued*)
Sample Healthcare Data Set

Year	Cause Name	Cause Name	State	Deaths	Age-Adjusted Death Rate
2008	Chronic lower respiratory diseases (J40-J47)			703	48.2
2006	Chronic lower respiratory diseases (J40-J47)			644	46.9
2005	Chronic lower respiratory diseases (J40-J47)			717	54.1

Courtesy: NCHS—Leading Causes of Death: United States.

6.4 NEED OF BIG DATA ANALYTICS IN HEALTHCARE

Expenses are a lot higher than they ought to be, and they have been ascending for as long as two decades. Basically, we need some savvy, data-driven reasoning here. What's more, current motivating forces are changing too: numerous insurance agencies are changing from charge-for-benefit intends to plans that organize understanding results [6].

Healthcare suppliers have no immediate motivating force to impart understanding of data to each other, which had made it harder to use the intensity of investigation. Since a greater amount of them are getting paid dependent on patient results, they have a money-related impetus to share data that can be utilized to enhance the lives of patients while cutting expenses for insurance agencies.

At last, doctor choices are winding up increasingly proof based, implying that they depend on huge swathes of research and clinical data instead of exclusive tutoring and professional assessment. As in numerous different enterprises, data social affair is getting bigger, and professionals require help in the issue. This new treatment state of mind implies that there is a more noteworthy interest for big data investigation in healthcare offices than at any time in recent history and the ascent of SaaS (Software as a service) devices is additionally noted.

Table 6.2 shows that the Kids' Inpatient Database (KID) is a set of pediatric hospital inpatient databases included in the Healthcare Cost and Utilization Project (HCUP) family. These databases are created by Agency for Healthcare Research and Quality (AHRQ) through a Federal-State–Industry partnership. The Nationwide Readmissions Database (NRD) is a unique and powerful database designed to support various types of analyses of national readmission rates for all payers and the uninsured. The NRD includes discharges for patients with and without repeat hospital visits in a year and those that have died in the hospital. Repeat stays may or may not be related. The criteria to determine the relationship between hospital admissions are left to the analyst using NRD. This database addresses a large gap in healthcare data—the lack of nationally representative information on

TABLE 6.2
HCUP Summary Statistics Report: KID 2016—Core File Means of Continuous Data Elements

Variable/Label	N	N Miss	Min.	Max.	Mean	Std. Dev.
HOSP_KID: KID hospital number	3,117,413	0	10,001	40,944	27,305.14	10,147.02
AGE: Age in years at admission	3,117,413	0	0.00	20.00	5.88	7.69
AGE_NEONATE: Neonatal age (first 28 days after birth) indicator	1,681,361	1,436,052	0.00	1.00	0.87	0.34
AMONTH: Admission month	3,116,797	616	1.00	12.00	6.51	3.48
AWEEKEND: Admission day is a weekend	3,117,403	10	0.00	1.00	0.21	0.41
DIED: Died during hospitalization	3,113,802	3,611	0.00	1.00	0.01	0.07
DISCWT: KID discharge weight	3,117,413	0	0.86	24.44	2.01	2.37
DISPUNIFORM: Disposition of patient (uniform)	3,113,802	3,611	1.00	99.00	1.34	1.87
DQTR: Discharge quarter	3,116,797	616	1.00	4.00	2.50	1.13
DRG: DRG in effect on discharge date	3,117,413	0	1.00	999.0	660.53	250.43
DRGVER: DRG grouper version used on discharge date	3,117,413	0	33.00	34.00	33.25	0.44
DRG_NoPOA: DRG in use on discharge date, calculated without POA	3,117,413	0	1.00	999.0	660.37	250.42
DXVER: Diagnosis version	3,117,413	0	10.00	10.00	10.00	0.00
ELECTIVE: Elective versus nonelective admission	3,108,222	9,191	0.00	1.00	0.12	0.33
FEMALE: Indicator of sex	3,116,216	1,197	0.00	1.00	0.52	0.50
HCUP_ED: HCUP emergency department service indicator	3,117,413	0	0.00	4.00	0.42	0.86
HOSP_REGION: Region of hospital	3,117,413	0	1.00	4.00	2.66	1.01
I10_HOSPBRTH: ICD-10-CM indicator of birth in this hospital	3,117,413	0	0.00	1.00	0.42	0.49
I10_NDX: ICD-10-CM number of diagnoses on this record	3,117,413	0	0.00	30.00	5.33	3.91
I10_NECAUSE: ICD-10-CM number of external cause codes on this record	3,117,413	0	0.00	4.00	0.12	0.49

Courtesy: www.hcup-us.ahrq.gov/db/nation/kid/kidsummarystats.jsp#2016.

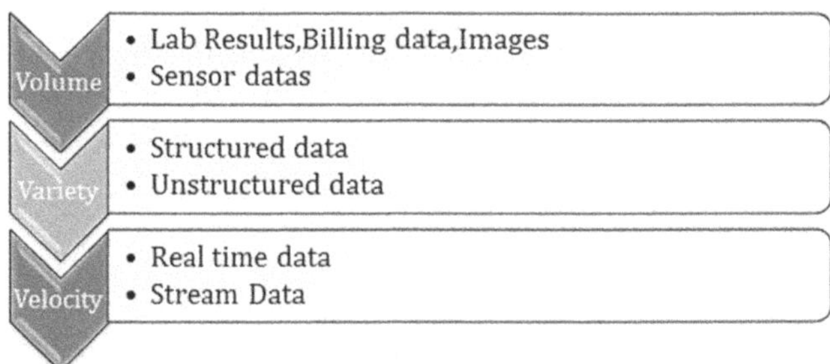

FIGURE 6.4 Common big data challenges in the three V's.

hospital readmissions for all ages. (Courtesy: www.hcup-us.ahrq.gov/db/nation/kid/kidsummarystats.jsp#2016 [7].)

6.5 CHALLENGES IN BIG DATA ANALYSIS IN HEALTHCARE

For social insurance associations that effectively coordinate information-driven bits of knowledge into their clinical and operational procedures, the prizes can be tremendous. More beneficial patients bring down consideration costs and greater deceivability into execution, and higher staff and shopper fulfillment rates are among the numerous advantages of transforming information resources into information experiences. The common big data challenges in the three V's are listed in Figure 6.4.

The earlier illustration explains the characteristics of big data in healthcare, which reflects the definition of the three V's through Volume, Variety, and Velocity. In order to infer the three V's, the common challenges have been listed as follows.

6.5.1 Capture

All information originates from some place, however, deplorably for some medicinal services suppliers, it doesn't generally originate from some place with pure information administration propensities. Capturing information is spotless, finished, precise, and arranged effectively, for their use in different frameworks is a continuous fight for associations. Poor EHR, ease of use, tangled work processes, and an inadequate comprehension of why enormous information is essential to capture well would all be able to add to quality issues that will torment information all through its life cycle.

6.5.2 Cleaning

Healthcare service suppliers are personally familiarized with the significance of orderliness in the facility. Ungraded information can rapidly crash a major information examination venture, particularly when uniting dissimilar information sources that may record clinical or operational components in marginally unique configurations. Information cleaning that is otherwise called purging or scouring guarantees that data sets are exact, right, reliable, applicable, and not undermined at all. While most information cleaning forms are still performed physically, some IT merchants do offer mechanized scouring apparatus that utilize rationale standards to think about, and differentiate huge data sets to guarantee large amounts of precision and respectability in human service information distribution centers.

6.5.3 Storage

As the volume of human services information increases exponentially, a few suppliers are not anymore ready to deal with the expenses and effects of on-start server farms. Distributed storage is turning into an undeniably well-known alternative as costs drop and unwavering quality develops. Nearly 90 percent of human services associations are utilizing a type of cloud-based well-being IT foundation, including capacity and applications. The cloud offers agile calamity recuperation, brings down advance expenses, and demands less extension, despite the fact that associations must be to a great degree cautious about picking accomplices that comprehend the significance of Health Insurance Portability and Accountability Act (HIPAA) and other social insurance explicit consistence and security issues.

6.5.4 Security

Data securities are the main need for social insurance associations, particularly in the wake of a quick fire arrangement of prominent breaks, hackings, and delivering product scenes. From phishing assaults to malware to workstations incidentally left in a taxi, social insurance data is liable to an almost vast cluster of vulnerabilities. The HIPAA Security Rule incorporates an extensive rundown of specialized shields for associations putting away Protected Health Information (PHI), including transmission security, validation conventions, and powers over access, uprightness, and examination. By and by, these shields convert into a sound judgment security methodology [8], for example, utilizing something like date hostile to infection programming, setting up firewalls, encoding delicate data, and utilizing multifaceted validation. Social insurance associations should oftentimes help their staff individuals to remember the basic idea of data security conventions and reliably survey [9–10] who approaches high-esteem data advantages by keeping noxious gatherings from causing harm.

6.5.5 Stewardship

Medical services data, particularly on the clinical side, has a long time frame of realistic usability. Even it is not required to keep data open for something like 6 years, suppliers may wish to use derecognized data sets for research ventures, which make progressing stewardship and curation a vital concern. Data may likewise be reused or rethought for different purposes, for example, quality estimation or execution benchmarking. It is difficult to understand when the data was made, by whom, and for what reason, and additionally, who has recently utilized the data, and why, or how. Creating a complete, precise, and up coming metadata is a key part of an effective data administration plan. Metadata enables examiners to precisely imitate past inquiries, which is crucial for logical investigations and exact benchmarking, and keeps the making of "data dumpsters," or confined data sets that are restricted in their handiness.

6.5.6 Querying

Additionally, hearty metadata and solid stewardship conventions make it simpler for associations to question their data and to find the anticipated solutions. The capacity to enquire data is primary for detail and investigation, yet human services associations do commonly conquer various difficulties, though they can participate in significant examination of their huge data resources. Right off the bat, they should defeat data siloes and interoperability issues that keep questioning devices about getting to the association's whole storehouse of data. On the off chance that distinctive segments of a data set are held in various walled-off frameworks or in various arrangements, it may not be conceivable to create an entire picture of an association's status or an individual patient's well-being. Numerous associations utilize Structured Query Language (SQL) to jump into substantial data sets and social databases, yet it is viable when a client would first be able to confide in the precision, culmination, and institutionalization of the current data [11].

6.5.7 Reporting

After suppliers have nailed down the inquiry procedure, they should produce a report that is clear, brief, and open to the intended interest group. By and by, the exactness and honesty of the data has a basic downstream effect on the precision and unwavering quality of the report. Poor data at the start will deliver suspect reports toward the finish of the procedure, which can hinder the clinicians attempting to utilize data to treat patients. Suppliers should likewise comprehend the distinction between "examination" and "revealing." Reporting is regularly essential for the investigation of data that must be extricated before it was very well inspected, yet announcing can likewise remain without anyone else as a final result. While a few reports might be equipped towards featuring a specific pattern, arriving at a novel resolution to make an explicit move, others must be introduced in a way that enables the peruse to draw his or her very own inductions about what the full range of data implies. Suppliers have various alternatives for meeting these different necessities, including qualified vaults, revealing apparatuses incorporated with their electronic well-being records, and web-based interfaces facilitated by Centers for Medicare & Medicaid Services (CMS) and different gatherings.

6.5.8 Visualization

For consideration purposes, spotless drawing of data perception can make it a lot less demanding for a clinician to retain and utilize data properly. Shading coding is a well-known data representation procedure that ordinarily delivers a quick reaction—for instance, red, yellow, and green are all comprehended to mean stop, alert, and go.

Associations should likewise consider great data introduction rehearses, for example, graphs utilize an appropriate extent to represent differentiating figures and the right marking of data to diminish the potential disarray [12]. Tangled flow charts, cramped or covering content, and low-quality designs can baffle and bother beneficiaries, driving them to disregard or confound data.

6.5.9 Updating

Healthcare data isn't static, and most components will require moderately visited refreshes, with the end goal to stay current and important. For a few data sets, similar to persistent imperative signs, these updates may happen like clockwork. Other data, such as personal residence or conjugal status, may just change a couple of times amid a person's whole lifetime. Understanding the instability of huge data, or how regularly and to what degree it changes, can be a test for associations that don't reliably screen their data resources. Suppliers must have an unmistakable thought of which data sets require manual refreshing, which can be computerized, how to finish this procedure without downtime for end clients, and how to guarantee that updates can be directed without harming the quality or uprightness of the data set. Associations ought to likewise guarantee that they are not making superfluous copy records from refresh to a solitary component, which may make it troublesome for clinicians to get to important data for the patient's basic leadership [13]. The challenges of big data in healthcare are listed in Table 6.3.

6.5.10 Sharing

Hardly any suppliers work in a void, and less patients get the majority of their consideration at an unaccompanied area. This implies that offering data to outer accomplices is basic, particularly, as the business moves towards populace well-being of the executives and esteem-based consideration. Data interoperability is a perpetual worry for associations of numerous types, sizes, and positions along the data development range. Major contrasts in the manner in which electronic well-being records are planned and executed can extremely reduce the capacity to move data between different associations, regularly leaving clinicians without data as they have to settle on key choices, catch up with patients, and create techniques to enhance by and large results. The business is presently endeavoring to enhance the sharing of data crosswise over specialized and hierarchical boundaries [14].

6.6 COLLECTION OF HEALTHCARE DATA

Data collection is nothing but gathering and sharing data across the healthcare system. Healthcare regards to be a heterogeneous arrangement of open and private information accumulation framework, adding health reviews, managerial enlistment and charging records, and medicinal records, which is utilized by different elements, including doctor's facilities, doctors, and health designs [15].

TABLE 6.3
The Challenges of Big Data in Healthcare

Domain	Challenges
Capture	Spotless, finished, precise, and arranged effectively for use in different frameworks is a continuous fight for associations
Cleaning	Purging or scouring guarantees that data sets are exact, right, reliable, applicable, and not undermined at all
Storage	Distributed storage is turning into an undeniably well-known alternative as costs drop and unwavering quality develops
	Near 90% of human services associations are utilizing a type of cloud-based well-being IT foundation, including capacity and applications
Security	Main need for social insurance associations, particularly in the wake of a quick fire arrangement of prominent breaks, hackings, and delivery of product scenes
Stewardship	A data steward can guarantee that all components have standard definitions and arrangements, are archived fittingly from creation to erasure, and stay helpful for the jobs that need to be done
Querying	The capacity to inquiry data is primary for detailing and investigation, yet human services associations should commonly conquer various difficulties previously they can participate in significant examination of their huge data resources
Visualization	Tangled flowcharts, cramped or covering content, and low-quality designs can baffle and bother beneficiaries, driving them to disregard or confound data. Regular instances of data perceptions incorporate warmth maps, bar outlines, pie graphs, scatterplots, and histograms, all of which have their very own particular use to show ideas and data
Updating	Understanding the instability of huge data, or how regularly, and to what degree it changes can be a test for associations that don't reliably screen their data resources.
Sharing	This implies offering data to outer accomplices is basic, particularly as the business moves towards populace, well-being of the executives and esteem based consideration
Reporting	Reporting is regularly essential for investigation—the data must be extricated before it may be inspected

6.6.1 Importance in Healthcare Data Collection

The healthcare is growing continuously and rapidly because of its expected "huge data," showing up as the new pattern in the business [16]. This has transformed into the spine of the whole framework to give various sources to haul out data that drives a more elevated amount of experiences, patterns, and other fundamental issues. Created over an assortment of sources, data collection in healthcare can likewise energize productive correspondence between specialists and patients in terms of care. Regardless of the merits with new innovations, there is still a need to deal with an effective healthcare data. In addition, the advanced security of patients close to home data is a tough task to resolve. Figure 6.5 shows the relative importance in healthcare data collection.

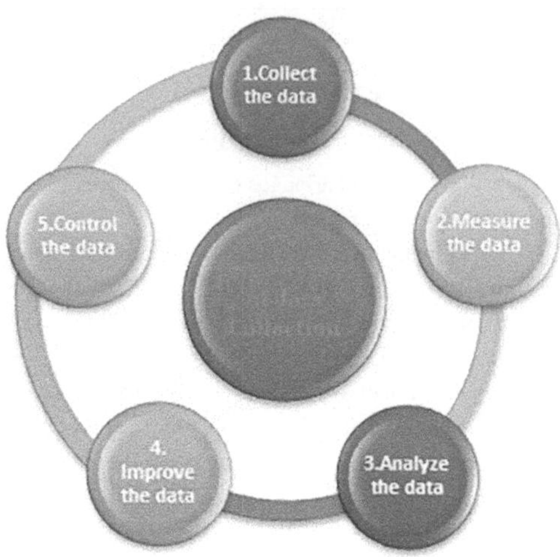

FIGURE 6.5 Importance in healthcare data collection.

6.6.2 Complications and Clarifications of Healthcare Data Collection

Questions enlightening security collaboration still requires an extreme arrangement. Here, we have three huge prevention types that keep us from creating data viably [17].

- **Deprived Data Quality**: Data realization is an underlying issue. We should take Medical Record Number (MRN), for instance. Most medicinal service foundations utilize Spreadmart to produce essential data; notwithstanding, this procedure requires affirmation that the got data coordinates the MRNs. Then again, there is dependably a shot for a mistype in any of the patient MRNs, when keeping in touch with them down physically in the spreadsheet. Additionally, some MRNs may change occasionally in the source framework. Such potential outcomes ought to be likewise considered.
- **Complex Relationship**: Envision a human services foundation that utilizes Enterprise Data Warehouse (EDW), with the point of diminishing the measure of data storehouses. Then again, Excel spreadsheets are as yet the best way to gather and keep up self-made data; however, this causes another data storehouse despite this technique. Accordingly, the procedure for collaboration of the given data storehouse transforms into an intense test. For instance, we have two individuals in our foundation who handle self-made data and record it in a spreadsheet. They have some available time toward the end of the working day to enter that data. We send them spreadsheets requesting that representatives enter refreshed records, and the two will make notes in their restrictive individual way. As it were, the data will be given in two unique adaptations.

- **Data Security**: Effective data security has dependably been a tremendous concern with regards to data gathering in nursing and other human service fields. Generally speaking, data gathering instruments in human services are mostly put away on a "Personal Computer," which frequently prompts data security issues. Utilizing cloud-based stages for putting away data might be a decent answer for this issue; however, it will require some redone instruction for staff individuals. Table 6.4 lists the healthcare data collection complications and clarifications in a detailed manner.

6.6.3 Current Data Collection Methods

Some of the current data collection methods are shown in Figure 6.6 with highlights.

1. **Sample versus evaluation**—some data are gathered for the whole populace to which they apply; such data are sometime alluded as enumeration data. One model is the real decennial statistics, which intends to acquire tallies by geographic area and essential statistic qualities for the whole occupant populace of the United States. Be that as it may, the term statistics might be used to allude to any data accumulation for each unit in the number of inhabitants in. On the other hand, numerous data can't be gathered for the whole populace without extreme expense as well as the weight on respondents. Rather, the data is gathered from a subset of the populace, or an example that is chosen in a way that makes it illustrative of the whole populace; along these lines, evaluations can be determined from the example that estimated those for the whole populace.

TABLE 6.4
Healthcare Data Collection Complications and Clarifications

Domain	Complications	Clarifications
Deprived data quality	• Data realization is the underlying issue • To be less productive with regard to creating genuine data	• Utilize Spreadmart to produce essential data; withstanding this procedure requires affirmation that the got data coordinates the MRNs
Complex relationship	• Excel spreadsheets are the best way to gather and keep up self-made data • This causes another data storehouse despite this technique	• Locking the records and giving access to one and just the individual capable may appear • As each individual depends on a specific calendar, this may result in wrong utilization of working time
Data security	• Depends on data gathering in nursing and other human service fields	• Utilizing cloud-based stages • Require redone instruction for staff individuals

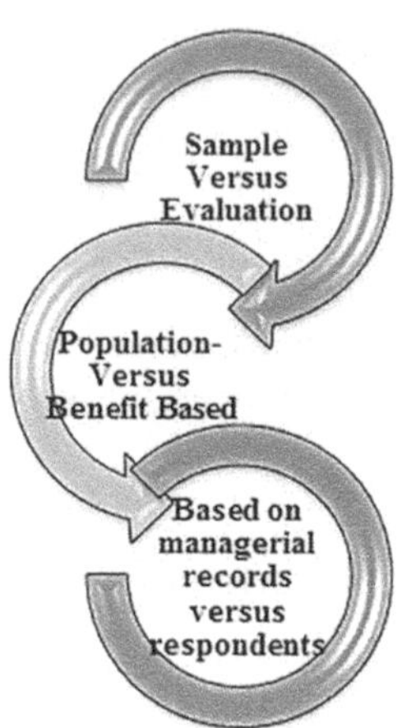

FIGURE 6.6 Current data collection methods.

2. **Population-versus benefit based**—some data accumulation endeavors center around an all-inclusive community characterized just by expansive statistic attributes, for example, all kids under age 6 or every single youthful young lady.
3. **Based on managerial records versus respondents**—some data are separated from records that as of now exist since they are essential for the organization of a program or intercession. Models are government records, business records, and restorative records. Other data are gathered specifically from respondents, for instance, by meeting people about their encounters. On account of youngsters, most respondent-based data are gathered from intermediary respondents. A third class to consider is that relating to clinical data, for example, observational examinations.

6.6.4 Advanced Data Collection Tools

- **360° Patient Views**: This tool covers health plan, care manager, provider, caregiver, and patient details. This is for the patient's past medical history and current lifestyle to enhance decision-making and provide optimal treatment effectively. This also supports proactive rather than reactive care, resulting in reduction in cost and higher quality of care with better quality of life. As the world continues to develop, the whole social insurance industry is swimming in huge amounts of data, and just getting access to that data isn't sufficient for getting a total image of the patient's outing. Establishments can be profited by officially existing advertising arrangements, similar to social insurance CRM, that give reports, estimations, and examination on different issues [18].

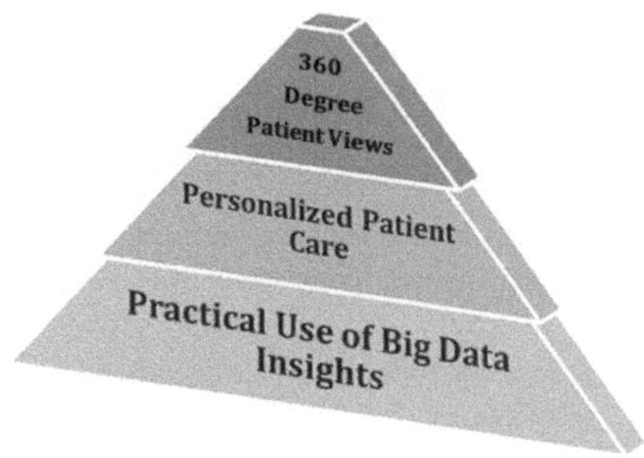

FIGURE 6.7 Advanced data collection tools.

- **Personalized Patient Care**: This tool incorporates every patient's confidential data by storing it electronically and is referred to the physician during examination of diagnosis at any time despite meeting in person. Late measurements enable human service associations in utilizing EHR to profit by even a more profound way to deal with patient bits of knowledge.
- **Practical Use of Big Data Insights**: This tool deals with the patient's data for continuous monitoring and analysis to make it practically available while using big data. Producing important data isn't sufficient, rather utilizing it astutely for the patient is significant with the aim of conveying care administration to another dimension. Tolerant data is quickly transforming into a vital resource for all medicinal service associations. Figure 6.7 states the advanced data collection tools.

6.6.5 Healthcare Data Standards

The term healthcare data standards refers to techniques, conventions, phrasings, trade, stockpiling, recovery of medical records, prescriptions, radiology pictures, managerial procedures, installation of restorative gadgets, and observing frameworks. Institutionalizing healthcare data includes the following things:

- **Definition of data components**—assurance of the data substance to be gathered and traded.
- **Data exchange groups**—standard configurations for electronically encoding the data components and data models that characterize the connections among data components in a message.
- **Terminologies**—the medicinal terms and ideas are used to depict, order, and code the data components.
- **Knowledge representation**—standard techniques for electronically speaking to therapeutic writing, clinical rules, and so forth for help.

Some of the examples of health data standards are listed in Table 6.5. (Courtesy: www.searo. who.int/entity/health_situation_trends/topics/health_data_standrads/.)

TABLE 6.5
Some of the Examples of Health Data Standards

Standards	Definition
International classification of diseases (ICD)	• International standard diagnostic classification for all general, epidemiological, many health management purposes, and clinical use
LONIC—Logical observation identifiers names and codes	• A universal code system for identifying laboratory and clinical observations • LOINC has standardized terms that enable exchange and aggregation of electronic health data from many independent systems and database that contains the usual categories of chemistry, hematology, serology, microbiology, toxicology, as well as categories for drugs and cell counts, antibiotic susceptibilities, and more
HL7—Health level seven	• HL7 is an international community to provide a framework for exchange, integration, and sharing and retrieval of electronic healthcare information • It is a coordinated message-based connection between two systems that allow information exchange reliably between application programs, and it includes patient demographics
SNOMED—Systematized nomenclature of medicine	• It designed as a comprehensive nomenclature of clinical medicine for the purpose of accurately storing and/or retrieving records of clinical care in human and veterinary medicine • It provides the core general terminology for the EHR and contains active concepts with unique meanings and formal logic-based definitions organized into hierarchies • It offers a standard-based foundation for different functions, e.g., collection of a variety of clinical information, linked to clinical knowledge bases, information retrieval, data aggregation, analyses, exchange, etc. • It contributes to the improvement of patient care by underpinning the development of systems that accurately record healthcare encounters and to deliver decision support
SDMX—Statistical data and metadata exchange	• SDMX provides standard formats for data and metadata, together with content guidelines, and an IT architecture for exchange of data and metadata • SDMX-HD (Health Domain) is a WHO implementation of the SDMX standard to allow medical facilities to share and exchange medical indicators and metadata between medical organizations

6.6.6 Inferences of Patient Data Collection in Healthcare

To interrogate the general health on a worldwide scale, computerized health data are changing the manner in which we see medicinal care. On the off chance that it does, there's an all likelihood that somebody in some place endeavor to locate a superior way [19].

At the present time, however, something other than the drug evolves. The manner in which we collaborate with our specialists and the manner in which our specialists associate with one another is reclassifying what tolerant care resembles. Specialists are accessible on interest, therapeutic records can be gotten to without sitting tight for a dispatch, and the measure of data accessible to a general society is developing constantly [20]. Worries about security exist paired up with the advantages of expanded access, and therapeutic offices can be defenseless against an assortment of digital assaults.

- **General Health Opportunities**: Expanded availability in the medicinal world makes data more accessible to more individuals. It's less demanding than at any other time to follow infection episodes, analyze national insights, and recognize drifts in worldwide health. We have less transferable sickness pestilences now than whenever it was written previously. Conclusion and treatment conventions are more complex, yet as a general rule, we don't need to stress over smallpox, measles, or rubella because of immunizations. With the data recorded by wellness trackers, electronic nourishment journals, and other health-centered gadgets, general health specialists can portray the way of life decisions that lead to sickness. General health crusades can turn out to be more focused and on an accentuation put on development and plan adherence through close to home innovation.
- **Office Implications**: While the worldwide stage absolutely gives a chance to use enormous measures of patient data, the suggestions are cozier at an office level. For clinic frameworks with different divisions and areas, we have an incorporated EHR changes quiet care. Quiet care results remain to profit more specialists by having similar data and making educated, helpful choices. Notwithstanding when it's not a client blunder, however, data isn't constantly sheltered. Tolerant records can be captured amid electronic transmission also, possibly trading off patient protection or uncovering significant data like government disability numbers [21].
- **Individual Impacts**: Having an individual data uncovered is surely a potential danger of electronic health databases; however, there isn't generally a choice to tell your specialist that you just need a paper diagram. In any case, patients do have the choice to choose whether or not to utilize wearable gadgets or exploit telehealth alternatives. The dangers don't really exceed the advantages, however. For people attempting to get fit or increment work out, wellness trackers give incredible criticism and can even be set to send updates when it's an ideal opportunity to go ahead. Telehealth enables patients to connect with medical attendants or different suppliers through talk or video call, possibly deciding whether a stumble into the specialist's office is justified, despite all the trouble. Keen pill bottles serve a comparable capacity, however, they can likewise ensure that prescription isn't mishandled, conceivably making a weapon against the mind-boggling narcotic pandemic [22].
- **Complete Accountability**: Technology and the patient data it gathers give a staggering number of research open doors for worldwide and national

health. In spite of the fact that talking in such expansive terms can influence the procedure to appear to be far evacuated, proof-based restorative treatment just stands to profit by having more data illuminating prescribed procedures. As the database develops [23], the impacts will stream down to singular patient results through groups of coordinated care suppliers. Be that as it may, as the data database develops, it likewise turns out to be more defenseless, and that duty is at present in transition.

6.7 ANALYSIS OF HEALTHCARE DATA

In general, healthcare domain has four categories of analytics: descriptive, diagnostic, predictive, and prescriptive analytics; a brief description of each one of them is given as follows [24].

Figure 6.8 shows the categories of healthcare data analysis, and the explanation of each type has been elaborately explained with relevant examples.

- **Descriptive Analytics**: It comprises depicting current circumstances and their investigation. A few strategies are utilized to play out this dimension of examination. For example, graphic measurement devices like histograms and diagrams are among the strategies utilized in engaging examination.
- **Diagnostic Analysis**: It means to clarify why certain occasions happened and what are the variables that activated them. For instance, symptomatic examination endeavors to comprehend the explanations for the customary readmission of a few patients by utilizing a few techniques, for example, grouping and choice trees [25].
- **Predicative Analytics**: It mirrors the capacity to anticipate future occasions; it likewise helps in distinguishing patterns and deciding probabilities of questionable results. An outline of its job is to foresee whether a patient can get intricacies or not. Prescient models are frequently constructed utilizing machine learning procedures.
- **Prescriptive Analytics**: It will probably propose reasonable activities by prompting ideal basic leadership. For example, prescriptive investigation may propose dismissing a given treatment on account of a hurting

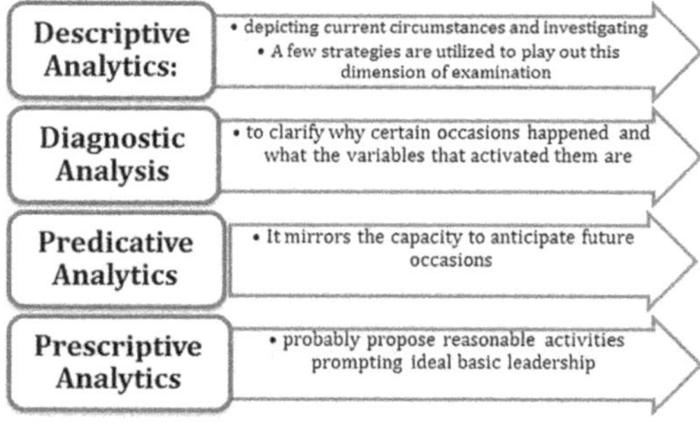

FIGURE 6.8 Categories of healthcare data analysis.

symptom high likelihood. The incorporation of huge data advances in healthcare examination may prompt better execution of medicinal frameworks [26].

The three other primary zones of big data analytics apart from the earlier ones are in healthcare data:

- **Image Processing**: Medical images are a critical wellspring of data that are as often as possible utilized for analysis, treatment appraisal, and arrangement. Computed tomography (CT), magnetic resonance imaging (MRI), X-beam, subatomic imaging, ultrasound, photograph acoustic imaging, fluoroscopy, positron emission tomography-CT (PET-CT), and mammography are portions of instances of imaging procedures that are entrenched inside clinical settings. Restorative image data can run anyplace from a couple of megabytes for a solitary report to many megabytes per examination. Such data require substantial capacity limits whenever put away for long haul. It likewise requests quick and precise calculations if any choice-helping computerization were to be performed utilizing the data. Moreover, if different wellsprings of data obtained for every patient are additionally used amid determination, visualization, and treatment forms at that point, the issue of giving firm stockpiling and creating proficient strategies equipped for embodying the expansive scope of data turns into a test.
- **Signal Processing**: Similar to restorative signals, therapeutic flags likewise present volume and speed snags particularly amid high goals and capacity from a huge number of screens associated with every patient. Notwithstanding the data estimate issues, physiological flags additionally present a multifaceted nature of a spatiotemporal sort. Examination of physiological signs is frequently more significant when introduced alongside situational setting mindfulness, which should be implanted into the improvement of persistent checking and prescient frameworks to guarantee its adequacy and strength [27].
- **Genomics**: The expense to grouping the human genome is quickly diminishing with the improvement of high-throughput sequencing innovation. With suggestions for current general health approaches and conveyance of care, examining genome-scale data for creating noteworthy proposals in a convenient way is a critical test to the field of computational science. Cost and time to convey suggestions are vital in a clinical setting. Activities handling this mind-boggling issue incorporate 100,000 subjects more than 20–30 years utilizing the prescient, preventive, participatory, and customized health, alluded to as P4, prescription worldview and additionally an integrative individual profile. The P4 activity is utilizing a framework approach for (i) dissecting genome-scale data sets to decide malady states, (ii) moving towards blood-based indicative instruments for consistent checking of a subject, (iii) investigating new ways to deal with medication target disclosure, creating apparatuses to manage big data

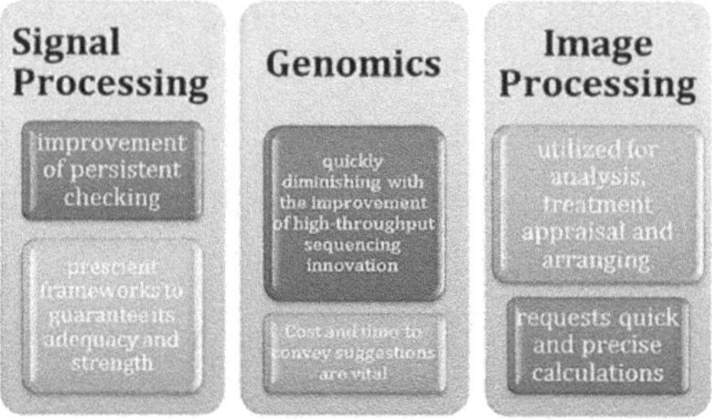

FIGURE 6.9 The primary zones of big data analytics.

difficulties of catching, approving, putting away, mining, coordinating, and lastly (iv) displaying data for every person. The integrative Personal Omics Profile (iPOP) joins physiological observing and different high-throughput techniques for genome sequencing to produce a point-by-point health and illness conditions of a subject. At last, acknowledging significant proposals at the clinical dimension remains a terrific test in this field. Using such high thickness data for investigation, revelation, and clinical interpretation requests novel big data approaches and analytics. Figure 6.9 shows the three other primary zones of big data analytics.

6.8 HEALTHCARE DATA MANAGEMENT

Healthcare data management is the process of storing, protecting, and analyzing data dragged from assorted bases. Dealing with the abundance of accessible medicinal services information enables well-being frameworks to make comprehensive perspectives of patients, customize medications, enhance correspondence, and upgrade well-being results [28].

6.8.1 Big Data and Care Management

ACOs (accountable care organization) center oversaw care and the need to keep individuals at home and out of the healing facility. Sensors and wearables will gather health data on patients in their homes and push the majority of that data into the cloud. Electronic scales, BP (Blood Pressure) screens, SpO2 (Blood oxygen saturation) sensors, nearness sensors like iBeacon, and destined to-be-designed sensors will shoot data from a huge number of patients persistently. Healthcare foundations and care supervisors, utilizing refined apparatuses will screen this huge data stream and the IoT to keep their patients healthy [29].

Besides, a majority of different sensor data will come into healthcare associations at an extraordinary volume and speed. In a healthcare future predicated on keeping individuals out of the doctor's facility, a health framework's capacity to deal with this data will be essential. These volumes of data are best overseen as streams coming into a big data group. As the data streams in, associations should have the capacity to distinguish any potential health issues and alarm a care supervisor to intercede. For instance, if a patient's pulse spikes, the framework will send a caution progressively to a care administrator, who would then be able to cooperate with the patient to recover his circulatory strain into a healthy range. Big data is the main seek after dealing with the volume, speed, and assortment of this sensor data.

6.8.2 Advantages of Healthcare Data Management

The Advantages of Healthcare Data Management are as follows

- Create 360° perspectives of patients and convey customized communications by incorporating quiet data from every single accessible source.
- Improve understanding commitment with precise display and investigation dependent on healthcare data.
- Improve populace health results by following current health drifts and foreseeing the forthcoming ones.
- Make educated, high-affect business choices dependent on data bits of knowledge.
- Understand doctor action and adjust them to the association's objectives.

6.9 BIG DATA IN HEALTHCARE

At the point when healthcare associations imagine the fate of big data, they frequently consider utilizing it for breaking down content-based notes. Current investigation innovations generally make utilization of discrete data and battle to profit by a majority of significant clinical data caught in doctors' and medical caretakers' notes. Big data ordering procedures and a portion of the new work discovering data in literary fields could surely increase the value of healthcare investigation later on.

6.9.1 Big Data and IoT

Big data will truly end up profitable to healthcare in what's known as the IoT.

The IoT is a developing system of ordinary items from modern machines to purchaser products that can share data and finish errands while you are occupied with different exercises, similar to work, rest, or exercise. Before long, our vehicles, our homes, our real machines, and even our city lanes will be associated with the Internet—making this system of items known as IoT for short. Composed of a large number of sensors and gadgets that create perpetual surges of data, the IoT can be utilized to enhance our lives and organizations from numerous point of views.

Any gadget that creates healthcare data about a man's health and sends it into the cloud will be a piece of this IoT. Wearable may be the most recognizable case of such a gadget. Numerous individuals presently wear a wellness gadget that tracks what number of steps they've taken, their heart rate, their weight, and how everything is drifting. Applications are accessible on PDAs (Personal Digital Assistants) that track how often and how seriously a client works out. There are likewise restorative gadgets that can likewise send data into the cloud: circulatory strain screens, beat oximeters, glucose screens, and a whole lot more [30].

6.9.2 Patient Prophecies for Upgraded Staffing

Healthcare data investigation helps to anticipate the quantity of patients to enhance staffing. One of the key data sets is one decade of doctor's facility affirmation records, which data researchers crunched utilizing "time arrangement investigation" procedures. These investigations enabled the scientists to see important examples in confirmation rates. At that point, they could utilize a machine to figure out how to locate the most precise calculations that anticipated future affirmation patterns. The outcome is an internet browser-based interface that is intended to be utilized by doctors, medical attendants, and clinic organization staff inexpert in data science to figure visit and confirmation rates for the following 15 days. Additional staff can be drafted in when high quantities of guests are normal, prompting diminished sitting tight occasions for patients and better nature of care.

6.9.3 Electronic Health Records

EHR is the most used across a board utilization of big data in prescription. Each patient has his very own computerized record that incorporates socioeconomics, therapeutic history, hypersensitivities, lab test results, and so forth. Records are shared by means of secure data frameworks and are accessible for suppliers from both open and private area. Each record is involved as one modifiable document, which implies that specialists can actualize changes after sometime with no printed material and no peril of data replication. EHRs can likewise trigger alerts and updates when a patient ought to get another lab test or track solutions to check whether a patient has been following the specialists' requests. In spite of the fact that EHR is an extraordinary thought, numerous nations still battle to completely execute them.

6.9.4 Real-Time Warning

Different instances of big data examination in healthcare share one urgent usefulness, which reminds constant caution. In doctor's facilities, CDS software investigates medicinal data on the spot, furnishing health professionals with exhortation as they settle on prescriptive decisions. Be that as it may, specialists need patients to avoid healing centers to maintain a strategic distance from expensive in-house medications. Individual investigation contraptions, as of now inclining as business insight trendy expressions, can possibly turn out to be a piece of another technique. Wearable devices will gather patients' health data constantly and send this data to the cloud. Moreover, this data will be gotten to the database on the condition of health of the overall population, which will enable specialists to look at this data in a financial setting and alter the conveyance techniques in a like manner. Organizations and care directors will utilize advanced instruments to screen this enormous data stream and will respond each time the outcomes will exasperate. For instance, if a patient's circulatory strain increments alarmingly, the framework will send a caution progressively to the specialist who will at that point make a move to achieve the patient and direct measures to bring down the weight. Another precedent is that of Asthmapolis, which has begun to utilize inhalers with Global Positioning System (GPS)-empowered trackers with the end goal to recognize asthma patterns, both on an individual dimension and taking a gander at bigger populaces. This data is being utilized related to data from the CDC, with the end goal to provide better treatment for asthmatics.

6.9.5 Augmenting Patient Engagement

Numerous buyers and, consequently, potential patients as of now have an enthusiasm for brilliant gadgets that record each progression they take, their pulses, dozing propensities, and so on on a lasting premise. This fundamental data can be combined with other identifiable data to recognize the hiding potential health dangers. A ceaseless sleep deprivation and a hoisted pulse can flag a hazard for future coronary illness, for example. Patients are straightforwardly associated with the checking of their own health, and motivations from health protections can push them to lead a healthy way of life. Another in-progress approach is to accompany a wearable that follows explicit health drifts and hands-off them to the cloud where doctors can screen them. Patients experiencing asthma or pulse could profit by it and turn into more autonomous with decreased pointless visits to the specialist.

6.9.6 Using Health Data for Informed Strategic Planning

The utilization of big data in healthcare takes into account key arrangement because of better experiences in individuals' inspirations. Care troughs can examine registration results among individuals in various statistic gatherings and recognize what factors dishearten individuals from taking up treatment.

6.9.7 Extrapolative Analytics in Healthcare

The objective of healthcare business insight is to enable specialists to settle on data-driven decisions in seconds and to enhance patients' treatment. This is especially valuable if there should be an occurrence of patients with complex restorative narratives experiencing numerous conditions. New instruments would likewise have the capacity to anticipate, for instance, which is in danger of diabetes, and accordingly be educated to utilize extra screenings or wait for the board [31].

6.9.8 Diminish Fraud and Enrich Security

A few investigations have demonstrated that this specific industry is to encounter data ruptures than some other industry. The reason is straightforward: individual data is to a great degree important and profitable in illegal businesses. What's more, any rupture would have emotional results [32]. In light of that, numerous associations began to utilize investigation to help counteract security dangers by distinguishing changes in system activity, or whatever other conducts that mirror a digital assault. Obviously, big data has characteristic security issues and many feel that utilizing it will make the associations more helpless than they are as of now. In any case, propels in security, for example, encryption innovation, firewalls against infection software, and so forth, answer that requirement for greater security, and the advantages brought to a great extent overwhelm the dangers. In a like manner, it can help anticipate extortion and mistaken cases in a fundamental, repeatable manner. Investigation helps to streamline the handling of protection claims, empowering patients to show signs of improvement returns on their cases and caregivers are paid quicker.

6.9.9 Telemedicine

Telemedicine term alludes to conveyance of remote clinical administrations utilizing innovation. It is utilized for essential counsels and introductory conclusion, remote patient observation, and restorative training for health professionals. Some more explicit uses incorporate telesurgery specialists to perform activities with the utilization of robots and fast constant data conveyance without physically being in a similar area with a patient. Clinicians utilize telemedicine to give customized treatment designs and avoid hospitalization or reaffirmation. Such utilization of healthcare data examination can be connected to the utilization of prescient investigation as observed beforehand. It enables clinicians to foresee intense restorative occasions ahead of time and avoid crumbling of patient's conditions. By fending off patients from healing facilities, telemedicine diminishes costs and enhances the nature of administration. Patients can abstain from holding up lines, and specialists don't sit idle for pointless meetings and printed material. Telemedicine additionally enhances the accessibility of care as patients' state can be checked and counseled anyplace and whenever.

6.9.10 Assimilating Big Data per Medical Imaging

Medical imaging is crucial, and every year, 670 million imaging techniques are performed. Breaking down and putting away these pictures physically are time consuming and costly, as radiologists need to inspect each picture exclusively, while clinics need to store them for a few years. Medical imaging clarified how big data examination for healthcare could change and the manner in which the pictures are perused: calculations created by investigating countless pictures could distinguish explicit examples in the pixels and convert it into a number to assist the doctor with the finding. They even go further, saying that it could be conceivable that radiologists will never again need to take a gander at the pictures, yet rather investigate the results of the calculations that will unavoidably examine and recall a larger number of pictures than they could in a lifetime. This would without a doubt affect the job of radiologists, their training, and the required range of abilities.

6.9.11 A Method to Avert Pointless ER (Emergency Room) Visits

Sparing time, cash, and vitality utilizing huge information examination for medicinal services is essential. As an example case, a patient is alluded to three diverse substance misuse facilities and two distinctive emotional wellness centers dealing with lodging. It is not just awful for the patient, it is also a misuse of valuable assets for the two healing centers. In some hospitals, doctor faculties meet up to make a program called PreManage ED, which shares records during crisis divisions.

6.10 FUTURE FOR BIG DATA IN HEALTHCARE

Similarly, as officials in trade and mechanical segments announce that their enormous information activities have been effective and transformational, the viewpoint for social insurance is much more energizing. The following are a couple of regions where huge information is bound to change medicinal services [33].

Exactness medication, as imagined by the National Institutes of Health (NIH), tries to select one million individuals to volunteer their well-being data in the All of us examine program. This program is a piece of the NIH Precision Medicine Initiative. As per the NIH, the activity expects to see how a man's hereditary qualities, condition, and way of life can help decide the best way to counteract or treat illness. The long-haul objectives of the Precision Medicine Initiative spotlight convey the exactness medication to all regions of well-being and social insurance on a vast scale.

Wearable devices and IoT sensors, officially noted earlier, can possibly change medicinal services for some patient populaces and to enable individuals to stay solid. A wearable gadget or sensor may one day give an immediate, constant feed to a patient's electronic well-being records, which enables therapeutic staff to screen and after that counsel with the patient, either eye to eye or remotely [34].

Machine learning, a part of man-made brainpower, and one that relies upon enormous information are as of now helping doctors to enhance persistent consideration. IBM with its Watson Health PC framework has just banded together with Mayo Clinic, Convenience, Value, and Service (CVS) Health, Memorial Sloan Kettering Cancer Center, and others. Machine learning, together with human services, extends information investigation and duplicate guardians' capacity to improve tolerant consideration.

6.11 CONCLUSION

The effective analysis of management of big data in healthcare has been presented in this book chapter. The procedure for collecting healthcare data has been discussed in the initial sections of the article. The four categories of big data analysis parts have also been illustrated, and the second part includes descriptive, diagnosis, predicate, and prescriptive types. Of course, the third part of this chapter has been dedicated to study the management of big data in healthcare and its integration with IoT in big data healthcare, which includes volume, speed, and data. This article also incorporates a variety of applications like EHRs, augmenting patient engagement, extrapolative analysis in healthcare, and telemedicine. In addition, the challenges of big data in healthcare have also been analyzed by means of clearing, storage, security, stewardship, querying, reporting, visualizing, updating, and sharing. As a conclusive remark, the utilization of machine learning could be the best solution to improve the tolerance along with human service information.

REFERENCES

1. N. Elgendy and A. Elragal, "Big data analytics: A literature review paper", Lecture Notes in Computer Science, Industrial Conference on Data Mining, Springer International Publishing, Vol. 8557, 2014, pp. 214–227.
2. M. D. Anto Praveena and B. Bharathi, "A survey paper on big data analytics," *2017 International Conference on Information Communication and Embedded Systems (ICICES)*, 2017, pp. 1–7, Chennai, India
3. C. Lakshmi and V. V. Nagendra Kumar, "Survey paper on big data," *International Journal of Advanced Research in Computer Science and Software Engineering*, Vol. 6, No. 8, 2016, pp. 368–381.
4. R. Hermon and P. A. Williams, "Big data in healthcare: What is it used for?" 2014.
5. Sample healthcare data set (Courtesy: NCHS - Leading Causes of Death: United States). www.hcup-us.ahrq.gov/db/nation/kid/kidsummarystats.jsp#2016.
6. M. U. Sarwar, et al., "A survey of big data analytics in healthcare," *International Journal of Advanced Computer Science and Applications*, Vol. 8, No. 6, 2017, pp. 355–359.
7. www.searo.who.int/entity/health_situation_trends/topics/health_data_standards.
8. J. R. Alam, A. Sajid, R. Talib, and M. Niaz, "A review on the role of big data in business," *International Journal of Computer Science and Mobile Computing*, Vol. 3, No. 4, 2014, pp. 446–453.
9. E. A. Archenaaa and M. Anitab, "A survey of big data analytics in healthcare and government," *Procedia Computer Science*, Vol. 50, 2015, pp. 408–413.
10. Q. Memon, "Smarter health-care collaborative network," *Building Next-Generation Converged Networks: Theory and Practice*, Editors: Al-Sakib K. Pathan, Muhammad M. Monowar, Zubair M. Fadlullah, 2013, pp. 451–476, CRC Press
11. B. Ristevski and M. Chen, "Big data analytics in medicine and healthcare", *Journal of Integrative Bioinformatics*, 2018, Vol. 15, No. 3, pp. 1–5. doi:10.1515/jib-2017-0030.
12. W. Raghupathi and V. Raghupathi, "Big data analytics in healthcare: Potent potential," *Health Information Science and Systems*, Vol. 2, No. 23, 2014.
13. Q. K. Fatt and A. Ramadas, "The usefulness and challenges of big data in healthcare," *Journal of Healthcare Communications*, Vol. 3, No. 2:21, 2018, pp. 1–4.
14. Big Data Technologies in Healthcare. Needs, opportunities and challenges, TF7 Healthcare subgroup,Report, 2016, pp. 1–31.
15. M. S. Islam, M. M. Hasan, X. Wang, H. D. Germack, and M. Noor-E-Alam, "A systematic review on healthcare analytics: Application and theoretical perspective of data mining," *Journal of Healthcare (Basel)*, Vol. 6, No. 2, p. 54, 2018. doi:10.3390/healthcare6020054.
16. J. R. Alam, A. Sajid, R. Talib, and M. Niaz, "A review on the role of big data in business," *International Journal of Computer Science and Mobile Computing*, Vol. 3, No. 4, 2014, pp. 446–453.
17. I. Barbier-Feraud and J. B. Malafosse, "Big data and prevention from prediction to Demonstration", Report, 2016, pp. 1–80.
18. S. Kumar and M. Singh, "Big data analytics for healthcare industry: Impact, applications, and tools," *Journal of Big Data Mining and Analytics*, Vol. 2, No. 1, 2019, pp. 48–57.
19. P. A. Vineet Kumar, "The use of big data analytics in information systems research," 2018, https://ssrn.com/abstract=3185883 or http://dx.doi.org/10.2139/ssrn.3185883.
20. S. Patel and A. Patel, "A big data revolution in health care sector: Opportunities, challenges and technological advancements," *International Journal of Information Sciences and Techniques (IJIST)*, Vol. 6, No. 1/2, 2016, pp. 155–162.
21. P. Groves, B. Kayyali, D. Knott, and S. Van Kuiken, "The big data revolution in healthcare," *McKinsey Quarterly*, Vol. 2, 2013.

22. T. L. Coelho da Silva and R. P. Magalhaes, et al., "Big data analytics technologies and platforms: A brief review," *LADaS 2018 - Latin America Data Science Workshop*, 2018, pp. 25–32, Rio de Janeiro, Brazil
23. T. Nizam and S. I. Hassan, "Big data: A survey paper on big data innovation and its technology," *International Journal of Advanced Research in Computer Science*, Vol. 8, No. 5, 2017, pp. 2173–2179.
24. S. Kumari and K. Sandhya Rani Dr., "Big data analytics for healthcare system," *2018 IADS International Conference on Computing, Communications & Data Engineering (CCODE)*, 2018, India
25. L. Taylor, R. Schroeder, and E. Meyer, "Emerging practices and perspectives on Big Data analysis in economics: Bigger and better or more of the same?" *Big Data & Society*, Vol. 1, No. 2, 2014, pp. 1–10.
26. M. Sarnovsky, P. Bednar, and M. Smatana, "Big data processing and analytics platform architecture for process industry factories," *Big Data and Cognitive Computing*, Vol. 2, No. 1, 2018, pp. 2–18.
27. R. Sonnati, "Improving healthcare using big data analytics," *International Journal Of Scientific & Technology Research*, Vol. 6, No. 03, 2017, pp. 142–147.
28. N. El Aboudi and L. Benhlima, "Big data management for healthcare systems: Architecture, requirements, and implementation," *Journal of Advanced Bioinformatics*, Vol. 2018, 2018. doi:10.1155/2018/4059018.
29. Gesundheit Österreich Forschungs and und Planungs GmbH, "Study on big data in public health, telemedicine and healthcare," Final Report, December 2016.
30. L. Wang and C. A. Alexander, "Big data in medical applications and health care," *Current Research in Medicine*, Vol. 6, No. 1, 2015, pp. 1–8.
31. A. Belle and R. Thiagarajan, et al., "Big data analytics in healthcare", *BioMed Research International Journal*, Vol. 2015, 2015, pp. 1–16.
32. E. A. Archenaaa and M. Anitab, "A survey of big data analytics in healthcare and government," *Procedia Computer Science*, Vol. 50, 2015, pp. 408–413.
33. McKinsey, "Big data, big Transformations", McKinsey Global Survey, Minding Your Digital Business, 2012.
34. Big Data is the Future of Healthcare, Cognizant 20-20 insights | September 2012.

7 Healthcare Analytics
A Case Study Approach Using the Framingham Heart Study

Carol Hargreaves
National University of Singapore

CONTENTS

7.1 INTRODUCTION AND BACKGROUND TO THE CASE STUDY: FRAMINGHAM HEART STUDY

In the early 1990s, heart disease became the leading cause of death. The effect of smoking, cholesterol, and obesity on heart disease and stroke was unknown. High blood pressure was seen as an inevitable consequence of aging. The Framingham Heart Study aimed to unravel the underlying causes of heart disease. Through the

Framingham Heart Study, hypertension treatment, cholesterol reduction, and smoking cessation have contributed to a 50-year decline in cardiovascular deaths.

The primary objective of this chapter is to educate clinicians and healthcare professionals on the value of healthcare analytics. This chapter will demonstrate how the analysis of health data, such as blood cholesterol, blood pressure, smoking, and obesity, can identify high-risk heart attack patients, and how the proactive changes in these high-risk patient lifestyles and the use of medication can prevent a heart attack from taking place. We will use "The Framingham Heart Study" as a case study for this chapter.

This chapter is largely nonmathematical but technical, as statistical techniques are used for the analysis of the Framingham Heart Study data.

The reader will gain a hands-on experience in analyzing healthcare data, and at the same time, also learn what questions can be asked, what techniques may be used, and how to interpret the results from an analytical evaluation and integrating a clinical approach for preventing or reducing the risk of a heart attack. A case study approach, using a data analytics framework, is provided so that the reader can easily understand and follow the analysis.

The following healthcare questions arise: Who are the high-risk heart attack patients? Which patients are obese and have a high risk of a heart attack? Does smoking increase the risk of a heart disease?

Clinical prediction is one of the most important branches of healthcare data analytics. While the linear regression and the logistic regression models are basic and are widely used for clinical prediction, more advanced methods such as decision trees and neural networks have also been successfully used in clinical classification and prediction applications.

Machine-learning models, such as the logistic regression model, are built to identify the risk factors for cardiovascular heart disease. The reader will learn how to identify high-risk patients for cardiovascular heart disease using the data analytics framework. Once the training model is built, the performance of the model will be tested on new patient data.

Metrics such as the overall accuracy of the model, the Confusion Matrix, Recall, Specificity, area under the curve (AUC) will be used to evaluate the performance of the training model.

Next, actionable insights are derived from the analytics model results. These insights drive the clinical treatment recommendations for different patient risk groups.

Clinicians and healthcare professionals will learn how data analytics can improve clinical care, predict which patients will have cardiovascular disease (CVD), and enable clinical decision support (CDS) tools for screening patients with cardiovascular risk.

7.2 LITERATURE REVIEW

Today's healthcare industries are moving from volume-based business into value-based business, which requires an overwork from doctors and nurses to be more productive and efficient. Efficient healthcare industries improve healthcare practice; change individual life styles; drive patients to live longer; and prevent diseases, illnesses, and infections [1].

CVD is one of the most common causes of death globally. Analytics is the way of developing insight through the efficient use of data and application of quantitative and qualitative analysis [2]. It can generate fact-based decisions for "planning, management, measurement, and learning" purposes. The most important risk factors identified for coronary heart disease (CHD) were age, gender, blood pressure, blood glucose, and the number of cigarettes per day. Other studies on CHD show similar results: age, smoking habit, history of hypertension, family history, and history of diabetes [3].

A big data analytical framework that utilizes a ubiquitous healthcare system was established [4]. The framework analyzes vital signs extracted from accelerometers to provide healthcare services. Vital signs are continuous time-series data that are unstructured in nature and have inadequacy to store in the traditional databases. Electrocardiogram signal (ECG), respiration, and motion data have been accounted as vital signs.

For the past two years, predictive analysis has been recognized as one of the major business intelligence approaches, but its real-world applications extend far beyond the business context. Parkland hospital in Dallas, Texas has launched a predictive system that scans all patients' details and information to identify potential risks and outcomes. As a result, the hospital has saved more than half a million dollars, especially in heart failure and disease predictions in terms of performing patients' monitoring and avoiding future complications [5].

The CDS aims to increase the quality of healthcare services by enhancing the outcomes. The primary focus of the CDS system is to provide right information to the right people and proper customized healthcare management process not limited to clinical guidelines, documentations, and diagnosis [6].

By analytics on data, the current state of the health of patients provides insight to them to take more ownership of their healthcare. The information sharing mechanism increases productivity and reduces overlapping of data. Thereby, it enhances the coordination of care. Big data will further personalize medicine by determining the tests and treatments needed for each patient. The provision of earlier treatment can reduce health costs and can eliminate the risk of chronic diseases [7].

7.3 INTRODUCTION TO THE DATA ANALYTICS FRAMEWORK

In today's digital world, patient data is easily available in many forms, such as electronic health records, biomedical images, sensor data, genomic data, biomedical signals, data gathered from social media, and clinical text. Once the patient data is available, medical organizations need to not only store the patient data but also use the patient data to track key healthcare analytics metrics. The healthcare analytics metrics help clinicians to understand the effectiveness of patient treatments and to be more proactive when the patient treatment proves to be ineffective. Healthcare analytics will help clinicians to build individualized patient profiles that can accurately compute the likelihood of an individual patient to suffer from a medical complication or condition in the near future.

One of the reasons for the development of healthcare analytics as a tool is that, with the advancement of technology, healthcare analytics applications in hospitals, clinics, and medical consulting practices may take place in a timely manner and enable emergency and proactive critical decision-making, resulting in many lives being saved. Healthcare analytics applications are used for a variety of reasons, ranging from improving patient service to improve the hospital's capability to predict patient heart attacks and to offer valuable real-time patient insights on hospital computer monitoring dashboards.

Today, these real-time dashboards may be used to effectively track key performance clinical indicators with the most current information and patient data available. These dashboards help the clinician to make key decisions in a timely manner, thereby increasing the number of early diagnosis for a particular disease condition, reducing patient mortality, and overall increasing the efficiency of the medical organization.

Today, many medical organizations are exploiting healthcare analytics to enable proactive clinical decision-making; in other words, they are switching from reacting to healthcare situations to anticipate them.

For whatever purpose the healthcare analytics is applied, the key outcome is the same: The process of solving a healthcare problem using relevant data and turning it into insights is by providing the clinician with the knowledge he or she needs to make clinical decisions. So what is healthcare analytics? Essentially, healthcare analytics is a seven-step process, outlined as follows.

7.3.1 Step 1. Defining the Healthcare Problem

This stage involves understanding what the clinician would like to improve or investigate. Sometimes, the goal is broken down into smaller goals. Relevant data needed to solve these healthcare goals are decided by the hospital stakeholders, expert clinicians, and healthcare analysts. At this stage, key questions such as, "which patients are at high risk of dying?," "how can we reduce patient mortality?," "what are the factors that contribute to patient mortality?," "what healthcare and patient data are available?," "how can we use this healthcare and patient data?," and "do we have sufficient patient data?" are considered.

7.3.2 Step 2. Explore the Healthcare Data

This stage involves cleaning the healthcare and patient data, making statistical computations for handling missing healthcare and patient data, removing outliers, and transforming combinations of input data variables to form new data variables. We usually plot time-series graphs, as they are able to indicate any patterns or outliers in the data. The removal of outliers from the healthcare dataset is a very important task as outliers often affect the accuracy of the model results. If the analyst does not remove the outliers, we may have a case of "Garbage in, garbage out (GIGO)"!

Once the analyst has cleaned the healthcare data, the analyst will then visualize the data as data visualizations that make it easy to understand the results of the patient profiles and treatments. Data visualization also helps the analyst to identify outliers and patterns in the patient data that can point to appropriate data analytics techniques for analysis and patient risk modeling.

Next, the analyst will summarize the patient data using descriptive statistics (such as mean, standard deviation, range, mode, and median) that will help provide a basic understanding of the patient data and the different patient profile groups. Often, to derive actionable insights from the patient data, it is at this stage that the analyst uses statistical hypothesis testing, to compare different patient groups using different assumptions.

To identify possible correlation between factors, the analyst will plot the patient data using scatterplots and apply statistical methods such as correlation analysis to identify relationships between factors.

The analyst will also perform simple regression analysis to see whether simple predictions can assist with clinical decision-making. At this stage, the analyst is already looking for general patterns and actionable insights to achieve the healthcare analytics goal.

Two very important checks that need to be done before the modeling stage are as follows:

1. Check whether two or more variables are highly correlated with each other (multicollinearity). If there are two or more variables highly correlated with each other, only one of these variables need to be included in the model. If the highly correlated variables are not removed, some of the model coefficients will have the wrong coefficient sign or extremely high values or extremely low values. In other words, some of the model coefficients will likely be unreliable if multicollinearity is not removed.
2. Check whether the target variable classes are balanced. Class imbalance problems are quite common in clinical applications. Let us consider an example that demonstrates the class imbalance problem where the clinical dataset has 5% of patients with heart disease and 95% of patients without heart disease. The model may have an overall accuracy of close to 100% and the specificity () metric close to 100%, while the sensitivity (measures the ratio of the number of actual heart disease patients that are correctly identified) metric is less than 20%. This clearly demonstrates that the clinical model is not good because we are more interested in a model that can accurately predict the heart disease cases (which is the minority class in this example).

To overcome the imbalance problem, the bagging and oversampling techniques may be used.

Once the multicollinearity and imbalance have been removed, depending on the healthcare analytics goal, the analyst may choose to perform predictive or classification modeling.

7.3.3 Step 3. Predict What Is Likely to Happen; or Perform Classification Analysis

Healthcare analytics is about being proactive in decision-making. At this stage, the analyst will model the healthcare data using machine-learning techniques that include decision trees, neural networks, and logistic regression. These techniques will uncover insights and patterns that highlight relationships and "hidden evidences" of the most influential variables.

7.3.4 Step 4. Check the Modeling Results

At this stage, the analyst will then evaluate how good the machine-learning model is, using metrics such as overall accuracy, recall, specificity, and AUC. Usually, the analyst will explore several machine-learning models and then select the best performing model based on the model evaluation results.

Check the modeling results stage involves checking that the model results are reasonable and then to evaluate the performance of the clinical prediction models. This is an important stage because if the following checks are not performed, it is highly likely that the model coefficients are not at all acceptable, and taking action using incorrect model coefficients can result in grouping patients in wrong patient profile groups, and therefore, treating patients with the wrong treatment.

1. To check whether the model results are reasonable, we need to check the following:
    - Are all the input variables statistically significant? The input variables are statistically significant when their p-values are less than 0.05.
    - Do the model coefficients have the right signs? If a coefficient has a positive sign and the expected sign should be negative, further investigation needs to be made. For example, the analyst needs to check whether there are outliers in the data or whether there are variables in the model that are highly correlated.
    - Is the size of the coefficients realistic? For example, if some coefficients are extremely small, almost zero, and some coefficients are extremely large, this will indicate that the input data needs to be scaled. This can be done by standardizing the data to z-scores (rescale the data to have a mean of 0 and standard deviation of 1 (unit variance). Another method is to normalize the data (rescales the values into a range of [0, 1]).
    - Is the standard deviation of the model coefficient acceptable? We sometimes find that the standard deviation of some model coefficients is extremely large. When we have this situation, it indicates that there are probably a few outliers for that variable, and we may need to remove the outliers so that the coefficient estimate of the variable is more accurate and reliable because the standard deviation will reduce and will not be so large.
2. To check whether the model performs well, we need to use the confusion matrix. The confusion matrix is a tool that is commonly used to evaluate the accuracy of the prediction model.

Once the clinical prediction model is built, it is important for the analyst to evaluate how good the prediction model is by testing the prediction model on unseen data (validation dataset). Using the validation dataset, the analyst needs to compute the overall accuracy, sensitivity, and specificity of the prediction model.

As a rule of thumb, the overall accuracy should be at least 70%.

From the confusion matrix, the analyst computes the sensitivity of the model, also known as the *true positive rate (TPR) or recall*. The sensitivity of the model measures the ratio of the number of patients with heart disease correctly identified. The sensitivity measure should be at least 70%.

From the confusion matrix, the analyst computes the specificity, also known as the *true negative rate (TNR)* and measures the ratio of the number of patients without heart disease that are correctly identified. The specificity measure should be at least 70%.

If the overall accuracy, sensitivity, and specificity of the model are at least 70%, this indicates the acceptance of the clinical predictive model and confirms that the model is performing well.

7.3.5 Step 5. Optimize (Find the Best Solution)

At this stage, the analyst will work with the clinician and apply the machine-learning model coefficients and outcomes and run "what-if" scenarios, to determine the best solution, with the given constraints and limitations. The analyst will select the optimal solution and model based on the lowest error, clinical targets, and his intuitive recognition of the model coefficients that are most aligned to the clinical goal.

7.3.6 Step 6. Derive a Clinical Strategy for Patient Care and Measure the Outcome

The clinician will apply the machine-learning tool to guide his clinical diagnosis/strategy and the appropriate treatment for his patients. After an appropriate time period, the patient's health condition is assessed to determine whether the machine-learning model and the clinical decision was effective or not, and whether the clinician should continue to be guided by the machine-learning tool and continue to use the clinical decision-making system.

7.3.7 Step 7. Update the CDS System

In the final stage, the analyst updates the patient outcome results in the CDS system, and the machine-learning tool derives new insights in real time. Information such as, "was the clinical decision and patient treatment effective?," "how did the patient treatment group compare with the control group?," and "what was the accuracy of the machine-learning tool?" are uploaded into the clinical database.

Over time, the clinic's database is continuously evolving with new patient treatment allocations, new patient outcomes, and real-time machine-learning tool accuracy information. It is important that the analyst reviews and monitors the CDS system to determine when the machine-learning tool is no longer accurate or to determine when the clinical treatment is no longer effective, so that steps can be taken to modify the CDS system appropriately.

For example, if the initial accuracy of the clinical model is 95%, over time this accuracy gets reduced to 90% (assuming a reduction of more than 5% indicates that action needs to be taken), the analyst should get an alert to investigate whether the model needs to be modified due to patient clinical changes and model coefficient changes.

Every year, the clinical model may need to be modified because patient group's lifestyles have changed over the years, and so the clinical model needs to adapt accordingly with more relevant input variables.

If the CDS tool is not monitored over time, it is possible that newer important input variables for clinical decision-making become available over time, and if these variables are not used for clinical modeling, the CDS tool will become less accurate over time, and eventually clinical treatment will become ineffective over time.

For example, if the initial accuracy of the clinical model is 95%, over time, this accuracy reduces to 85%, and the analyst should get a serious message to investigate whether the model needs to be revamped completely and built from scratch.

It may happen that every 3 years, a fresh clinical model may need to be built as the number of changes in the patient lifestyle is much greater, and the accuracy of the clinical model has decreased considerably where a small modification to the model is not sufficient.

It is highly recommended that the analyst builds a data-driven clinical decision-making system that includes trigger alerts to the analyst when unacceptable changes to the model accuracy occur. Changes to the clinical model accuracy must be communicated to all clinicians and healthcare professionals. The clinicians and healthcare professionals need to consult with their patients to better understand the patient lifestyle changes and health changes so that the analyst has better insights on how to modify or revamp the clinical model. This communication will likely improve the clinical model and patient treatment, resulting in better patient health.

The benefit of monitoring the CDS system is that it allows the analyst to be proactive in investigating the clinical models when the accuracy of the model reduces considerably. This way, the patients will receive the right treatment when their lifestyle and health conditions change.

So, in a nutshell, an automated CDS system will enable clinicians and healthcare professional to be proactive and to enlighten the analyst through the decision support system when patient lifestyle changes take place. The automated CDS system will improve the communication between the analyst and clinicians, resulting in a more effective patient care system. Analysts will receive patient treatment and health condition updates in a timely manner, allowing the CDS system to automatically trigger to the clinician the right treatment for the patient, at the right time, resulting in an effective patient care system.

7.4 DATA EXPLORATION AND UNDERSTANDING OF THE HEALTHCARE PROBLEM

The objective of the Framingham Heart Study was to unravel the underlying causes of heart disease using patient blood cholesterol, blood pressure, smoking, and obesity information to identify high-risk heart attack patients.

The R programming software was used to perform our statistical analysis. R is a programming language and free software environment for statistical computing and graphics supported by the R Foundation for Statistical Computing. The R language is widely used among statisticians and data miners for developing statistical software and data analysis.

An initial data analysis using 4,240 patients of the framingham.csv [8] dataset was undertaken to understand the quality of the patient health data and to correct any errors if found. Using frequency counts, it was found that 582 (13.7%) patients had some missing data. As missing data was only 13.7%, patients who had missing data were excluded from the study analysis. The study now consisted of 3,658 patients with 2,035 (56%) females and 1,623 (44%) males.

Further, the predictive model input data was standardized to enable all input variables in the model to have a fair chance of being significant. Predictive modeling is a process that uses data mining and probability to forecast outcomes. Each model is made up of a number of risk factors, which are variables that are likely to influence Ten Year CHD, the target variable of interest [9]. Once data has been collected for the relevant risk factors, a statistical model is formulated.

The dataset had four types of risk factors:

- Demographic Risk Factors
- Behavior Risk Factors
- Medical History Risk Factors
- Risk Factors from First Examination

The Demographic Risk Factors included:

- Male: sex of the patient (44%)
- Age: age in years at first examination (32–70 years)
- Education:
  - Some high school (1) (42%),
  - High school/GED (General Education Development) (2) (30%),
  - Some college/vocational school (3) (17%),
  - College (4) (11%)

The Behavior (Smoking Behavior) Risk Factors included:

- Current smoker (49%)
- cigsPerDay (1–70 cigarettes per day)

Medical History Risk Factors included:

- BPmeds: On blood pressure medication at the time of first examination (3%)
- prevalentStroke: Previously had a stroke (1%)
- prevalentHyp: Currently Hypertensive (31%)
- diabetes: Currently has diabetes (3%)

Risk Factors from First Examination included:

- totChol: Total cholesterol (mg/dL) (113–600 mg/dL)
- sysBP: Systolic blood pressure (83.5–295)
- diaBP: Diastolic blood pressure (48–142.5)

- BMI: Body Mass Index (weight (kg)/height (m*m)) (15.5–56.8)
- heartrate: Heart rate (beats/minute) (44–143)
- glucose: Blood glucose level (mg/dL) (40–394)

Ten-Year CHD: Currently have CHD (15%)

After performing a frequency count analysis, we identified which risk factors were significant using chi-square tests for categorical variables and independent two-sample T-test for the continuous variables.

There were eight continuous risk factors. Using the two-sample T-test, seven (totchol, sysBP, diaBP, BMI, glucose, cigsPerDay, age) of the eight risk factors were identified as significant factors for CHD. Only heart rate was not a significant factor for CHD in this dataset.

Of the seven categorical variables, six (male, education, BPMeds, prevalentStroke, prevelentHyp, diabetes) were significant factors for CHD. Only currentSmoker was not a significant factor for CHD.

Next, we checked to see if multicollinearity existed amongst continuous variables and found that diaBP and sysBP are highly correlated with a correlation value of 0.787 and a *p*-value of 0.000. This means that we need to exclude one of these two variables before running our predictive model. It does not matter which one we exclude. We have decided to exclude the diaBP variable.

To overcome the imbalance (85:15) of the presence of CHD in the target classes, we will adjust the threshold from 0.5 (the default) to 0.85.

7.5 MACHINE-LEARNING MODEL APPLICATION

The logistic regression model was selected for use in this study because it is the most basic and robust classification algorithm. The logistic regression is a special type of regression where the binary response variable is related to a set of explanatory predictor variables that can be continuous or discrete. The logistic regression model is perfect for situations where the aim is to predict the presence or absence of a characteristic or outcome based on the values of a set of predictor variables [10]. The relationship between the target and input variables is not always a straight line, and so a nonlinear or logistic regression model is used. Furthermore, the logistic regression model was chosen as it required little running time compared with other complicated machine-learning algorithms and its output is also easy to interpret. As the dependent variable of the logistic regression model is binary, the dependent variable selected was "TenYearCHD" and was used to classify the patients.

The main assumption for building a logistic regression model is that the independent variables do not have significant multicollinearity [11]. Initially, there were 15 input variables. After multicollinearity was handled, we were left with 14 input variables for our logistic regression model. As "diaBP" was highly correlated with "sysBP," with a correlation value of 0.787 and a *p*-value of 0.000, it was necessary to remove one of the two variables ("diaBP," "sysBP"), the diaBP variable was removed to overcome multicollinearity.

Our logistic regression model was carried out using the backward stepwise regression method. Variables with a p-value greater than 0.05 were deemed as insignificant to "TenYearCHD" and were removed from the model. The process stopped when the model was left with only variables with p-values less than 0.05.

7.6 EVALUATION OF THE MACHINE-LEARNING MODEL RESULTS

Insights about the input variables and their effect on "TenYearCHD" were gained by understanding the signs and size of the coefficients of the logistic regression model results and the resulting odds ratios.

The size of the coefficient indicates the main drivers of "TenYearCHD." The larger the coefficient, the higher the significance of the variable to "TenYearCHD." A positive coefficient indicated that the predictor variable had a positive relationship with "TenYearCHD." The CHD predictive model has five significant risk factors: male, age, cigsPerDay, sysBP, and glucose. See Table 7.1 later.

Age (0.614) is the most important risk factor, followed by sysBP (0.389), male (0.303), cigsPerDay (0.222), and glucose (0.207), as they have the highest positive relationship with "TenYearCHD" and were the only significant variables in the logistic regression model.

An added advantage of using the logistic regression is that it calculates the logarithmic odds of a "TenYearCHD" event. The odds ratio is calculated to study the effect of each variable in affecting the odds of the "TenYearCHD" [2]. The odds ratio for "male," "age," "cigsPerDay," "sysBP," and "glucose" was 1.354, 1.848, 1.248, 1.475, and 1.229, respectively.

For example, the odds ratio for males means that holding all other variables constant, and patients who are "male" have a 35% higher risk of "TenYearCHD" compared with females.

Other than understanding the association of different variables to a "TenYearCHD," there is a need to evaluate the training model to find out its predictive power. This is done by measuring the accuracy of the model. Accuracy measures are computed from the confusion matrix. The confusion matrix measured the ability of the model to classify the patients who are at risk of a CHD. See Table 7.2 later.

TABLE 7.1
Significant Risk Factors for CHD

Coefficient	Estimate	Std. Error	Z-Value	p-Value	Odds Ratio
Intercept	−2.02182	0.069	−29.216	0.000	0.132
Male	0.30334	0.063	4.797	0.000	1.354
Age	0.61417	0.662	9.276	0.000	1.848
cigsPerDay	0.22166	0.058	3.819	0.000	1.248
sysBP	0.38852	0.056	6.954	0.000	1.475
Glucose	0.20659	0.047	4.376	0.000	1.229

TABLE 7.2
Confusion Matrix for Test Data

Reference	Prediction		
	0	1	Total
0	654	276	930
1	58	109	167
Total	712	385	1,097

From the confusion matrix, 654 plus 109 patients were classified correctly as not having a risk of CHD or having a risk of CHD, respectively. Hence, from Table 7.2, the overall accuracy of the CHD risk factor model is 70%. However, 276 patients who did not have CHD were classified as having CHD, and 58 patients who were classified as having CHD were predicted as having no CHD.

$$\text{The "}\textbf{True Positive Rate/Recall/Sensitivity}\text{" value} = \frac{\text{True positive}}{\text{True positive} + \text{False negative}}$$

$$= \frac{109}{(109+58)}$$

is 65%, which means for every 100 patients who have a TenYearCHD risk, the model is able to classify 65 of them correctly.

$$\text{The "}\textbf{Specificity}\text{" value} = \frac{\text{True positive}}{\text{True positive} + \text{False negative}} = \frac{654}{(654+276)} \text{ is } 70\%,$$

which means for every 100 patients without TenYearCHD, the model is able to classify 70 of them correctly.

Sensitivity and specificity are inversely proportional to each other. So, when we increase sensitivity, specificity decreases and vice versa.

$$\text{The } \textbf{False Positive Rate} = 1 - \text{Specificity}.$$

The AUC is plotted with the TPR on the y-axis against the false positive rate on the x-axis [12]. See Figure 7.1 later.

The AUC is a performance measurement for a classification problem at various threshold settings. Our classification problem was to classify patients who were at risk of a "TenYearCHD" and those who were not at risk. The AUC tells us how much our model is capable of distinguishing between the "TenYearCHD" class and the "Non-TenYearCHD" class. The higher the AUC value, the better the model is at distinguishing between patients with disease and without disease. An excellent model has an AUC value near to 1, which means it has good measure of separability. A poor model has AUC near to 0, which means it has a poor measure of separability.

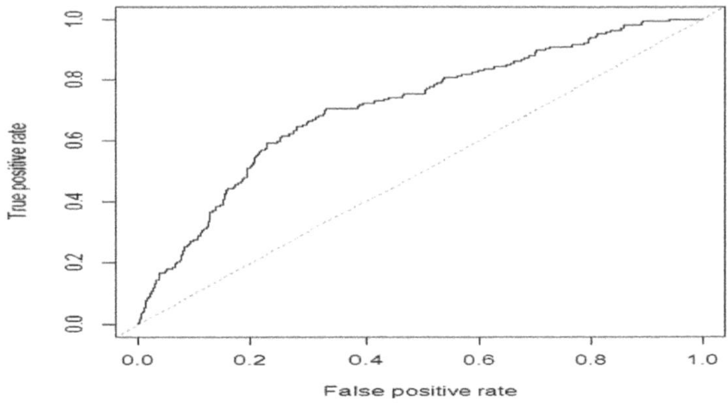

FIGURE 7.1 AUC for test data.

Further, the AUC for our classification model was 71%, demonstrating a good measure of how well our classification model's performance is.

The logistic regression model results were considered good, and hence, the training CHD risk factor model was accepted for predicting patients at risk of CHD.

7.7 CONCLUSION

The logistic regression model was used to identify which patients are likely to be at high risk of CHD. The outcome from this study confirms first that machine-learning techniques, such as logistic regression technique, can be used to classify patients as high-risk CHD patients.

Further, from 15 risk factors, five risk factors were identified as CHD indicators. Age was the most important risk factor, the older the patient, the more likely they were at risk of a CHD. Male patients were found to have a higher risk of CHD than females. Systolic blood pressure was a significant factor, indicating that diastolic blood pressure can also be seen as a significant factor as systolic blood pressure and diastolic blood pressure were highly correlated. So, patients with high blood pressure (hypertension) have a high risk for CHD. Lastly, cigsPerDay was also a significant risk factor for CHD, the more cigarettes a patient smoked, the more likely they were at risk of CHD.

7.8 FUTURE DIRECTION

Data analytical techniques, particularly machine-learning techniques, can be very valuable for clinicians, as these techniques can accurately identify which patients are high-risk CHD patients and require treatment that can reduce their risk of CHD or prevent them from getting a heart attack. The coronary risk factor model can be used by clinicians as a CDS tool (CDST). The CDST can serve as a good guide for the clinicians when diagnosing patients by providing the required support and confidence to the clinician when decisions about the patient's risk of a heart attack have to be made.

With the advancement of technology, clinicians can quickly update patient information in the CDST, and the patient CHD risk level will be immediately informed to the clinician. The use of a CDST will assist clinicians in making important clinical decisions faster; hence, administration of patient treatment will be earlier, preventing or reducing the coronary heart risk of the patient early.

ACKNOWLEDGEMENTS

The Framingham Heart Study is supported by Contract No. HHSN268201500001I from the National Heart, Lung, and Blood Institute (NHLBI) with additional support from other sources. This manuscript was not approved by the Framingham Heart Study. The opinions and conclusions contained in this publication are solely those of the author and are not endorsed by the Framingham Heart Study or the NHLBI and should not be assumed to reflect the opinions or conclusions of either.

REFERENCES

1. Alkhatib M. A., Talaei-Khoei A., Ghapanchi A. H. Analysis of Research in Healthcare Data Analytics. *Australasian Conference on Information Systems*, Sydney, 2015.
2. Simpao A.F., Ahumada L.M., Gálvez J.A., Rehman M.A. A review of analytics and clinical informatics in health care. *J. Med. Syst.* 2014;38:45. doi:10.1007/s10916-014-0045-x.
3. Karaolis M., Moutiris J.A., Papaconstantinou L., Pattichis C.S. Association rule analysis for the assessment of the risk of coronary heart events, *Proceedings of the Annual International Conference of the IEEE Engineering in Medicine and Biology Society*, Minneapolis, MN, USA. 3–6 September 2009, pp. 6238–6241.
4. Kim, T.W., Park, K.H., Yi, S.H., Kim, H.C. A big data framework for u-healthcare systems utilizing vital signs. *Proceedings – 2014 International Symposium on Computer, Consumer and Control*, IS3C, 2014, pp. 494–497. doi:10.1109/IS3C.2014.135.
5. Jacob, S. Young parkland physician makes a splash with predictive modeling software, *D Healthcare Daily*, Dallas, 2012, http://healthcare.dmagazine.com/2012/12/10/young-parkland-physician-makes-asplash-with-predictive-modeling-software/, Retrieved 09 August, 2015.
6. Sonnati, R. Improving healthcare using big data analytics. *Int. J. Sci. Tech. Res.* March 2017;6(03), ISSN 2277-8616.
7. Big Data Offers Big Opportunities in Healthcare, Retrieved from www.villanovau.com/resources/bi/big-data-healthcare-opportunities/#.VnfRArZ95kg.
8. https://courses.edx.org/asset-v1:MITx+15.071x_2a+2T2015+type@asset+block/framingham.csv.
9. https://en.wikipedia.org/wiki/Predictive_modelling.
10. Menard, S.W., NetLibrary, I. *Applied Logistic Regression Analysis.* Thousand Oaks, CA: Sage Publications.
11. Hosmer, D.W., Lemeshow, S., Sturdivant, R.X. *Applied Logistic Regression* (third ed.). Hoboken, NJ: Wiley.
12. https://towardsdatascience.com/understanding-auc-roc-curve-68b2303cc9c5.

8 Bioinformatics Analysis of Dysfunctional (Mutated) Proteins of Cardiac Ion Channels Underlying the Brugada Syndrome

Carlos Polanco
Universidad Nacional Autónoma de México

Manlio F. Márquez
Instituto Nacional de Cardiología Ignacio Chávez

Vladimir N. Uversky
University of South Florida

Thomas Buhse
Universidad Autónoma del Estado de Morelos

Miguel Arias Estrada
Instituto Nacional de Astrofísica, Óptica y Electrónica

CONTENTS

8.1 INTRODUCTION

This work illustrates a possible physicochemical trait that is unique to mutated proteins that underlie a malignant disease known as Brugada syndrome (BrS). We speculate that this information could be incorporated into a bionanosensor to identify the related proteins by this notorious physicochemical trait.

The electrical activity of the heart is based on the proper function of ion channels. There are specific pores located mainly on their cytoplasmic membranes that permit the passage of ions through it. In subjects who have BrS [1], specific defects in some of these channels have been reported, resulting in a characteristic electrocardiographic phenotype and the possibility to generate malignant ventricular arrhythmias [2]. Mutations in BrS-related proteins affect the physical form of these ion channels. Currently, the only treatment that can effectively prevent this syndrome is the insertion of an automatic implantable cardiac defibrillator, or, in subjects who already have such a device, the recurrence of ventricular arrhythmias can be prevented with a quinidine treatment [3].

Medical studies on close relatives of subjects diagnosed with BrS have determined that this syndrome can be inherited [4], as some of these relatives also have mutations in BrS-related proteins. In the UniProt database [5], there are 4,388 overrepresented (redundant) BrS mutated proteins from 36 proteins [2] that underlie this abnormality. This overrepresentation or duplicity is not exclusive of this database, since it often comes from the multiple functions found for the same protein.

To develop a new method for the molecular diagnosis of the syndrome, based on parameters other than the detection of the mutation, this work introduces a highly discriminating metric based on the concept of electronegativity in Pauling's work [6] called "polar profile" [7]. Several protein groups were used in the evaluation of the BrS mutated proteins that were selected based on one of the following factors: (i) they are peptides or proteins previously used by this team to carry out analysis of functional identification, (ii) there is a large number of corresponding proteins found in nature, (iii) they have a leading role in diseases that affect humans, and (iv) they are closely related to the BrS mutated protein, as it is the case of BrS proteins. These protein groups are (i) a set of BrS proteins from which the BrS-related mutated proteins derive; (ii) a set of antibacterial peptides named selective cationic amphipathic antibacterial peptides (SCAAP) [8], and this group stands out for being highly toxic to bacteria but causes almost no harm to human cells; (iii) three sets of antimicrobial proteins: bacteria, fungi, and viruses extracted from UniProt and APD2 [9] databases; (iv) six sets of lipoproteins extracted from UniProt database [5] and associated with coronary and heart diseases; (v) two sets of proteins with different degrees of structural disorder named intrinsically disordered (unfolded) and partially folded proteins, taken from the work of Oldfield and coworkers [10]. This group is particularly important because it is used as a reference on the degree of disorder [11] for other protein groups; and (vi) a set of 557,713 "reviewed" proteins from UniProt database [5] that includes all the proteins whose annotation is confirmed as correct. This group was not used in the calibration of the Polarity Index Method® (PIM), but in the search of proteins with a "polar profile" similar to that of BrS mutated proteins.

To obtain the "polar profile" of BrS mutated proteins, a computational system called PIM [7] was used. It was designed and programmed to perform two functions: to obtain the BrS "polar profile" of the mutated proteins and to compare that profile with the polar profiles of other protein groups. The PIM is a nonsupervised system, with a set of programs written in Fortran 77 [12] and scripts in Linux [13] that extensively evaluate a single physicochemical property, the "polar profile." The PIM system is trained in evaluating different groups of proteins and making automatic changes on the "polar profile" of the target protein group, to reach the "polar profile" that best characterizes this target. The process is independent of the number of protein groups participating. Its metric only uses the linear representation of the protein, in FASTA format, reading one pair of amino acids at a time from one end to the other to find the polarity type that represents each pair.

The discriminatory efficiency of the PIM system was verified by comparison of the proportion of accepted/rejected proteins from the BrS mutated protein group and BrS protein group with respect to the real proportion of proteins in the same protein groups. This analysis was performed using the nonparametric two-sided Kolmogorov–Smirnov test (Section 8.4.4). An analysis of the same protein groups was carried out to obtain the characterization of their "degree of disorder" using the set of supervised per-residue disorder predictors, such as PONDR® FIT, PONDR® VLXT, PONDR® VSL2, PONDR® VL3, FoldIndex, IUPred, and TopIDP (Section 8.4.2). This analysis was aimed at verifying the efficiency of PIM when identifying and discriminating the "polar profile" of the set of BrS mutated proteins and getting a "fingerprint" of the degree of disorder of this group.

This article comprises three sections: (i) A computational analysis of the degree of disorder of each BrS mutated protein as well as its association with other diseases, (ii) A bioinformatic characterization (through the PIM®) of the BrS mutated proteins, and their contrast with a large and diverse set of protein groups of different structural and functional types, with the objective to obtain a "fingerprint" of the BrS mutated protein set, and (iii) the use of this "fingerprint" on all of the reviewed proteins listed in the largest known public primary database UniProt containing protein sequence and functional information. We speculate that this characterization could be used for the rapid identification of this syndrome at early stages, potentially even before the development of major symptoms.

8.2 RESULTS

We provide a workflow of the PIM® (see Figure 8.1) to clarify and make the results and procedures of this nonsupervised computational algorithm understandable.

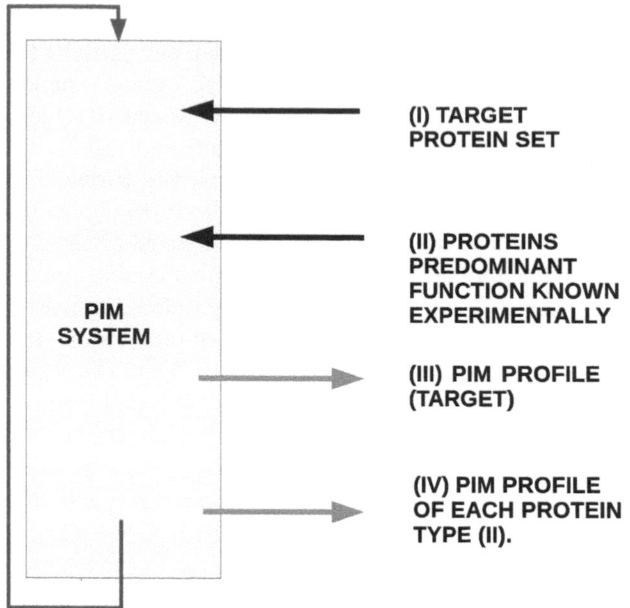

FIGURE 8.1 Workflow of PIM®. (i) Assembly of the target set whose polar profile will be obtained by the PIM system. (ii) Assembly of protein sets whose functions and/or structures are known experimentally. (iii) Extraction of the polar profile characteristic of the target set. (iv) Polar profile of each protein in protein sets, whose functions and/or structures are known. (v) Comparison and reacquisition of the polar profile of the target set, based on the calculation of similarity of the target set and the group of protein type (ii) (see Section 8.4.1).

8.2.1 Brief Description of Unique BrS-Related Proteins

Later, we provide a brief description of the 20 unique proteins associated with BrS and also a list of other diseases linked to mutations in these proteins.

Hairy/enhancer-of-split related with YRPW motif protein 2 (HEY2, UniProt ID: Q9UBP5); *HEY2* gene serves as the BrS susceptibility gene (PMID: 28637782).

Natriuretic peptides A (NPPA, UniProt ID: P01160); mutations in *NPPA* gene are associated with atrial standstill 2 (ATRST2) and familial atrial fibrillation 6 (ATFB6).

Ankyrin-3 (ANK3, UniProt ID: Q12955); mutations in *ANK* gene are associated with mental retardation, autosomal recessive 37 (MRT37); ANK3 protein also plays a role in BrS via interaction with the cardiac sodium channel Na(v)1.5 (a product of the *SCN5A* gene).

Calmodulin-1 (CALM1, UniProt ID: P0DP23); mutations in *CALM1* gene are associated with ventricular tachycardia, catecholaminergic polymorphic 4 (CPVT4), and Long QT syndrome 14 (LQT14).

Voltage-dependent L-type calcium channel subunit beta-2 (CACNB2, UniProt ID: Q08289); mutations in *CACNB2* gene are associated with BrS 4 (BRGDA4).

Myosin-7 (MYH7, UniProt: P12883); mutations in *MYH7* gene are associated with familial hypertrophic cardiomyopathy 1 (CMH1), myosin storage autosomal dominant myopathy, (MSMA), Scapuloperoneal myopathy MYH7-related (SPMM), dilated cardiomyopathy 1S (CMD1S), Myopathy, distal, 1 (MPD1), myosin storage autosomal recessive myopathy, (MSMB), and left ventricular noncompaction 5 (LVNC5).

Potassium/sodium hyperpolarization-activated cyclic nucleotide-gated channel 4 (HCN4, UniProt ID: Q9Y3Q4); mutations in *HCN4* gene are associated with Sick sinus syndrome 2 (SSS2) and BrS 8 (BRGDA8).

Potassium voltage-gated channel subfamily H member 2 (KCNH2, UniProt ID: Q12809); mutations in *KCNH2* gene are associated with Long QT syndrome 2 (LQT2) and Short QT syndrome 1 (SQT1).

Potassium voltage-gated channel subfamily E member 3 (KCNE3, UniProt ID: Q9Y6H6); mutations in *KCNE3* gene are associated with BrS 6 (BRGDA6).

Potassium voltage-gated channel subfamily KQT member 1 (KCNQ1, UniProt ID: P51787), mutations in *KCNE3* gene are associated with long QT syndrome 1 (LQT1), Jervell and Lange-Nielsen syndrome 1 (JLNS1), familial atrial fibrillation 3 (ATFB3), short QT syndrome 2 (SQT2), and non-insulin-dependent diabetes mellitus, (NIDDM).

Potassium voltage-gated channel subfamily D member 3 (KCND3, UniProt ID: Q9UK17), mutations in *KCND3* gene are associated with Spinocerebellar ataxia 19 (SCA19) and BrS 9 (BRGDA9).

Voltage-dependent L-type calcium channel subunit alpha-1C (CACNA1C, UniProt ID: Q13936), mutations in *CACNA1C* gene are associated with Timothy syndrome (TS) and BrS 3 (BRGDA3).

Sodium channel protein type 5 subunit alpha (SCN5A, UniProt ID: Q14524), mutations in *SCN5A* gene are associated with progressive familial heart block 1A (PFHB1A), long QT syndrome 3 (LQT3), BrS 1 (BRGDA1), sick sinus syndrome 1 (SSS1), familial paroxysmal ventricular fibrillation 1 (VF1), sudden infant death syndrome (SIDS), atrial standstill 1 (ATRST1), dilated cardiomyopathy, 1E (CMD1E), and familial atrial fibrillation 10 (ATFB10).

Sodium channel protein type 10 subunit alpha (SCN10A, UniProt ID: Q9Y5Y9), mutations in *SCN10A* gene are associated with episodic pain syndrome, familial, 2 (FEPS2); SCN10A gene is considered by some groups as a major susceptibility gene for BrS (BRGDA).

Transforming growth factor beta-3 (TGFB3, UniProt ID: P10600), mutations in *TGFB3* gene are associated with arrhythmogenic right ventricular dysplasia, familial, 1 (ARVD1) and Loeys-Dietz syndrome 5 (LDS5).

Sodium channel subunit beta-2 (SCN2B, UniProt ID: O60939), mutations in *SCN2B* gene are associated with familial atrial fibrillation, 14 (ATFB14). Also, defects in this gene can be a cause of BrS (BRGDA) and sudden infant death syndrome (SIDS).

ATP (Adenosine triphosphate)-sensitive inward rectifier potassium channel 8 (KCNJ8, UniProt ID: Q15842), mutations in *KCNJ8* gene are associated with sudden infant death syndrome (SIDS), hypertrichotic osteochondrodysplasia (HTOCD), variation in the *KCNJ8* gene are associated with BrS (BRGDA).

Sodium channel subunit beta-3 (SCN3B, UniProt ID: Q9NY72), mutations in *SCN3B* gene are associated with BrS 7 (BRGDA7) and familial atrial fibrillation, 16 (ATFB16).

Sodium channel subunit beta-1 (SCN1B, UniProt ID: Q07699), mutations in *SCN1B* gene are associated with generalized epilepsy with febrile seizures plus 1 (GEFS+1), BrS 5 (BRGDA5), familial atrial fibrillation, 13 (ATFB13), and epileptic encephalopathy, early infantile, 52 (EIEE52).

Glycerol-3-phosphate dehydrogenase 1-like protein (GPD1L, UniProt ID: Q8N335) and mutations in *GPD1L* gene are associated with BrS 2 (BRGDA2).

8.2.2 PIM-Based Analysis of the Unique BrS-Related Proteins

When the PIM system was calibrated with the unique BrS mutated protein group, the method was able to identify 70% of them (Table 8.1). The graphic interpretation of the "polar profiles" (Figure 8.2) of the unique BrS proteins and the unique BrS mutated proteins showed that these profiles differ in all interactions in less than 15%, except for the polar interaction [NP, NP], whose difference was greater than 70% (Figure 8.2a).

In addition, when the PIM system was calibrated with the unique BrS mutated proteins (MutBRUGADA), its "polar profile" was compared with the 557,713 "polar profiles" of the "reviewed" proteins from UniProt database, and it was found that 35,944 *(35,944/557,713≈6.4%) of the "reviewed" proteins from the* UniProt *database have their "polar profiles" similar to the "polar profile" of* the unique BrS mutated proteins.

From these results, it can be concluded that the "polar profile" of BrS mutated proteins is *not* similar to any of the "polar profiles" studied here, indicating that BrS mutated proteins are characterized by a highly specific "polar profile" according to the percentages mentioned in these tables.

8.2.3 Intrinsic Disorder Analysis of the BrS-Related Proteins

This important observation indicates that the peculiarities of the "polar profile" of the BrS proteins and their mutated proteins can be used for computational identification and differentiation of the BrS mutations in large protein datasets. To further illustrate the power of the polar profile approach in discriminating the proteins of interest, we analyzed 20 unique BrS-related proteins with a set of commonly used algorithms for prediction of protein intrinsic disorder predisposition, such as PONDR® VLXT, PONDR® VSL2, PONDR® VL3, PONDR® FIT, IUPred_short, and IPred_long. Table 8.2 represents an overview of the results of this analysis, showing for each query protein a mean disorder score evaluated by averaging the per-residue disorder profiles generated by the individual disorder predictors as well as a mean predicted percentage of intrinsic disorder (PPID, which is a content of residues predicted to be disordered; i.e., those with the disorder scores 0.5 threshold) calculated using the mean per-residue disorder profile generated for each query

TABLE 8.1
Analysis of Proteins by PIM System

Groups	SCAAP	Bacteria UniProt	Fungi UniProt	Virus UniProt	Bacteria APD	Fungi APD	Virus APD	Completely Disordered
Brugada proteins	0	3	12	13	2	0	0	1
Brugada mutated proteins	0	4	5	3	1	0	0	0
Groups	Partially disordered	HDL	LDL	VLDL	Chylomicrons	Atherosclerosis	Brugada proteins	Brugada protein mutations
Brugada proteins	3	16	6	18	7	0	64	11
Brugada mutated proteins	0	5	8	0	7	0	25	67

Similarities (%) found by PIM system in the protein groups. The score represents the percentage of proteins (column) with a similar "polar profile" (row). For example, the PIM system calibrated with the **BRUGADA** group has a similarity of 1% with the "polar profile" of the completely disordered protein group. See Section 8.4.1.

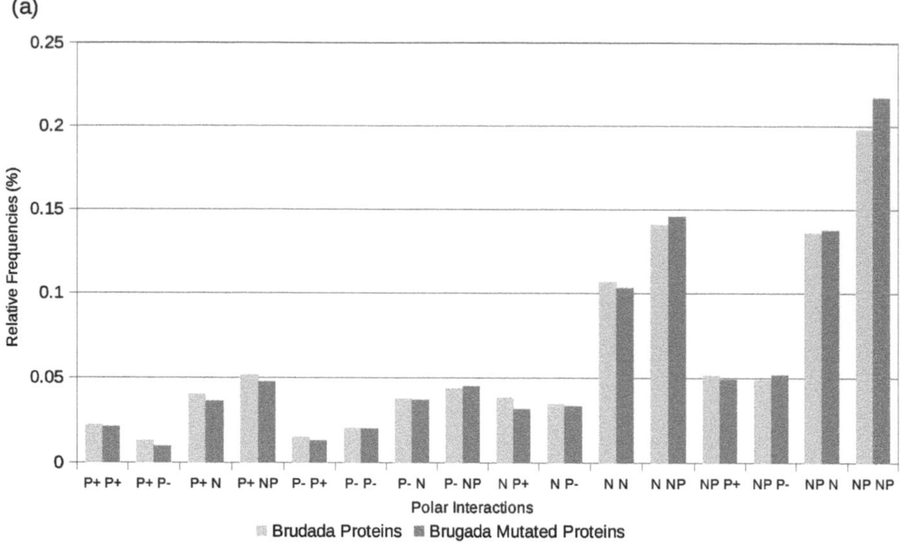

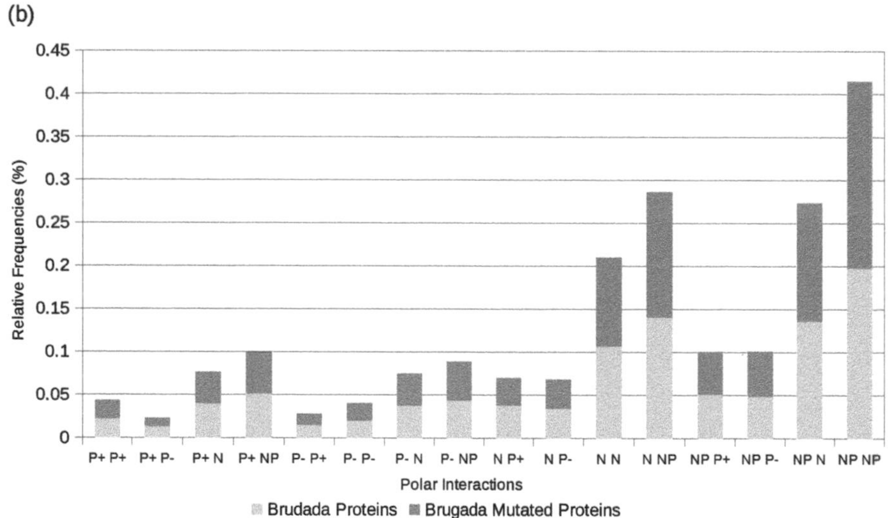

FIGURE 8.2 Graphical representation of the BrS protein and BrS protein mutations. (a) Column-Normal Excel software. (b) Column-Stacked Excel software. The *X*-axis represents the 16 polar interactions.

protein and PONDR® VSL2-based PPID values. Some of the data from Table 8.2 was used to generate Figure 8.3, which provided a compelling overview of the overall intrinsic disorder predisposition of these proteins.

In fact, Table 8.2 and Figure 8.3 clearly show that all BrS proteins analyzed in this study contain noticeable levels of intrinsic disorder, and a vast majority of them clearly belong to the category of hybrid proteins containing ordered domains/regions and intrinsically disordered regions of different lengths. This conclusion is

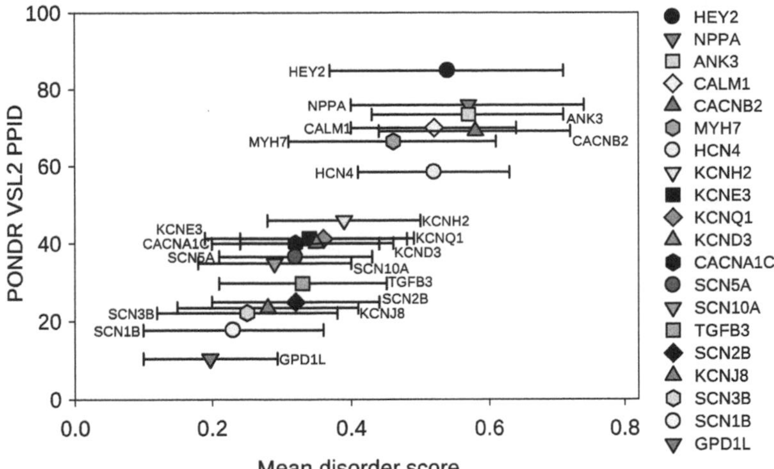

FIGURE 8.3 Quantification of the overall disorder predisposition of 20 unique BrS-related proteins. Plot shows the dependence of PONDR® VSL2-based PPID values on the mean disorder scores evaluated for individual proteins by averaging the outputs of six disorder predictors. Error bars show standard deviations evaluated using the mean disorder profile calculated by averaging six per-residue disorder profiles generated for each protein by individual disorder predictors.

further illustrated by the mean disorder profiles for these proteins provided in the Supplementary Materials section, where the presence of intrinsically disordered regions in almost all these proteins is clearly seen. The conclusion on the highly disordered nature of BrS proteins is further supported by grouping these proteins using the accepted classification of intrinsic disorder-containing proteins based on their PPID values, where proteins are considered as highly ordered, moderately disordered, or highly disordered, if their PPID < 10%, 10% ≤ PPID < 30%, or PPID 30%, respectively [14]. This analysis based on the investigation of the mean disorder profiles revealed that only three proteins from this dataset, SCN3B, SCN1B, and GPD1L, were predicted as highly ordered, whereas 6 and 11 BrS proteins were classified as moderately and highly disordered (see Table 8.2). According to a similar classification based on the analysis of PONDR® VSL2 profiles, all BrS proteins are either moderately or highly disordered (6 or 14 proteins, respectively). Furthermore, seven BrS proteins were predicted to have PPID values exceeding 50% by both approaches. Since the intrinsic disorder is crucial for the functionality of many proteins [14–22], and since high levels of intrinsic disorder are commonly found in proteins associated with various diseases [23–29], the results of this analysis suggest that structural plasticity is crucial for functionality and pathological implications of BrS-related proteins. Despite these important conclusions, disorder-based profiles of BrS proteins do not contain any common features that could be used for their classification and differentiation from other disorder-containing proteins.

Similarly, analysis of the effects of BrS-related mutations on the disorder propensity of carrier proteins did not show any repeated peculiarity that can be used for the

TABLE 8.2
Peculiarities of Intrinsic Disorder Analysis of the Unique BrS-related Proteins

Protein (Disease)	UniProt ID	Length	pI (Charges)	Mean Disorder	Standard Deviation	Mean PPID (%)	PONDR® VSL2 PPID (%)
HEY2 (BRGDA)	Q9UBP5	337	8.31 (+29/–27)	0.54	0.17	57.86	84.91
NPPA (ATFB6; ATRST2)	P01160	153	6.59 (+17/–17)	0.57	0.17	71.90	75.97
ANK3 (BRGDA; MRT37)	Q12955	4,377	6.07 (+544/–608)	0.57	0.14	59.36	73.46
CALM1 (LQT14; CPVT4)	P0DP23	149	4.09 (+14/–38)	0.52	0.12	60.40	70.00
CACNB2 (BRGDA4)	Q08289	660	8.11 (+94/–92)	0.58	0.14	65.45	69.14
MYH7 (CMH1; MSMA; SPMM; CMD1S; MPD1; LVNC5)	P12883	1,935	5.63 (+316/–358)	0.46	0.15	54.99	66.43
HCN4 (BRGDA8; SSS2)	Q9Y3Q4	1,203	9.07 (+113/–99)	0.52	0.11	50.21	58.47
KCNH2 (SQT1; LQT2)	Q12809	1,159	8.20 (+120/–115)	0.39	0.11	37.62	45.95
KCNE3 (BRGDA6)	Q9Y6H6	103	8.82 (+12/–10)	0.34	0.15	22.33	41.35
KCNQ1 (ATFB3; LQT1; JLNS1; SQT2; NIDDM)	P51787	676	9.88 (+86/–53)	0.36	0.12	30.92	41.21
KCND3 (BRGDA9; SCA19)	Q9UK17	655	8.56 (+69/–62)	0.35	0.11	34.05	40.09
CACNA1C (BRGDA3; TS)	Q13936	2,221	6.33 (+225/–240)	0.32	0.12	29.81	40.05
SCN5A (BRGDA1; LQT3; ATFB10; SSS1; VF1; PFHB1A; ATRST1; CMD1E)	Q14524	2,016	5.34 (+187/–229)	0.32	0.11	30.06	36.64
SCN10A (BRGDA; FEPS2)	Q9Y5Y9	1,956	5.67 (+189/–218)	0.29	0.11	23.26	34.95
TGFB3 (ARVD1; LDS5)	P10600	412	8.31 (+54/–50)	0.33	0.12	20.15	29.78
SCN2B (ATFB14; BRGDA; SIDS)	O60939	215	5.98 (+25/–28)	0.32	0.12	16.74	25.00

(*Continued*)

TABLE 8.2 (*Continued*)
Peculiarities of Intrinsic Disorder Analysis of the Unique BrS-related Proteins

Protein (Disease)	UniProt ID	Length	pI (Charges)	Mean Disorder	Standard Deviation	Mean PPID (%)	PONDR® VSL2 PPID (%)
KCNJ8 (SIDS; HTOCD; BRGDA)	Q15842	424	9.38 (+52/–40)	0.28	0.13	15.09	23.53
SCN3B (BRGDA7; ATFB16)	Q9NY72	215	4.66 (+19/–32)	0.25	0.13	8.83	22.22
SCN1B (BRGDA5; EIEE52; GEFS+1; ATFB13)	Q07699	218	4.86 (+21/–33)	0.23	0.13	4.13	17.81
GPD1L (BRGDA2)	Q8N335	351	6.61 (+40/–41)	0.197	0.097	2.85	10.51

Some basic properties of the unique BrS-related proteins are shown. We list proteins names and associated diseases, UniProt IDs for proteins, length of their sequences (as corresponding number of residues), pI, and the number of positively and negatively charged residues provided by ProtParam tool (https://web.expasy.org/protparam/) of ExPaSy Bioinformatics Resource Portal. For each query protein, we also list a mean disorder score evaluated by averaging the per-residue disorder profiles generated by individual disorder predictors and the corresponding standard deviation as well as a mean PPID (which is a content of residues predicted to be disordered; i.e., those with the disorder scores 0.5 threshold) calculated using the mean per-residue disorder profile generated for each query protein and also PONDR® VSL2-based PPID values.

identification of such mutations. To illustrate this point, Figure 8.4 shows the mutation effect on intrinsic disorder profiles of one of the most disordered proteins in the dataset, NPPA (Figure 8.4a), whereas Figure 8.4b shows the disorder profiles of the wild-type and mutated forms of one of the most highly ordered proteins, SCN1B. It can be seen that, in both these cases, BrS-related mutations induce some changes (increases or decreases) in local intrinsic disorder propensity of regions in close proximity to mutations. Again, there are no features in the corresponding disorder profiles that can discriminate BrS-related mutations from any other mutations in these mutated proteins. This is in contrast to the PIM-based analysis, which generates specific polar profiles containing characteristic features that can be used for finding computable differences to identify BrS mutated proteins.

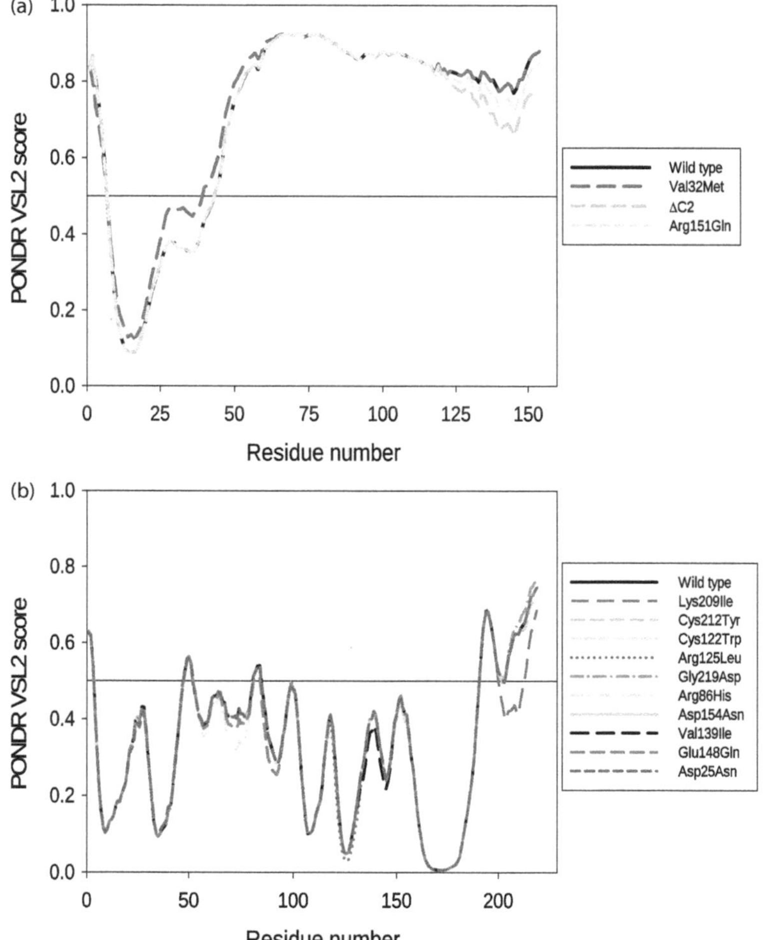

FIGURE 8.4 Illustrative examples of the effect of BrS-related mutations on the intrinsic disorder propensity of a highly disordered protein, NPPA (a) and a mostly ordered protein, SCN1B (b). In both cases, wild-type proteins are shown by black lines, whereas different lines represent disorder profiles of several BrS-related mutated proteins.

8.2.4 Kolmogorov–Smirnov Test

Motivation: There are several statistical tests (parametric and nonparametric) that respond to the correlation between two or more samples. In particular, nonparametric tests do not presuppose a normal distribution. This is an advantage for the analysis, where it is not known what distribution would follow the "polar profile" of a protein. Within the nonparametric tests, two tests are equivalent and robust: Mann–Whitney U test and Kolmogorov–Smirnov test, but the last one is less demanding. The statistical two-sided test confirmed (with alpha = 0.01) that the proportion of proteins accepted/rejected by the PIM system correlates with the actual proportion of the groups of BrS proteins and BrS mutated proteins. This also provided support to the conclusion that the "polar profile" of each one of these groups contains group-specific features.

8.3 DISCUSSION

The nonsupervised PIM® can be implemented in parallel mode. Since it uses the master-slaves scheme, its processing time is drastically reduced to one-tenth of the corresponding CPU time of the methods utilizing the monoprocessor scheme. We were able to implement a version in CUDA (Compute Unified Device Architecture) language to be executed on NVIDIA microprocessors. This implementation is viable because the only element that resides in memory is the representative vector of the polar profile of the target protein and the protein whose function and/or structure is known from experimental studies. We believe that the PIM system constitutes a fortunate computational innovation, because its high-performance computing scheme corresponds to master-slaves, which substantially reduces the processing time, and from a biological point of view, its metrics use only a linear representation of the protein and not its three-dimensional structure, which effectively reduces the complexity in the programming in comparison with the metrics that acts in three-dimensional space. Another important consideration is the physicochemical nature of the only property assessed by the PIM system, the polarity, which represents the electromagnetic balance of a protein. This is one of the four major forces that affect matter.

A distinctive "polar profile" was obtained for the set of BrS mutated proteins that discriminated the other protein groups analyzed in this study, including a set of 36 BrS proteins the BrS mutated proteins came from. When comparing the "polar profile" of BrS mutated proteins with the "reviewed" proteins from the UniProt database [5], it was identified that 6.4% of them share the same "polar profile".

The metrics of the PIM system named "polar profile" only takes the polarity of a peptide or a protein and expresses that measure with 16 numbers as an incidence matrix, equivalent to the 16 possible polar interactions from $\{P^+, P^-, N, NP\} \times \{P^+, P^-, N, NP\}$. A typical algorithm evaluating protein polarity (which is not PIM-based) generates a single real number to characterize the polarity status of a query protein. However, we assume that it is more precise when the algorithm uses 16 real numbers instead of a single one. It is possible that this particularity is what makes the PIM system effective; using the polar interaction with 16 values gives more representability to the amino acids and its physical properties at the molecular level. Therefore, our method can effectively identify a specific protein or find computable differences to identify specific proteins, e.g., mutations associated with BrS, as shown in this work. On the other hand, it is important to differentiate such mutants from other protein groups and search for other proteins that can possibly be associated with the syndrome by extensively checking different databases and individually evaluating all the proteins found in the UniProt database. A special feature of the PIM system is its sensitivity in the detection of differences when comparing two "polar profiles". Although the graphs of the "polar profile" (Figure 8.2a) of the BrS protein and BrS mutated protein groups are very similar, except for the polar interaction [NP, NP], the PIM system maximizes the differences that at first sight would not be possible to recognize.

An alternative approach would be the construction and analysis of the 20^n proteins of "n" length, yet this approach would be extremely expensive computationally. For instance, to analyze proteins with a length of 11 residues, the number of possible combinations would be $20^{11} \approx 204$ trillions of proteins. This would require parallel computing with a computational power of tera- or petaFLOPS (Floating point operations). Fortunately, current computational resources make both approaches feasible. Knowledge of the "polar profile" of this group of mutated proteins enables the construction of synthetic peptides that could be used for the neutralization of the ion channel imbalance. It could also be used in biochips that identify and measure the presence of BrS mutated proteins in the blood. Having a proved computational method will make the implementation of a biochip much compact and relatively straightforward, compared with a processor-based implementation. These biological microdevices would have a very low cost for the consumer and will enable the population to use it massively as a preventive measure. It would not be necessary to program the PIM system in the biochip, rather only the "polar profile" of the BrS mutated proteins may be needed, unlike other bioinformatics methods with several metrics that require the program directly in the biochip.

8.4 MATERIALS AND METHODS

The computational nonsupervised PIM system [7] has been used to identify different protein groups and its metrics named "polar profile" has been previously explained in detail [7]. However, it will be useful to illustrate the metrics with the next example, and by adding some considerations related to the computational scheme.

The PIM system only stores the polar profile of the representative protein group that is to be characterized (target protein set), and the polar profile of one of the proteins that comes from the set (or sets), whose function and/or structure are defined by experimental methods (Figure 8.1). This approach substantially reduces the memory demand, thus reducing processing time, but more importantly, it allows to take advantage of supercomputer resources, whose computational scheme is master-slave.

8.4.1 Evaluation of Polar Profile

The linear representation of a protein is taken as an orderly succession of amino acids that can be selected from this group of 20: H, K, R, D, E, C, G, N, Q, S, T, Y, A, F, I, L, M, P, V, and W. These amino acids are subclassified into four polar groups {P^+, P^-, N, NP} based on their polar charge affinity: P^+ (polar positively charged) = *{H, K, R}*, P^- (polar negatively charged) = *{D, E}*, *N* (polar neutral) = *{C, G, N, Q, S, T, Y}*, and *NP* (nonpolar) = *{A, F, I, L, M, P, V, W}*.

The PIM algorithm starts by replacing each amino acid of the protein with its numeric equivalent according to this rule: if the amino acid is P^+ = {H, K, R}, the equivalent would be 1; if the amino acid is P^- = {D, E}, the equivalent would be 2; if the amino acid is N = {C, G, N, Q, S, T, Y}, the equivalent would be 3; and if the amino acid is NP = {A, F, I, L, M, P, V, W}, the equivalent would be 4. For instance, if the proteins studied were (i) MSWQSYVDDHLMCDVE and (ii) *FLPIEDGTY*, their numeric equivalent would be (i) 4343334221443242, and (ii) 444422333, respectively.

From this numeric representation of the protein, a 4 × 4 **A**[*i*, *j*] incidence matrix is built, where *i* is the row and *j* is the column, each row and column representing one of the four types of polarity [7]. To illustrate this process, the protein (i) MSWQSYVDDHLMCDVE will be taken with its numerical equivalent 4343334221443242. The first step is to set the matrix to zero **A**[*i*, *j*] = 0 for all *(i, j)* = (1, 4). Then, the protein is read in numerical notation by pairs, one digit at a time, from left to right until the end of the protein. In this example, the first pair is "43," which means that to element (*i*, *j*) = (4, 3) in the matrix **A**[*i*, *j*] will be added 1. Then, taking one digit to the right is pair "34," therefore, to element (3, 4) in the matrix, **A**[*i*, *j*] will be added 1 (Table 8.3, a). If this procedure is repeated with protein (b) *FLPIEDGTY, the **B**[i, j] incidence matrix will be obtained (*Table 8.2, *b*). These two examples show how to get the "polar profile" of a protein.

To obtain the "polar profile" of a group of proteins, this procedure will be applied to each protein of the group, but the polar incidences will be accumulated in the same matrix. As a result, the incidence matrix of the group of proteins studied, in this case, the BrS protein mutations, will be unique, and it will be called the "target incidence matrix" **A**[*i*, *j*].

The PIM system only requires the "target" protein group and the other protein groups it will be compared to. With this information, the PIM system calculates the incidence matrix representative of the target **A**[*i*, *j*], normalizes it, and compares it

TABLE 8.3
Evaluation of Polar Profiles

(a)		P^+	P^-	N	NP
$\mathbf{A}_{MSWQSYVDDHLMCDVE}[i,j]$ =	P^+	0	0	0	1
	P^-	1	1	0	1
	N	0	1	2	2
	NP	1	1	3	1
(b)		P^+	P^-	N	NP
$\mathbf{B}_{FLPIEDGTY}[i,j]$ =	P^+	0	0	0	0
	P^-	0	1	1	0
	N	0	0	2	0
	NP	0	1	0	3

Incidence matrix of sample proteins (see Section 8.4.1).

with the incidence matrix of each protein of the other protein groups. The incidence matrix of each of these proteins will be called incidence matrix $\mathbf{B}_k[i, j]$, where "*k*" is the identifier of each protein. Now it will be explained in detail how the incidence matrices $\mathbf{A}[i, j]$, *y* $\mathbf{B}_k[i, j]$ are compared.

8.4.1.1 Weighting of Polar Profiles

To illustrate this stage, it will be assumed that the protein MSWQSYVDDHLMCDVE (Section 8.4.1) has the property studied, and it is required to verify whether the protein FLPIEDGTY (Section 8.4.1) has a similar "polar profile." Their incidence matrices will be $A[i, j]$ and $B[i, j]$, respectively. Once the matrices are determined, they are normalized, matrix $B[i, j]$ is weighted with matrix $A[i, j]$; thus, $B[i, j] = B[i, j]+A[i, j]$ (Table 8.4, a, b).

8.4.1.2 Comparison of Polar Profiles

The PIM system compares the matrices $\boldsymbol{A}[i, j]$ and $\boldsymbol{B}[i, j]$ (Table 8.5) checking if the highest relative frequency from $\boldsymbol{A}[i, j]$ corresponds to the same position in $\boldsymbol{B}[i, j]$ (Table 8.5). Then, the first highest relative frequency will be checked in both matrices (Table 8.5, position 1), and if the position matches in both matrices, the

TABLE 8.4
Weighting of Polar Profiles

(a)		P^+	P^-	N	NP
$\mathbf{B}_{FLPIEDGTY}[i,j] = \mathbf{A}_{MSWQSYVDDHLMCDVE}[i,j]+\mathbf{B}_{FLPIEDGTY}[i,j] =$	P^+	0.0000	0.0000	0.0000	0.0370
	P^-	0.0370	1.0370	1.0000	0.0370
	N	0.0000	0.0370	2.0741	0.0741
	NP	0.0000	1.0741	0.1111	3.0370
(b)	P^+	P^+	P^-	N	NP
$\mathbf{A}_{MSWQSYVDDHLMCDVE}[i,j] =$	P^-	0.0000	0.0000	0.0000	0.3333
	N	0.3333	0.3333	0.0000	0.6666
	NP	0.0000	0.6666	1.0000	0.3333

Weighting the incidence matrix of sample proteins (see Section 8.4.2).

TABLE 8.5
Comparison of Polar Profiles

Position	**1**	**2**	**3**	**4**	**5**	**6**	**7**	**8**	**9**	**10**	**11**	**12**	**13**	**14**	**15**	**16**
A[*i*,*j*]	15	14	12	11	16	10	8	6	5	4	13	9	7	3	2	1
B[*i*,*j*]	16	11	14	6	7	15	12	10	8	5	4	13	9	3	2	1
Similarity	×	×	×	×	×	×	×	×	×	×	×	×	×	✓	✓	✓

Comparison of sample proteins by position (see Section 8.4.1.2). (✓) The position matches in the matrices. (×) The position does not match in the matrices.

level of similarity would be 1 in 16, i.e., (1/16 = 6.25%). This procedure continues until the lowest relative frequency in both matrices is determined (Table 8.5, position 16). When comparing two groups of proteins, the corresponding results will be referred to as the "level of similarity," which will be stated with the corresponding percentage. Getting back to the example, the *level of similarity would be 3 in 16, which means 18.75% (3/16 = 18.75%).*

8.4.1.3 Graphics of Polar Profiles

If the incidence matrix is represented as histograms (see Figure 8.5), their differences and similarities are clearly appreciated.

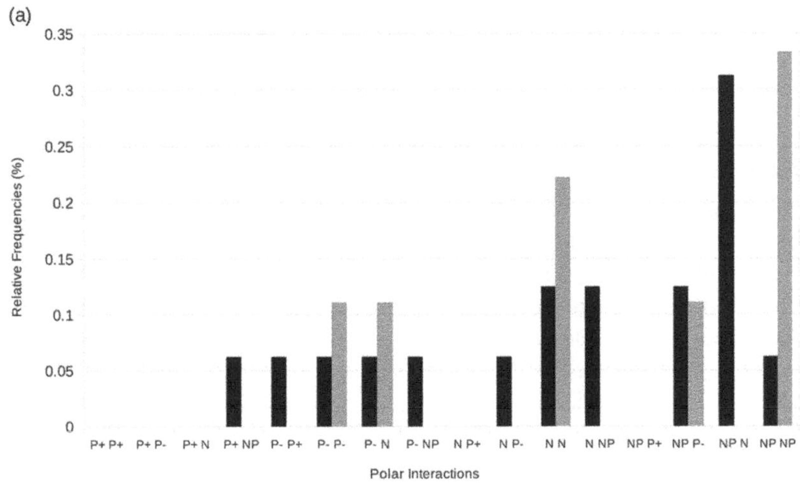

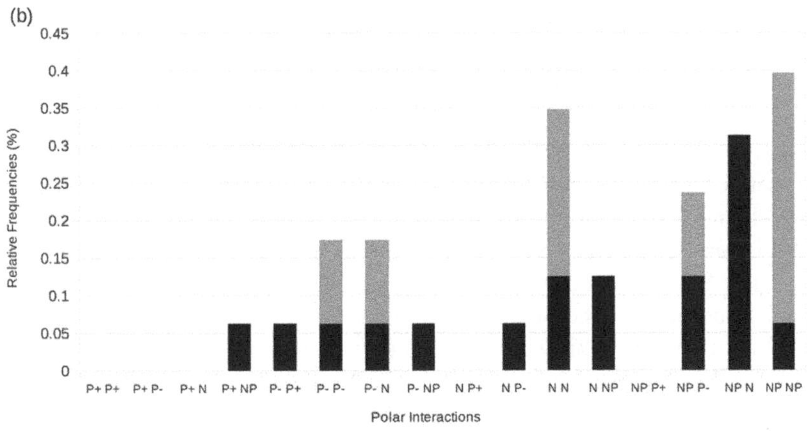

FIGURE 8.5 Graphical representation of sample proteins. (a): Column-Normal Excel software. (b) Column-Stacked Excel software. The *X*-axis represents the 16 polar interactions.

8.4.2 Evaluation of Intrinsic Disorder Predisposition

The peculiarities of intrinsic disorder distribution within the amino acid sequences of proteins associated with BrS were analyzed by a set of commonly used per-residue disorder predictors, such as PONDR® FIT [30], PONDR® VLXT [31], PONDR® VSL2 [32], PONDR® VL3 [33], and two forms of IUPred suitable for prediction of short and long intrinsically disordered regions, IUPred_short and IUPred_long, respectively [34]. These predictors were selected based on their specific sensitivities to different features associated with intrinsic disorder. PONDR® VSL2 is one of the more accurate stand-alone disorder predictors [35–37], PONDR® VL3 is characterized by high accuracy for predicting long intrinsically disordered regions [33], PONDR® VLXT is known to have high sensitivity to local sequence peculiarities and can be used for identifying disorder-based interaction sites [37], whereas a metapredictor PONDR® FIT is moderately more accurate than each of the component predictors [30], PONDR® VLXT [31], PONDR® VSL2 [32], PONDR® VL3 [33], FoldIndex [37], IUPred [38], and TopIDP [39]. IUPred was designed to recognize intrinsically disordered protein regions (IDPRs) from the amino acid sequence alone, based on the estimated pairwise energy content [38]. We also analyzed mean disorder propensity for these proteins, calculated by averaging disorder profiles of individual predictors. Use of the consensus for evaluation of intrinsic disorder is based on the empirical observations, showing that such approach usually increases the predictive performance compared with the use of a single predictor [30,36,40]. In these analyses, predicted disorder scores above 0.5 are considered to correspond to the disordered residues and regions.

8.4.3 Data Files

The files were formed based on the different groups of proteins. There are proteins and mutated proteins associated with BrS located in the UniProt database [5] that are duplicated. In each case (Table 8.6), they are highlighted in the "duplicated protein" column and in the mutated proteins evaluated by the PIM system. The PIM losses sensitivity when identical sequences are input, for that reason, the identical sequences were eliminated from all protein groups described here.

The sets of proteins, with which the BrS mutated proteins will be compared (Tables 8.8 and 8.9), intend to establish (or discard) their functional similarity with respect to these groups, in order to characterize (by association) the BrS mutated proteins. The PIM system automatically measures the similarity between the target protein set and each of the proteins that is part of the other groups, and for this reason, it is important to know the preponderant function of proteins in the groups used for comparison. We think that it is logical to consider possible associations of BrS mutated proteins with the group of lipoproteins, and in addition, to establish, in a similar manner, some similarity with proteins from fungi, virus, and bacteria groups. Particularly, the bacteria group contains the SCAAP group, which, together with the lipoprotein group, has a similar mechanism of action to the BrS mutated proteins.

TABLE 8.6
Whole Set of Brugada Proteins Found in UniProt

#	ID UniProt	Number of Unique BrS *Proteins*	Number of Unique BrS Mutated Proteins	Reference
1	ANK3	1	6	[41]
2	ARVD1	1	2	[42]
3	ATFB10	1	252	[37]
4	ATFB3	1	153	[43]
5	ATFB6	1	3	[44]
6	BRGDA1	0	0	[22]
7	BRGDA2	1	4	[45]
8	BRGDA3	1	30	[46]
9	BRGDA4	1	3	[47]
10	BRGDA5	1	8	[47]
11	BRGDA6	1	3	[48]
12	BRGDA7	1	9	[49]
13	BRGDA8	1	3	[46]
14	BRGDA9	1	8	[50]
15	CACNA1C	0	0	[46]
16	CACNB2	0	0	[47]
17	CALM1	0	0	[51]
18	GPD1L	0	0	[45]
19	HCN4	0	0	[46]
20	HEY2	1	2	[52]
21	KCND3	0	0	[50]
22	KCNE3	0	0	[48]
23	KCNH2	1	168	[53]
24	KCNJ8	1	4	[54]
25	KCNQ1	0	0	[43]
26	LQT14	1	0	[53]
27	LQT3	0	0	[51]
28	MYH7	1	137	[55]
29	SCN10A	1	8	[56]
30	SCN1B	0	0	[47]
31	SCN2B	1	4	[57]
32	SCN3B	1	4	[49]
33	SCN5A	0	0	[51]
34	SQT1	0	0	[53]
35	SSS1	0	0	[51]
36	VF1	0	0	[51]
	Total	**20**	**804**	

Unique BrS proteins and BrS mutated proteins extracted from UniProt database [5].

The protein groups taken are

i. 36 proteins (equivalent to 20 unique proteins) associated with BrS, extracted from the OMIM (www.ncbi.nlm.nih.gov/omim). This information is updated daily, and it identifies these proteins with the UniProt ID [5]. These proteins are: BRGDA8, BRGDA3, BRGDA6, BRGDA7, BRGDA4, BRGDA5, BRGDA2, BRGDA9, BRGDA1, SSS1, CACNB2, LQT3, LQT14, SQT1, GPD1L, SCN3B, SCN5A, KCND3, KCNE3, SCN1B, HCN4, CACNA1C, KCNJ8, SCN2B, KCNQ1, SCN10A, CALM1, VF1, KCNH2, ANK3, ATFB3, HEY2, MYH7, ARVD1, ATFB10, and ATFB6; (Table 8.6) from UniProt database [5]. Note: These proteins were taken in FASTA (Fast Adaptive Shrinkage Threshold Algorithm) format. A group of all unique BrS proteins (HEY2, NPPA, ANK3, CALM1, CACNB2, MYH7, HCN4, KCNH2, KCNE3, KCNQ1, KCND3, CACNA1C, SCN5A, SCN10A, TGFB3, SCN2B, KCNJ8, SCN3B, SCN1B, and GPD1L) was analyzed as the "BRU" set. The plain text file, and Excel file, related to these proteins can be found in the Supplementary Materials section.

ii. 4,388 mutated proteins (equivalent to 804 unique mutated proteins) of dysfunctional proteins from the 36 proteins were associated with BrS (Table 8.6) from UniProt database ([5], taken in FASTA format and extracted using Perl code varsplic.pl (from http://europepmc.org/ abstract/ MED/11159319). The overrepresentation of these proteins can be seen in Table 8.6.

iii. Four groups of antimicrobial peptides derived from fungi [4, Table 8.8], virus [4, Table 8.8], and bacteria [4, Table 8.8], as well as a subgroup of the bacteria group so-called SCAAP [8, Table 8.8]. The overrepresentation of the latter was 60% of all the protein sequences registered in the updated antimicrobial peptide database (APD2) (Table 8.8) [9]. The plain text file and Excel file related to these proteins can be found in the Supplementary Materials section.

iv. Two groups of unique intrinsically disordered proteins: 149 partially ordered and 50 completely disordered proteins [4, Table 8.7; 7, Supplementary Materials section]. The plain text file and Excel file, related to these proteins, can be found in the Supplementary Materials section.

v. Six groups of unique lipoproteins: High-density lipoprotein (HDL), low-density lipoprotein (LDL), intermediate-density lipoprotein (IDL), very-low-density lipoprotein (VLDL), chylomicrons, and atherosclerosis (the group formed of all lipoproteins related to atherosclerosis) (Table 8.9) [5]. The plain text file and Excel file, related to these proteins, can be found in the Supplementary Materials section.

vi. Finally, we used *557,713* "reviewed" proteins from UniProt database, extracted from [www.UniProt.org/UniProt/?query=&sort=score] accessed June 4, 2018. This file by size, is *not* provided as supplementary material, and if required, can be requested from the corresponding author.

TABLE 8.7
Analysis of the Overrepresentation of Brugada Groups

#	PIM System Trained with Protein/ Mutated Protein	Proteins^{P}/Mutated-Proteins^{M} with "Polar Profile" Similar
1	BRGDA8	HCN4^{PM}
2	BRGDA3	CACNA1C^{PM}
3	BRGDA6	KCNE3^{PM}
4	BRGDA7	SCN3B^{PM}
5	BRGDA4	CACNB2^{PM}
6	BRGDA5	SCN1B^{PM}
7	BRGDA2	GPD1L^{PM}
8	BRGDA9	KCND3^{PM}
9	BRGDA1	SSS1,^{PM} LQT3,^{PM} SCN5A,^{PM} VF1,^{PM} and ATFB10^{P}
10	SSS1	BRGDA1,^{PM} LQT3,^{PM} SCN5A,^{PM} VF1,^{PM} and ATFB10^{P}
11	CACNB2	BRGDA4,^{PM} BRGDA1,^{PM} SSS1,^{PM} KCND3,^{M} VF1,^{M} SCN5A,^{P} KCNH2,^{P} and ATFB10^{P}
12	LQT3	BRGDA1,^{PM} SSS1,^{PM}
13	LQT14	CALM1^{PM}
14	SQT1	KCNH2^{PM}
15	GPD1L	BRGDA2^{PM}
16	SCN3B	BRGDA7^{PM}
17	SCN5A	BRGDA1,^{PM} SSS1,^{PM} LQT3,^{PM} VF1,^{P} ATFB10,^{P} BRGDA1,^{PM} SSS1,^{PM} LQT3,^{M} CACNA1C,^{M} and ANK3^{M}
18	KCND3	BRGDA9^{PM}
19	KCNE3	BRGDA6^{PM}
20	SCN1B	BRGDA5^{PM}
21	HCN4	BRGDA8^{PM}
22	CACNA1C	BRGDA3^{PM}
25	KCNQ1	ATFB3^{PM}
27	CALM1	LQT14^{PM}
28	VF1	BRGDA1,^{PM} SSS1,^{PM} LQT3,^{PM} SCN5A,^{PM} and ATFB10^{PM}
29	KCNH2	SQT1^{PM}
31	ATFB3	KCNQ1^{PM}
35	ATFB10	BRGDA1,^{PM} SSS1,^{PM} LQT3,^{PM} SCN5A,^{PM} and VF1^{PM}

Overrepresentation of Brugada groups. For example, the PIM system identified that the "polar profile" of protein/mutation LQT3 is similar to the polar profile of *proteins* BRGDA1,^{P} SSS1,^{P} SCN5A,^{P} KCNH2,^{P} and ATFB10^{P}; and to mutated *proteins* BRGDA1,^{M} SSS1,^{M} KCND3,^{M} and VF1^{M}. The proteins not mentioned in this table have no similarity.

8.4.4 Kolmogorov–Smirnov Test

The Kolmogorov–Smirnov *two-sided* test (alpha = 0.01) [58] was applied to the BrS proteins (36 proteins) and the BrS mutated proteins (4,388 proteins), counting the number of matches and rejections produced by the PIM system. The Excel file and Kolmogorov–Smirnov test related to these proteins can be found in the Supplementary Materials section.

8.4.5 Test Plan

8.4.5.1 Polar Profile

The PIM system implements two procedures to characterize the mutated proteins associated with BrS. It registers the number of similarities (see Section 8.4.1.2) to identify the BrS protein group (Table 8.6) and compares it with the other groups (Tables 8.7 and 8.8). For instance, if the PIM system is calibrated with the unique BrS mutated proteins, it is tested with the unique groups of BrS proteins (Table 8.6), antimicrobial proteins (Table 8.8), disordered proteins, and lipoproteins (Table 8.9). This computational test generates false positives and false negatives. In addition, the "polar profiles" of unique proteins and mutated proteins associated with BrS are also compared (Figure 8.2). At this stage, the "polar profile" representative of the unique mutated proteins associated with BrS is obtained and, with this information, the PIM system looks for the coincidences of this "polar profile" with the "polar profiles" generated for the total reviewed proteins registered in the UniProt database [5].

TABLE 8.8
Antimicrobial Peptides

#	Number	Group	References
1	21	SCAAP	[8, Table 8.7]
APD2 Database			
2	469	Bacteria	[9]
3	86	Fungi	[9]
4	21	Virus	[9]
UniProt Database			
5	117	Bacteria	[5]
6	46,342	Fungi	[5]
7	1,104	Virus	[5]

Unique antimicrobial peptide groups [rows #2–4] taken from APD2 database (accessed December 2012), UniProt database (accessed August 16, 2017), and SCAAP [row #1] sequences identified [8, Table 8.7].

TABLE 8.9
Group of Lipoproteins Analyzed in This Study

#	Quantity	Description	Symbol	Search Engine
1	19	High-density	HDL	("high density lipoprotein" AND "homo sapiens") AND hdl AND reviewed:yes NOT "lipoprotein binding protein" NOT (disease:"high density lipoprotein")
2	0	Intermediate-density lipoprotein	IDL	("intermediate density lipoprotein" AND "homo sapiens") AND idl AND reviewed:yes NOT "lipoprotein binding protein" NOT (disease:"intermediate density lipoprotein")
3	51	Low-density lipoprotein	LDL	("low density lipoprotein" AND "homo sapiens") AND ldl AND reviewed:yes NOT "lipoprotein binding protein" NOT (disease:"low density lipoprotein")
4	17	Very-low-density lipoprotein	VLDL	("very low density lipoprotein" AND "homo sapiens") AND vldl AND reviewed:yes NOT "lipoprotein binding protein" NOT (disease:"very low density lipoprotein")
5	14	Chylomicron density lipoprotein	Chylomicrons	("chylomicrons" AND "homo sapiens") AND chylomicrons AND reviewed:yes NOT "lipoprotein binding protein" NOT (disease:"chylomicrons density lipoprotein")
6	11	Lipoproteins relevant to atherosclerosis	Atherosclerosis	(lipoproteins atherosclerosis AND organism:"Homo sapiens (Human) [9606]")

Unique lipoproteins extracted from UniProt database [5], and only human proteins are included that have the annotation "reviewed."

8.5 CONCLUSIONS

In this work, a bioinformatics algorithm PIM was tested and validated; its "fingerprint" enables the rapid identification of BrS mutated proteins. The polar profile of the BrS mutated proteins analyzed in this study is different from the polar profiles of proteins from the structural groups: intrinsically disordered proteins, or those of proteins from the functional groups: lipoproteins, or fungi, virus, or bacteria. On the other hand, the polar profile of each protein group associated with BrS mutated proteins serves as an effective discriminator. Therefore, it is concluded that it is possible to use the polar profile as a "fingerprint" that identifies, with high efficiency, those mutated proteins. The polar profile constitutes a metric that only measures the electromagnetic balance of the protein using its amino acid sequence, which makes it useful for database analysis.

Acknowledgments: The authors thank Concepción Celis Juárez for proofreading.

Author contributions: Theoretical conceptualization, and design: CP. Computational performance: CP, and VNU. Data analysis: CP, MFM, VNU, TB, and MAE. Results discussion: CP, MFM, VNU, TB, a MAE. **Competing interests**: We declare that we do not have any financial and personal interest with other people or organizations that could inappropriately influence (bias) our work. **Data and materials availability**: Copyright & Trademark. All rights reserved (México), 2018: PIM®, PONDR® FIT, PONDR® VLXT, PONDR® VSL2, PONDR® VSL2, PONDR® VL3, and PONDR® VSL2-based PPID values. **Software & Hardware**: *Hardware*: The computational platform used to process the information was HP Workstation z210—CMT—4 x Intel Xeon E3–1270/3.4 GHz (Quad-Core) —RAM 8 GB—SSD 1 x 160 GB—DVD SuperMulti—Quadro 2000—Gigabit LAN, Linux Fedora 14, 64-bits. Cache Memory 8 MB. Cache Per Processor 8 MB. RAM 8. *Software*: PONDR® FIT, PIM®, PONDR® VLXT, PONDR® VSL2, PONDR® VL3, FoldIndex, IUPred, and TopIDP, as well as PONDR® VSL2-based values. **Supplementary Materials**: The test files and the PIM ® system was supplied as support of the manuscript to the journal, but it can be requested from the corresponding author (polanco@unam.mx). The materials related to "Intrinsic disorder propensity in 20 unique BrS-related proteins," were supplied as support of the manuscript to the journal.

REFERENCES

1. P. Brugada, J. Brugada, Right bundle branch block, persistent ST segment elevation and sudden cardiac death: a distinct clinical and electrocardiographic syndrome. A multicenter report. *J Am Coll Cardiol* **20**, 1391–1396 (1992).
2. S. G. Priori, C. Blomström-Lundqvist, A. Mazzanti, N. Blom, M. Borggrefe, J. Camm, P. M. Elliott, D. Fitzsimons, R. Hatala, G. Hindricks, P. Kirchhof, K. Kjeldsen, K. H. Kuck, A. Hernandez-Madrid, N. Nikolaou, T. M. Norekvål, C. Spaulding, D. J. Van Veldhuisen, ESC Scientific Document Group. 2015 ESC Guidelines for the management of patients with ventricular arrhythmias and the prevention of sudden cardiac death: the task force for the management of patients with ventricular arrhythmias and the prevention of sudden cardiac death of the European Society of Cardiology (ESC). Endorsed by: Association for European Paediatric and Congenital Cardiology (AEPC). *Eur Heart J.* **36**, 2793–2867 (2015).
3. M. F. Márquez, A. Bonny, E. Hernández-Castillo, A. De Sisti, J. Gómez-Flores, S. Nava, F. Hidden-Lucet, P. Iturralde, M. Cárdenas, J. Tonet, Long-term efficacy of low doses of quinidine on malignant arrhythmias in Brugada syndrome with an implantable cardioverter-defibrillator: a case series and literature review. *Heart Rhythm.* **9**, 1995–2000 (2012).
4. J-M. J. Juang, M. Horie, Genetics of Brugada syndrome. *Journal of Arrhythmia.* **32**, 418–425 (2016).
5. UniProt Consortium. UniProt: a hub for protein information. *Nucleic Acids Res.* 43(database issue), D204–D212 (www.ncbi.nlm.nih.gov/omim), accessed May 17, 2016 (2015).

6. L. Pauling, *The Nature of the Chemical Bond and the Structure of Molecules and Crystals: An Introduction to Modern Structural Chemistry.* ISBN:9780801403330 (Cornell University Press, 1960).
7. C. Polanco, Polarity Index in Proteins-A Bioinformatics Tool. doi:10.2174/97816810826911160101, eISBN: 978-1-68108-270-7, 2016, ISBN: 978-1-68108-269-1 (Bentham Science Publishers, Sharjah, 2016).
8. C. Polanco, J. L. Samaniego, J. A. Castañon-González, T. Buhse, M. Leopold Sordo, Characterization of a possible uptake mechanism of selective antibacterial peptides. *Acta Biochim Pol.* **60**, 629–633 (2013).
9. G. Wang, X. Li, Z. Wang, APD2: The updated antimicrobial peptide database and its application in peptide design. *Nucleic Acids Res.* **37**, D933–D937 (2012).
10. C. J. Oldfield, A. K. Dunker, Intrinsically disordered proteins and intrinsically disordered protein regions. *Annu Rev Biochem.* **83**, 553–584 (2014).
11. C. Polanco, J. L. Samaniego, V. N. Uversky, J. A. Castañón-González, T. Buhse, M. Leopold-Sordo, A. Madero-Arteaga, A. Morales-Reyes, L. Tavera-Sierra, J. A. González-Bernal, M. Arias-Estrada, Identification of proteins associated with amyloidosis by polarity index method. *Acta Biochim Pol.* **62**, 41–55 (2014).
12. J. M. Claverie, A common philosophy and FORTRAN 77 software package for implementing and searching sequence databases. *Nucleic Acids Res.* **12**, 397–407 (1984).
13. M. Hanke, Y. O. Halchenko, Neuroscience Runs on GNU/Linux. *Front Neuroinf.* **5**, 8 (2011).
14. P. E. Wright, H. J. Dyson, Intrinsically unstructured proteins: re-assessing the protein structure-function paradigm. *J Mol Biol.* **293**, 321–331 (1999).
15. V. N. Uversky, A. K. Dunker, Understanding protein non-folding. *Biochim Biophys Acta.* **1804**, 1231–1264 (2010).
16. A. K. Dunker, C. J. Oldfield, J. Meng, P. Romero, Y. J. Yang, J. W. Chen, V. Vacic, Z. Obradovic, V. N. Uversky, The unfoldomics decade: an update on intrinsically disordered proteins. *BMC Genom.* 9(Suppl 2), S1 (2008).
17. V. U. Uversky, J. R. Gillespie, A. L. Fink, Why are "natively unfolded" proteins unstructured under physiologic conditions? *Proteins.* **41**, 415–427 (2000).
18. A. K. Dunker, J. D. Lawson, C. J. Brown, R. M. Williams, P. Romero, J. S. Oh, C. J. Oldfield, A. M. Campen, C. M. Ratliff, K. W. Hipps, J. Ausio, M. S. Nissen, R. Reeves, C. Kang, C. R. Kissinger, R. W. Bailey, M. D. Griswold, W. Chiu, E. C. Garner, Z. Obradovic, Intrinsically disordered protein. *J Mol Graph Model.* **19**, 26–59 (2001).
19. A. K. Dunker, Z. Obradovic, The protein trinity--linking function and disorder. *Nat Biotechnol.* **19**, 805–806 (2001).
20. A. K. Dunker, I. Silman, V. N. Uversky, J. L. Sussman, Function and structure of inherently disordered proteins. *Curr Opin Struct Biol.* **18**, 756–764 (2008).
21. V. U. Uversky, A decade and a half of protein intrinsic disorder: biology still waits for physics. *Protein Sci.* **22**, 693–724 (2013).
22. A. L. Darling, V. U. Uversky, Intrinsic disorder and posttranslational modifications: the darker side of the biological dark matter. *Front Genet.* **9**, 158 (2018).
23. V. U. Uversky, C. J. Oldfield, A. K. Dunker, Intrinsically disordered proteins in human diseases: introducing the D2 concept. *Annu Rev Biophys.* **37**, 215–246 (2008).
24. U. Midic, C. J. Oldfield, A. K. Dunker, Z. Obradovic, V. N. Uversky, Unfoldomics of human genetic diseases: illustrative examples of ordered and intrinsically disordered members of the human diseasome. *Protein Pept Lett.* **16**, 1533–1547 (2009).
25. V. N. Uversky, Intrinsic disorder in proteins associated with neurodegenerative diseases. *Front Biosci.* (Landmark Ed). **14**, 5188–5238 (2009).
26. V. N. Uversky, C. J. Oldfield, U. Midic, H. Xie, B. Xue. S. Vucetic, L. M. Iakoucheva, Z. Obradovic, A. K. Dunker, Unfoldomics of human diseases: linking protein intrinsic disorder with diseases. *BMC Genomics.* 10(Suppl 1), S7 (2009).

27. V. N. Uversky, Targeting intrinsically disordered proteins in neurodegenerative and protein dysfunction diseases: another illustration of the D(2) concept. *Expert Rev Proteomics.* **7**, 543–564 (2010).
28. V. N. Uversky, Intrinsically disordered proteins and their (disordered) proteomes in neurodegenerative disorders. *Front Aging Neurosci.* 7, 18 (2015).
29. A. Ludwig, X. Zong, J. Stieber, R. Hullin, F. Hofmann, M. Biel, Two pacemaker channels from human heart with profoundly different activation kinetics. *EMBO J.* **18**, 2323–2329 (1999).
30. B. Xue, R. L. Dunbrack, R. W. Williams, A. K. Dunker, V. N. Uversky, PONDR-FIT: a meta-predictor of intrinsically disordered amino acids. *Biochim Biophys Acta.* **1804**, 996–1010 (2010).
31. P. Romero, Z. Obradovic, X. Li, E. C. Garner, C. J. Brown, A. K. Dunker, Sequence complexity of disordered protein. *Proteins.* **42**, 38–48 (2001).
32. Z. Obradovic, K. Peng, S. Vucetic, P. Radivojac, A. K. Dunker, Exploiting heterogeneous sequence properties improves prediction of protein disorder. *Proteins.* **61**, 176–182 (2005).
33. Z. Obradovic, K. Peng, S. Vucetic, P. Radivojac, C. J. Brown, A. K. Dunker, Predicting intrinsic disorder from amino acid sequence. *Proteins.* **53**, 566–572 (2003).
34. Z. Dosztanyi, V. Csizmok, P. Tompa, I. Simon, IUPred: web server for the prediction of intrinsically unstructured regions of proteins based on estimated energy content. *Bioinformatics.* **21**, 3433–3434 (2005).
35. K. Peng, S. Vucetic, P. Radivojac, C. J. Brown, A. K. Dunker, Z. Obradovic, Optimizing long intrinsic disorder predictors with protein evolutionary information. *J Bioinform Comput Biol.* **3**, 35–60 (2005).
36. L. Kurgan, Comprehensive comparative assessment of in-silico predictors of disordered regions. *Curr Protein Pept Sci.* **13**, 6–18 (2012).
37. A. K. Dunker, J. D. Lawson, C. J. Brown, R. M. Williams, P. Romero, J. S. Oh, C. J. Oldfield, A. M. Campen, C. M. Ratliff, K. W. Hipps, J. Ausio, M. S. Nissen, R. Reeves, C. Kang, C. R. Kissinger, R. W. Bailey, M. D. Griswold, W. Chiu, E. C. Garner, Z. Obradovic, Intrinsically disordered protein. *J Mol Graph Model.* **19**, 26–59 (2001).
38. Z. Dosztanyi, V. Csizmok, P. Tompa, I. Simon, IUPred: web server for the prediction of intrinsically unstructured regions of proteins based on estimated energy content. *Bioinformatics.* **21**, 3433–3434 (2005).
39. A. Campen, R. M. Williams, C. J. Brown, J. Meng, V. N. Uversky, A. K. Dunker, TOP-IDP-scale: a new amino acid scale measuring propensity for intrinsic disorder. *Protein Pept Lett.* **15**, 956–963 (2008).
40. I. Walsh, M. Giollo, T. Di Domenico, C. Ferrari, O. Zimmermann, S. C. Tosatto, Comprehensive large-scale assessment of intrinsic protein disorder. *Bioinformatics.* **31**, 201–208 (2015).
41. E. Kordeli, S. Lambert, V. Bennett, Ankyrin. A new ankyrin gene with neural-specific isoforms localized at the axonal initial segment and node of Ranvier. *J Biol Chem.* **270**, 2352–2359 (1995).
42. D. P. Ten, P. Hansen, K. K. Iwata, C. Pieler, J. G. Foulkes, Identification of another member of the transforming growth factor type beta gene family. *Proc Natl Acad Sci USA.* **85**, 4715–4719 (1998).
43. C. Chouabe, N. Neyroud, P. Guicheney, M. Lazdunski, G. Romey, J. Barhanin, Properties of KvLQT1 K+ channel mutations in Romano-Ward and Jervell and Lange-Nielsen inherited cardiac arrhythmias. *EMBO J.* **16**, 5472–5479 (1997).
44. S. Oikawa, M. Imai, A. Ueno, S. Tanaka, T. Noguchi, H. Nakazato, K. Kangawa, A. Fukuda, H. Matsuo, Cloning and sequence analysis of cDNA encoding a precursor for human atrial natriuretic polypeptide. *Nature.* **309**, 724–672 (1984).

45. T. Nagase, N. Miyajima, A. Tanaka, T. Sazuka, N. Seki, S. Sato, S. Tabata, K. Ishikawa, Y. Kawarabayasi, H. Kotani, N. Nomura, Prediction of the coding sequences of unidentified human genes. III. The coding sequences of 40new genes (KIAA0081-KIAA0120) deduced by analysis of cDNA clones from human cell line KG-1. *DNA Res.* **2**, 37–43 (1995).
46. N. M. Soldatov, Molecular diversity of L-type Ca^{2+} channel transcripts in human fibroblasts. *Proc Natl Acad Sci USA.* **89**, 4628–4632 (1992).
47. A. I. McClatchey, S. C. Cannon, S. A. Slaugenhaupt, J. F. Gusella, The cloning and expression of a sodium channel beta 1-subunit cDNA from human brain. *Hum Mol Genet.* **2**, 745–749 (1993).
48. Y. F. Melman, A. Domenech, S. de la Luna, T. V. McDonald, Structural determinants of KvLQT1 control by the KCNE family of proteins. *J Biol Chem.* **276**, 6439–4644 (2001).
49. K. Morgan, E. B. Stevens, B. Shah, P. J. Cox, A. K. Dixon, K. Lee, R. D. Pinnock, J. Hughes, P. J. Richardson, K. Mizuguchi, A. P. Jackson, beta 3: an additional auxiliary subunit of the voltage-sensitive sodium channel that modulates channel gating with distinct kinetics. *Proc Natl Acad Sci USA.* **97**, 2308–2313 (2000).
50. W. Kong, S. Po, T. Yamagishi, M. D. Ashen, G. Stetten, G. F. Tomaselli, Isolation and characterization of the human gene encoding Ito: further diversity by alternative mRNA splicing. *Am J Physiol.* **V275**, H1963–H1970 (1998).
51. E. J. Wawrzynczak, R. N. Perham, Isolation and nucleotide sequence of a cDNA encoding human calmodulin. *Biochem Int.* **9**, 177–185 (1984).
52. C. Leimeister, A. Externbrink, B. Klamt, M. Gessler, Hey genes: a novel subfamily of hairy- and Enhancer of split related genes specifically expressed during mouse embryogenesis. *Mech Dev.* **85**, 173–177 (1999).
53. J. W. Warmke, B. Ganetzky, A family of potassium channel genes related to eag in Drosophila and mammals. *Proc Natl Acad Sci USA.* **91**, 3438–3442 (1984).
54. N. Inagaki, J. Inazawa, S. Seino, cDNA sequence, gene structure, and chromosomal localization of the human ATP-sensitive potassium channel, uKATP-1, gene (KCNJ8). *Genomics.* **30**, 102–104 (1995).
55. T. Jaenicke, K. W. Diederich, W. Haas, J. Schleich, P. Lichter, M. Pfordt, A. Bach, H. P. Vosberg, The complete sequence of the human beta-myosin heavy chain gene and a comparative analysis of its product. *Genomics.* **8**, 194–206 (1990).
56. D. K. Rabert, B. D. Koch, M. Ilnicka, R. A. Obernolte, S. L. Naylor, R. C. Herman, R. M. Eglen, J. C. Hunter, L. Sangameswaran, A tetrodotoxin-resistant voltage-gated sodium channel from human dorsal root ganglia, hPN3/SCN10A. *Pain.* **78**, 107–114 (1998).
57. J. Eubanks, J. Srinivasan, M. B. Dinulos, C. M. Disteche, W. A. Catterall, Structure and chromosomal localization of the beta2 subunit of the human brain sodium channel. *Neuroreort.* **8**, 2775–2779 (1997).
58. S. Siegel, Estadística no paramétrica aplicada a las ciencias (Trillas, pp. 155–165, 2 ed., 1985), Mexico

9 Discrimination of Healthy Skin, Superficial Epidermal Burns, and Full-Thickness Burns from 2D-Colored Images Using Machine Learning

Aliyu Abubakar, Hassan Ugail, and Ali Maina Bukar
University of Bradford

Kirsty M. Smith
Bradford Teaching Hospitals

CONTENTS

9.1 INTRODUCTION

Early identification of burn injuries and the precise evaluation of burn depth is of great importance to the patient towards management of an injury, thereby ensuring early decision-making on whether surgical intervention is required or not. Accurate assessment is important because it helps to decide the potential healing times of any identified burn wound that allows proper treatment to be performed as early as possible. Burns that take a long period to heal, such as superficial dermal and deep dermal burns, can be treated as early as possible if they are identified, which may minimize further complications due to assessment delay [1]. Although superficial epidermal and full-thickness burns are easy to be identified clinically, can machine learning (ML) algorithms be used to differentiate superficial burns and full-thickness burns?

According to a report released by the World Health Organization (WHO) in 2008, burn injury is ranked as the fourth most devastating injury after road accident and internal conflicts, with approximately 11 million [2,3] patients, while in United State alone, almost 450,000 patients are treated for burn injuries every year. Burns are ranked as the third most common injury affecting children below the age of 5 in England. This was reported in a study by Kandiyali et al. [4], and the investigation shows that more than 7,000 children have been reported at hospital in 2015 for burn management. Similar cases of burn incidences are rampant in low- and middle-income countries such as Africa. Report also shows that children are mostly affected by burn accidents in Africa, where sub-Saharan Africa has the highest rate of mortality cases with about 4.5 per 100,000 compared with 2.5 per 100,000 across 103 countries [5].

A large volume of medical images is being collated everyday due to the increase of imaging capturing devices. These images are processed by visual inspection by medical experts, but concern has been raised on the subjective nature (assessment inconsistency) associated with humans [6,7]. Furthermore, evaluation of burn injuries by visual inspection is subjected to two potential errors—underestimation and overestimation [8]. When burns are underestimated, patients are likely to spend a long time in hospitals, and further complications may arise due to assessment delay. Overestimation may subject patients to unnecessary surgical diagnosis such as skin grafting. In both cases, unnecessary hospital cost increases.

Due to proliferation of technology and the increase in imaging capturing devices, such devices are very expensive [8] and not user-friendly, and a huge volume of clinical burn images has been accumulating every day and are processed by human visual system. However, interpretation of such images by humans is prone to errors due to the fatigue nature, and above all, the assessment is highly subjective [6,7]. The subjectivity of this method has been identified with two common problems—overestimation and underestimation. Overestimation of burn may subject a patient to unnecessary skin grafting/surgery, whereas an increase in hospital delay and possibility of subjecting patient to a high risk of infection is associated with underestimation of burns [8].

Towards this end, this work is primarily intended to determine whether ML can be used to discriminate between burn and normal skin. The skin burn images comprise of superficial and full-thickness injuries. The rest of the chapter is organized as follows: Section 9.2 presents related literatures; Section 9.3 presents the overview of Convolutional Neural Network (ConvNet); goals and methodology are presented in Section 9.4; experimental result and discussion are presented in Section 9.4 while Section 9.5 concludes the work.

9.2 LITERATURE REVIEW

In this section, various causes of burn injuries and its classification are presented. Additionally, the section presented a review on some effective diagnostic procedure of addressing burn injuries and highlighted some associated limitations.

9.2.1 Skin Burns

According to WHO "*burn is an injury caused by heat (hot objects, gases, or flames), chemicals, electricity and lightening, friction, or radiation*" [9]. It is an injury that damages the body tissues, affecting people of different ages. It is one of the most disturbing injuries in the world affecting children and adults, though children are affected in most cases.

9.2.2 Causes of Burn Injuries

Burns are caused via various processes such as thermal, chemical, friction, and radiation [10]. Burns that are caused by flames, hot objects, hot liquid, and steams are referred to as thermal burns. Burns caused by hot liquid and steams are sometimes referred to as scald burns. Corrosive substances such as acid are also known to cause severe damage to the skin when they come in contact, and this category of injury caused by such substances are called chemical burns, mostly occurring due to accidental spillage in industries/workplaces. Frictional forces relative to each other generate heat and leads to mechanical commotion of the skin resulting in friction burn injury. Burns caused by radiation are mostly due to exposure to ultraviolet sun radiation. This is in addition caused to prolong exposure to therapeutic radiations in hospitals as well as industry workers dealing with radiation substances. The severity of the burn depends on how long the body gets exposed to any of the causes mentioned.

9.2.3 Burns Category

A burn is categorized into three stages depending on the severity or how deep the body tissue is affected:

- First-degree burn
- Second-degree burn
- Third-degree burn

First-degree burn is a situation where burn affects only the uppermost layer of the skin (epidermis) [11]. This type of burn is mostly common in Caucasian people, where sunlight is usually the causative agent. This category of burn does not raise much concern due to its ability to heal spontaneously within the first 7 days after injury. The most common feature that identifies this type of burn is reddish color, absence of swelling with no blistering, but slightly painful. Burn does more harm when its impact reaches the layer beneath the epidermis (dermis). This is because the demise housed most of the essentials organs that keep the skin in good shape. Most of the nutrients and oxygen supplied to the epidermis are contained in the dermis. Therefore, when burn affects this layer, this means that the body won't be able to regulate body temperature as well as vitamin D synthesis. In a nutshell, when burn destroyed the epidermis and affects the dermis layer, this category of burn is referred to as second-degree burn [12]. Burns extending deep to the subcutis layer, destroying the whole dermis, thereby exposing the subcutaneous layer including muscles and tendons is called third-stage (degree) burns [10,12]. In other words, this category of burn is called full-thickness burn since all the three layers of the skin are affected.

9.2.4 Burn Assessment Techniques

This section presents a succinct review of burn evaluation techniques.

9.2.4.1 Clinical Assessment

Clinically, burn assessment begins by capturing the history of injury, such as mode of injury, contact duration with the heat, and the type of first aid given to the patient. Afterwards, burn is assessed based on the physical appearance of the wound by considering color appearance, sensitivity to pain, and capillary refill. The features observation is conducted by human expert, and diagnosis is based on human judgment, which depends on the level of experience. Color is the most considered characteristic used by medical personnel when assessing burns visually and during blister formation. Burns that only affect the upper layer of skin (superficial) is known to be reddish, dry, and with no blistering. If the skin disruption and the reddish appearance of the burn wound is considered as a criterion to identify burn injury, there exist skin injuries that share similar physical characteristics with burn, such as bruises, as such a potential misdiagnosis may occur. Therefore, in a situation that involves similar characteristics shared by different skin injuries, probability of misdiagnosis is very high [13]. Misdiagnosis in this context can be likely of assessing and categorizing burns as something different, which eventually will lead to unwanted complications as a result of wrong treatment.

Additional means of diagnosing burns is through histological means (biopsy), which involves extracting sample from the patient's body and then analysing to determine the extent of injury. This technique is being used in hospitals as a gold standard; however, its invasive nature causes pains and leaves patients with scary marks. Moreover, a biopsy only covers a limited portion of the burn site, and in most cases, biopsy misses the most critical site due to the sampling variation [14].

Pain experienced by individuals who sustained burn injury are only a fraction of the total pains. Pains are experienced during the diagnostic processes such as identification of burn severity. Therapeutic procedures in assessing burn, especially those that affected dermal layer and subcutaneous tissues required pharmacological techniques such as administration of pain relievers (e.g., opioid analgesics). However, it is reported that, some individuals do not react to opioid, as such patients are subjected to experience unquantified pains during the debridement process [15]. In addition, there is likely that even those that react to opioid administration experience little pains. Unfortunately, the procedure of recognizing the burn severity is also questionable. Most of the diagnostic procedures in hospitals adopt visual inspection to determine the nature of burn.

9.2.4.2 Blood Perfusion Measurement

Within the tissue of living organisms, circulatory channels are responsible for transporting important materials such as oxygen, hormones, and other nutrients needed to keep the living tissues in perfect working condition. However, those substances are carried out by blood, and the flow of blood through the circulatory system is referred to as blood perfusion [16,17]. Measurement of blood flow by using noninvasive techniques, such as laser Doppler flowmetry and laser Doppler imaging (LDI), has provided a good and reliable procedure for assessing burn wounds [18] and has been recommended by the United States Food and Drug Administration (FDA).

9.2.4.2.1 Laser Doppler Imaging

The first study to propose the usage of LDI in burn assessment was in 1993 [19], even though the idea of using noninvasive method was coined since 1970s [20]. This new device was proposed to provide an efficient means of diagnosing tissue injuries such as burn wounds without contact, thereby minimizing pains experienced by patients. It works based on the principle that light traveling undergoes frequency shift when colliding with a moving object. What is the moving object in the body tissue? The moving objects in the body tissue are blood and its constituents. The architecture of this device is depicted in Figure 9.1, and it consists of a laser tube, four assembled arrays of photodiodes, scanning mirror, and a computer device. The laser tube that transmits laser light is mounted behind the photodiodes, the beam of light passes through the assembled photodiodes that is then reflected by the scanning mirror. The mirror directs the light on to the scanning area where the burn wound to be assessed is expected to be positioned. The mirror increases the total coverage area by the reflected light on the body surface. Light reflected by the scanned object is collected again by the mirror and back to the photodiodes.

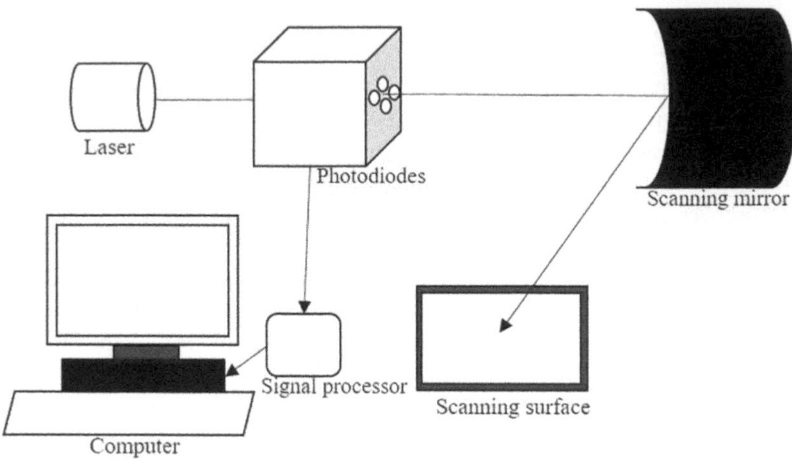

FIGURE 9.1 Architecture of an LDI.

The investigation was conducted by the authors in [19], where patients were recruited following their willingness and acceptance to participate, all of which sustained burns that are challenging to diagnose with visual assessment. Most of the burns sustained by these patients were within 24 h and were suspected to be deep dermal burns with less than 12% body surface area. To compare the effectiveness of this new technique, samples were collected using punch biopsies at 72 h after injury. The scanning was conducted, which captures images of the burn wounds lying 1.6 m from the scanner in a controlled room temperature of 25°C, and the scan images were generated by the computer. The result of the study shows that there was a 77% correlation in the assessment by both LDI and clinical assessment. Superficial dermal or epidermal depth that was correctly assessed clinically indicated high perfusion rate using LDI, where deep dermal and full-thickness burns indicate a low perfusion rate. However, 15.3% were assessed clinically as deep dermal wounds, but with LDI, there was an indication of high perfusion rate. This shows the overestimation by clinicians. Studies were conducted later by different scholars, ranging from its ability to predict healing times at early hours after injuries and assessment of burns in children.

Using LDI to diagnose burns in children was described as a challenging task, because scanning is expected to be carried out while the patient is still [21], and the restlessness is believed to affect the scanning process and render inaccurate outcome. As such, the study did investigate the prediction of burn wounds outcome in pediatrics using LDI. The study comprises 57 patients and was approved by the ethical committee of children's hospital at Westmead, Australia. Colored images of the burn wounds were captured using Olympus Camedia C-2500L digital camera. The scanning was conducted once in 36 to 72 h after injury. Each scan takes about 5–7 min for children who cooperated and stood still, while for the restless children the scanning time was reduced between 90 s and 3.5 min, although some generated images are noisy due to movement artifacts. The acquired images were then assessed and interpreted by a surgeon and recorded for comparison at subsequent assessment. After day 12, another assessment was conducted, and burn that does not heal within such period is categorized as deep partial or full-thickness burn. The LDI has effectively recognized a negative instance with 96% accuracy, while clinically, 71% was achieved. Similarly, true positive (TP) rate prediction by LDI is 90% as compared to the clinical outcome of 66% accuracy.

Following the innovation by Niazi et al. [19], investigations were conducted by different researchers to prove the efficiency of LDI in burn care unit. Researchers in [22] investigated various techniques of burn wound assessment, such as clinical evaluation, thermal imaging, and LDI. Clinical evaluation that was characterized by nonuniformity when conducted with different surgeons/dermatologists relied heavily on the physical characteristics of the wound and sensitivity to touch, and hence it is the easiest and costless means of burn assessment. However, with clinical evaluation, diagnosing superficial burns and very deep dermal burn wounds is reported to be optimal but imprecise for intermediate burns. In total, clinical evaluation is only accurate in two-thirds of all evaluations. In addition, the authors lamented that biopsy is more accurate than clinical assessment and is the most accepted means in determining burn depth. However, it has its own disadvantages such as sampling error, where the biopsied wound might not be an actual representation of the burn site, it leaves scar on the body due to specimen extraction, it is painful, and the interpretation of the result is subjective (nonuniformity) by different histopathologists. As such, biopsy is considered to be more of a research technique. Furthermore, the authors [22] highlighted the use of tissue perfusion measurement in burn care units, starting with thermal imaging. Thermal imaging operates based on the idea that deeper wounds are cooler than shallow wounds. it is stated that deeper burn wounds are more than 2°C cooler than normal/healthy skin. Despite being simple and efficient with up to 90% accuracy, thermal imaging technique is carried out by directly in contact with the wound area, and this makes it inefficient to measure the whole burn area, as it only assesses a sample portion and remains painful due to contact with burn wounds.

LDI techniques have been in used for more than two decades, and investigations were conducted to ascertain the reliability of using this device. In 2009, a study was conducted to determine the most effective assessment of burn wounds that require surgical intervention using both clinical assessment and LDI [23]. Moreover, the study provided an important information regarding the specific times burn assessment, when carried out, could yields good result. The authors similarly lamented the importance of early burn diagnosis to enhance early healing and recovery by the patients, but stressed the worrisome situation where differentiating superficial dermal burns from very deep dermal burns with clinical evaluation is very challenging. Superficial dermal burns are the type of burns believed to heal without surgical intervention, while deep dermal burns take a long time to heal and requires surgical intervention such as skin grafting. They also stated that visual assessment of burn wounds by experienced surgeons is also reliable in discriminating superficial and full-thickness burns. The accuracy of diagnosing dermal burns with clinical evaluation is between 50% and 75%, as reported by scholars. As such, they iterated the need for an alternative to improve the diagnostic accuracy using a more reliable technique. The authors proposed a study to determine the LDI accuracy in assessing burn depth and healing times on different days, starting from day 0. This study was aimed to provide vital information regarding on which day after the sustenance of burn injury does LDI assesses burn depth effectively when compared with clinical evaluation. They conducted the study on 40 patients (7 women and 33 men), all with intermediate burn wounds at the Burns center in University of Gent, Belgium. Both clinical and LDI techniques were used to determine the accuracy of assessments. The authors highlighted that, due to unforeseen condition, not all patients attended all the days. This is as a result of not arriving or reporting to the hospital on the day that the injury occurred, or due to restriction suggested by medical personnel for further scanning. The assessment was conducted on days 0, 1, 3, 5, and 8, As shown in Table 9.1, LDI is less accurate at day 0, and afterwards, the accuracy of LDI in assessing burn depth increases, with 100% accuracy at day 8 after burn incidence. Similarly, clinical evaluation shows inefficiency at day 0, increases at day 1 after burn, deteriorates at day 3, and regains accuracy at day 5, and shows its full effectiveness (100%) at day 8 after burn. Deteriorating or lack of good assessment at day 3 was reported to be as a result of dead tissue presence on the wound surface. However, the adherence of dead tissue on day 3 shows to have no effect on the assessment using LDI. The result also shows that LDI might not be required at day 8 after the burn, because clinical assessment is easy, less costly, and achieved a 100% accuracy.

It was stated that differentiating first-degree and third-degree burns is easier and effective with just clinical evaluation, but differentiating burns that include second-degree burn is the challenge faced by medical personnel even with experienced surgeons. A comparison of LDI and clinical assessments is presented in [24]. The authors' aim was to compare the performance efficiency of clinical assessment and LDI in discriminating superficial burn and intermediate burn wound (deep partial-thickness burns). The investigation was conducted at a burn unit in Mayo Hospital Lahore, Pakistan, where 34 patients were recruited, and a total of 92 burn wounds are considered from these participants within a period of 21 months (March 2015–November 2016). Laser Doppler scanning was done between days 3 and 5 after injury, because the burn depth is more reliable after the third day. Clinical evaluation was simultaneously performed by experienced surgeons but blinded by laser Doppler scanning outcome. The assessment has shown an accuracy of LDI outperforming clinical evaluation, with an LDI sensitivity of 92.75% against 81% for clinical assessment, and both achieved a specificity of 82%, positive predictive value of 94% for LDI against 93% for clinical assessment, negative predictive value of 79% for LDI against 59% for clinical assessment, and diagnostic accuracy of 90.12% for LDI against 81.52% for clinical assessment.

Moreover, another study was approved by the ethics committee at the Children's Hospital Westmead in Australia, in which 400 patients were included in the study over a period 12 months. This study served as a follow-up to a study presented in [21], where the effective performance of using LDI for the assessment of burn wound in children between 48 h and 72 h after injury was investigated. How effective is LDI in assessing burns in children before 48 h? This is the aim of the study in [25]. The authors found that the sensitivity and specificity using LDI are 78% and 74%, respectively, compared with those scans conducted after 48 h, which resulted in 75% and

TABLE 9.1
Performance Comparison of Clinical Assessment and LDI

Days	Number	Clinical Evaluation (%)	LDI (%)
0	31	40.6	54.7
1	39	61.5	79.5
3	40	52.5	95.0
5	34	71.4	97.0
8	25	100	100

85% for sensitivity and specificity, respectively. Statistically, its shows no significant difference. This indicates an inefficiency of LDI to detect burns at early hours, as reported by [23] and the need to have an efficient technique to do the job at early hours is of paramount importance.

The advantage of using LDI instead of visual assessment and histological means are as follows:

- highly experienced personnel can diagnose superficial burns and full-thickness burns, but there are burns in between, such as burns at the dermis layer, which are very challenging to diagnose. LDI can do that more accurately than humans.
- LDI can diagnose burns without any contact with the wound surface, thereby avoiding traumatic pains experienced by patients.

9.2.4.2.2 Challenges of Using LDI for Burn Assessment

However, despite the advantages offered by LDI over other techniques of burns assessment as highlighted in the previous sections, LDI has some underpinning factors. Some of these factors are summarized as follows:

i. Studies have shown that, in the first 24-h postburn, the LDI assessment is very poor, which is found to be good after 2–5 days. Late assessment of burn wounds is inaccurate due to the healing process and the formation of new tissue (also known as granulation tissue).
ii. The penetration power of laser light is limited by certain factors, such as blisters, cream, unremoved dead tissues, and coverings. Other factors that make burn evaluation so challenging include tattoos that were found absorb laser light [1,26] and natural skin pigment in a situation where it is bound to the dermis. However, natural skin pigmentation, in some cases, was found to be a noninfluencing factor because the pigment is detached along with the epidermis.
iii. Patient sickness and some vasoactive medications have been discovered to affect LDI measurement. Vasoactive medications reduce blood flow and as such affect burn wound assessment using LDI. Similarly, sickness affects skin blood flow.

iv. Skin appendages that are essential for restoration of a damaged tissue play a vital role when probing a burn wound with LDI. However, areas that were previously grafted due to a certain reason give an inaccurate LDI prediction.
v. In a situation where patient is restless or refuses to stand still, the movements were found to increase the LDI flux and tend to diminish the image appearance.
vi. Parts of the body, such as those at the edges of the limb and other parts that bent off from the LDI spot, have been found to receive limited LDI laser illumination, which requires to be rescanned from an appropriate direction.
vii. Another factor that hinders the use of LDI is the expensive modality reported by [6].

9.3 MACHINE LEARNING

ML is a subset of artificial intelligence (AI) that provides how to make machines intelligent and enable them to act like humans [27–29]. The terms AI and ML are sometimes used interchangeably; however, the terms differ but with a strong relationship. AI is a concept of making machines intelligent, while ML is a technique of how to achieve AI. ML basically can be grouped into two main categories: supervised ML and unsupervised ML. In supervised learning technique, machines are trained with labeled data so as to learn to map an input data with the corresponding output. The labeled data means a known data (i.e., a problem at hand with the known solution), and the idea is to enable machines to learn the relationship between data and output by learning unique representations that associated the input with the output. The goal is to enable accurate prediction of unseen data when presented to the machine without human intervention. The process of learning that involves only input data with no output information is called unsupervised learning. In this type of learning, machines are allowed to figure out and group data based on the similarity of representations. Unsupervised learning is considered more of true AI by some researchers than supervised learning approach, because during the learning process, there is complete absence of human intervention to guide the learning process. Some examples of supervised ML algorithms are support vector machines (SVMs) and artificial neural networks, while clustering is an example of unsupervised learning strategy.

Neural network has remained limited to not more than three layers, which consist of an input layer, a single hidden layer, and lastly the output layer till around 1990s. Subsequently, the evolution keeps growing to networks with multiple hidden layers. However, multiple layered networks were hindered by lack of training and evaluation datasets and unavailability of powerful computational machines to run the experiments. In 2010, the availability of a large database of annotated clean images called ImageNet [30] has revived the vison research. Since then, each year, vision communities have been participating in a challenge to train and evaluate their innovations.

9.3.1 Convolutional Neural Networks

ConvNets have recently achieved a remarkable success in the field of computer vision, such as image classification [31]. The idea was inspired by a study in 1950, but it was not fully recognized due to the unavailability of data to train the network and the lack of machines with computational power to train with. It became more popular in 2010 with the availability of a large database of annotated images comprising about 1,000 categories of different classes of images and were made publicly available to researchers who are willing to train and test their proposed models [30].

All the ConvNet models are composed of similar architectural layers, such as input layers, convolutional layers, pooling layers, and output layers or fully connected layers, but may differ in the layout they are presented. Input layer is the first layer that interfaces the outside world responsible for receiving input to be processed. However, input layer is not counted as part of the layer of a ConvNet model.

9.3.1.1 Convolution Layer

Convolutional layer in neural network is the most crucial part of the network. It is a layer of the CNN architecture that serves as a learning layer. It is a special layer of ConvNet architecture described in [32], and it consists of numerous neurons grouped together as filters. This layer performs the convolution (learning) operation on every portion of the input (usually image) and learns each pixel value [33], and the convolution operation means sliding a filter over an image while learning every single part of an input image.

A good way to understand the filter is by introducing a torchlight analogy. Assuming one is looking for a needle that falls in a room, say on a carpet in a sitting room that is very dim and makes searching a difficult task. One of the best means to look for it is to get a bright light into the searching environment. One might decide to use torchlight for the search, although the torchlight may not provide the desired brightness that can illuminate the entire region in the room. One may realize that the torchlight is brighter in a single focused area, and in ConvNet, this focused area is called receptive field. However, the filter size can be of different sizes: 3×3 filter, 5×5 filter, 7×7 filter, etc., depending on the required architecture defined by the developer. The operation is performed by a convolving (sliding) filter over the input image, where on every receptive field covered by the filter, the elements of the filter are multiplied with the receptive field's elements. The output of this operation is called feature map, and that should serve as the input to the next layer in the network. In addition, the convolving or sliding operation is guided by how many pixels a filter can move on at a time. This operation is called stride. For example, a stride of 1 means a filter moves from one receptive field to another by one pixel, a stride of 2 means moving a filter from a receptive field position onto another by skipping two pixels.

9.3.1.2 Pooling Layer

Thereafter, pooling layers are used to reduce the number of feature outputted by convolution layers so as to minimize computational cost and increase efficiency [34]. Pooling operations are performed similar to convolution operation by convolving against the features extracted by a convolution layer, and the operation is either maximum pooling operation or average pooling operation. Pooling operation reduces the number of features obtained through convolution operations into specific, small features. This is much like discarding unnecessary features, thereby minimizing the cost of computation and increasing the efficiency [34]. Examples of pooling

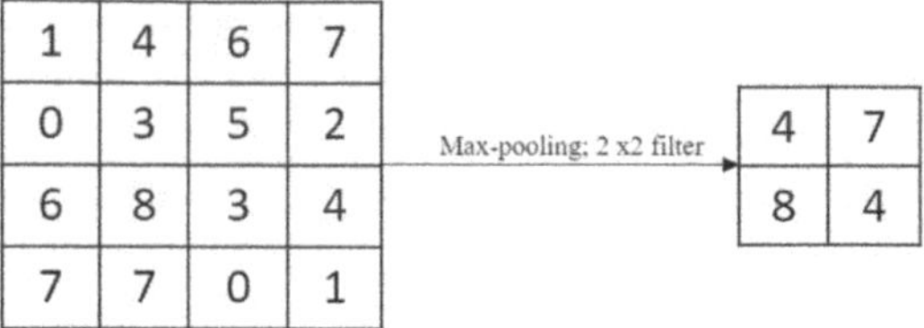

FIGURE 9.2 Example of a maximum pooling filter.

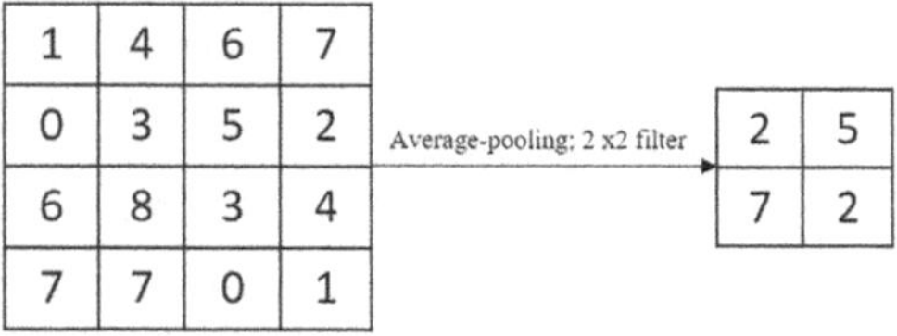

FIGURE 9.3 Example of an average pooling filter.

operations are maximum pooling (max-pooling) where the maximum value in the receptive field is selected to be the new pixel value in the condensed feature map, average pooling where the average of the features in the receptive fiedd is determined, the resulting average value serves as the new feature in the subregion of the new feature map. Examples of both maximum and average pooling filters are shown in Figures 9.2 and 9.3, respectively, both with 2×2 filter and a stride of 2.

9.3.1.3 Output/Classification Layer

The last layer of ConvNet model is the output layer, which outputs the information regarding the class to which each instance learned by the previous belongs. In a nutshell, the output layer is a classification layer.

9.3.2 Training a ConvNet

Training a ConvNet model requires a huge amount of data of very good quality. The presence of noise in the training dataset may hamper the performance effectiveness of the learning model and failed to achieve the desired expected output. Machines with high computational power are also a prerequisite for training deep learning models, as the emergence of a graphics processor unit greatly aids in wide popularity of deep learning. Training is normally performed by dividing the data into three subsets: training set used for the training; validation set used during the training to examine how accurate the model is on the unseen sample; and lastly, the testing set used after training to inspect the performance of the model. Thanks to the ImageNet Large Scale Visual Recognition Task (ILSVRC) introduced as an object recognition and classification benchmark in 2010 [35,36]. However, with limited database of images, it is now possible to train pretrained models via a concept called transfer earning. Transfer learning is a good strategy to train a ConvNet model with limited datasets, where the model that was trained on a large database can be employed to reuse the lower layers as feature extractors. This approach has been reported in many literatures [37,38] to be a good practice when dealing with insufficient data. A good example of the application of this strategy is in the health sector, not because the data is limited, but because the availability is hampered by ethical issues [33]. Therefore, the experiment provided in section 4 utilizes the concept of reusing a pretrained ConvNet model as a feature extractor while applying a classification algorithm for the classification task.

9.3.3 Common ConvNet Models

The illustrations of some ConvNet models that were trained on a huge database of annotated images containing 1,000 different categories are given as follows.

9.3.3.1 AlexNet

AlexNet is the first popular ConvNet architecture developed by Alex Krizhevsky et al. [39], which in 2012 participated in the ILSVRC and outperformed all other models presented in that year by achieving a top five error rate of 16%. The AlexNet has a total of eight layers, which consists of five convolutional layers (both convolution and pooling layers) and three fully connected layers. Convolution is performed using a 11 × 11 filter size, while the pooling operation is performed using 3 × 3 filters in the first layer with a stride of 2. In the second layer, the same operation is repeated but with a filter of size 5 × 5, while in all the remaining three layers (third, fourth, and fifth), 3 × 3 filters were used. Figure 9.4 gives a graphical view of the AlexNet model.

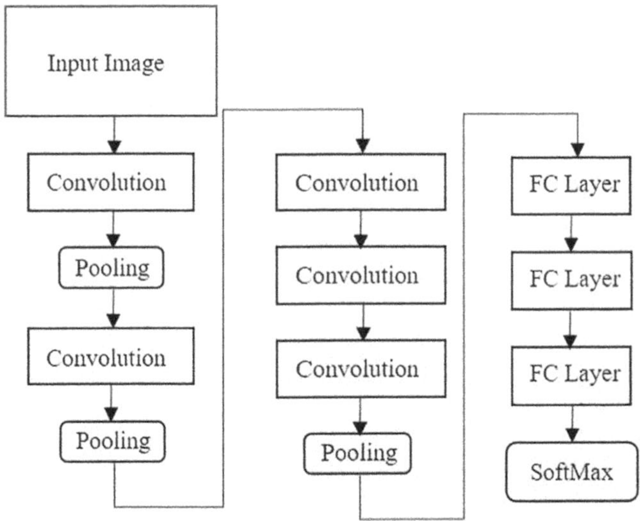

FIGURE 9.4 Graphical representation of AlexNet.

9.3.3.2 GoogleNet

GoogleNet by Szegedy et al. [40] from Google achieved a top five error rate of 6.67% in 2014. The GoogleNet model has multiple convolution layers in each layer arranged in parallel, which was proposed and incorporated as inception layers, as shown in Figure 9.5. These convolution layers are equipped with variable sizes, and the outputs are concatenated to form a single input to the next layer. GoogleNet has 22 layers, which makes it a deeper network than any model proposed before it.

9.3.3.3 VGGNet

Similarly, the Visual Geometry Group (VGG) at the University of Oxford proposed a model and won a second place in the ILSVRC challenge in 2014 [31]. Two variants of this network (VGG-16 and VGG-19) were the best performing models and are very deep with a similar architectural design but different numbers of layers and uses smaller convolution filters (i.e., 3 × 3) compared with 11 × 11 filter sizes in AlexNet. VGGNet differs with AlexNet in the placement of pooling layers—AlexNet is stacked with pooling layer every convolution, while VGGNet uses a pooling layer after two or three convolution layers.

9.3.3.4 Residual Network

In 2015, Microsoft proposed a deep network called Residual Network (ResNet) [41], which is almost eight times deeper than VGGNet and won the first position on the ImageNet ILSVRC challenge, with an error rate of 3.57%. The unique feature of this network is the residual connection (shortcut connection). In a nutshell, shortcut connection (or rather in some context, skip connection) is the addition of input of the previous layer with an output of a lower layer, and the resulting value then passes through an activation function that serves as the input for the next layer down the

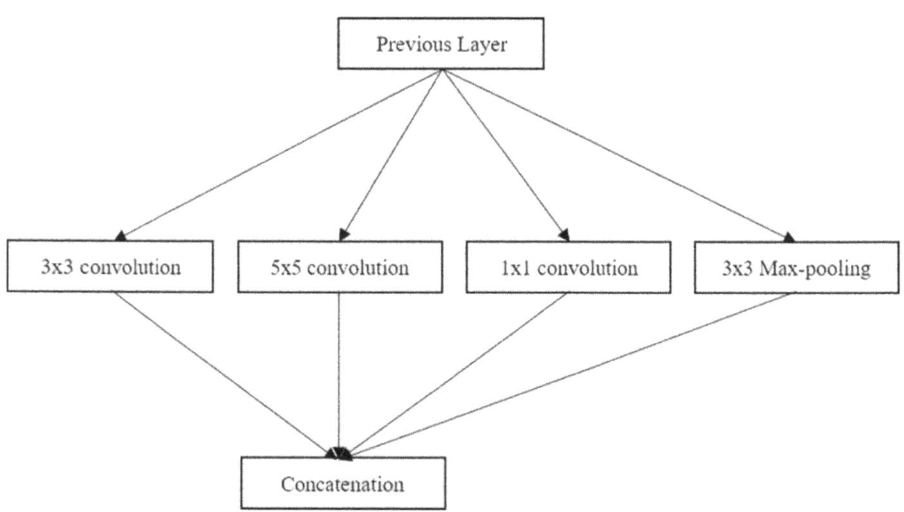

FIGURE 9.5 Inception layer.

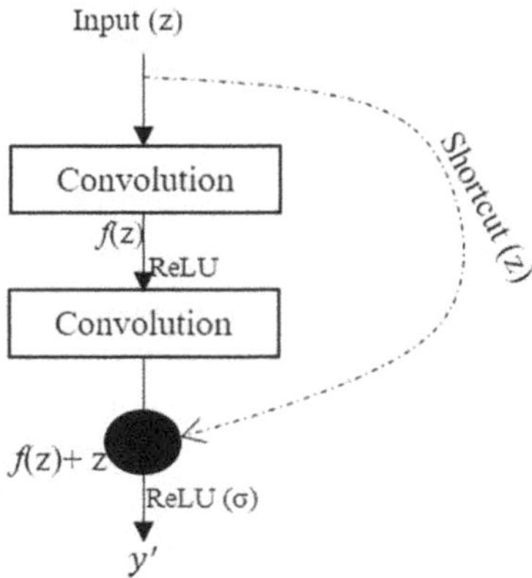

FIGURE 9.6 Description of a ResNet model.

network. This provides the possibility of a stacking network with more layers, with increasing accuracy as shown in Figure 9.6.

Skip connection was adopted after every few stacked convolution layers, which is represented by a shortcut path as shown in Figure 9.6. Mathematically, the resulting operation is represented by the following equation:

$$y = f(z) + (z) = f(z) + z \tag{9.1}$$

The (z) function, which is a shortcut presented in Figure 9.6, allows the gradient to be maintained and gives an opportunity to train the network quicker and with more layers stacked, and $f(z)$ is the output of the convolution layer (weight layer). The convolution layers may vary depending on the number of hyperparameters contained in each layer.

Figure 9.6 shows that ResNet has two parts: the main part and the shortcut part. The main part comprises the regular convolution (weight layer) and activation (ReLU) layers, while the shortcut path is the input of the previous layer directly concatenated to the output of the subsequent layers. After concatenation, nonlinearity is applied as represented in the following equation: $y' = \sigma(f(z) + z)$, where σ is the activation operation (nonlinearity) after the elementwise operation, as depicted in Figure 9.6, and y' is the output of the nonlinearity operation.

9.4 GOALS AND METHODOLOGY

The main goal is to investigate the use of ML techniques for recognition of burn wounds via the application of off-the-shelf learning model. The learning model is a pretrained model on objects classification, and it is proposed in this work to apply a particular model as feature extractor to extract discriminatory features contained in the images and classify them into healthy skin, superficial burn (first-degree burn), or full-thickness burn (third-degree burn) via the utilization of a classification algorithm.

9.4.1 Image Acquisition and Preprocessing

The burn images were directly captured from patients in Plastic Surgery and Burns Research Unit, Bradford Teaching Hospitals, with full consent, where a total of 34 images of different burn categories were obtained, which were subsequently augmented by flipping, zooming, and rotating, giving a total of 660 full-thickness burn images. Additionally, the superficial burn injuries (first-degree burns) were obtained via search on the internet, comprising mainly sunburn images. A total of 46 superficial burn images were successfully generated and then augmented to 660 images. Similarly, 660 healthy skin images were incorporated in the investigation, making a total of 1,980 images. All images were resized to 224 × 224 before feature extraction and classification, and in this way, all images will be of same size and correspond to a standard image size required by ResNet101. The samples of burn and healthy skin images were depicted in Figure 9.7.

9.4.2 Feature Extraction and Classification

The work presented here is inspired by the performance of ConvNet model (ResNet) during the ILSVRC challenge in 2015, and the choice of using a pretrained model is based on research conducted by researchers [37,42] who helped to achieve outstanding results. It was shown that, using a pretrained model to extract features from images, instead of training a ConvNet model from scratch with limited dataset, yields a remarkable result [37]. As such, to extract useful discriminatory features from images, lower layers (convolutional layers) of ResNet101 were used (excluding the topmost layer) as learning layers. These layers were convolved against the 2D images to extract features. In a nutshell, the approach is simple, yet powerful filters of fully trained state-of-the-art neural network are used to convert raw image pixels into dense features, and then, an SVM classifies the features into nonburns and burns of different degrees.

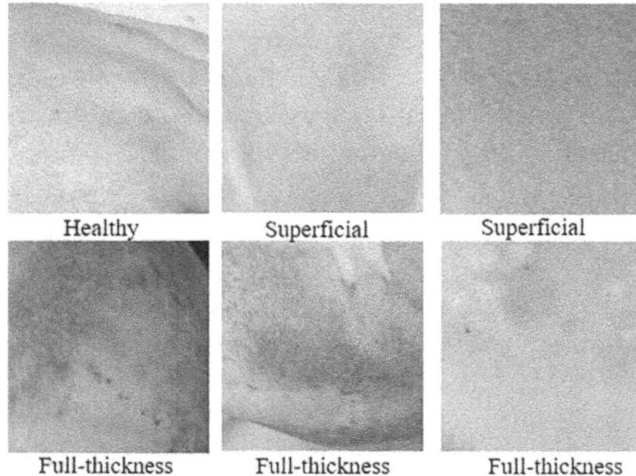

FIGURE 9.7 A sample of healthy and burn images.

Classiofcation of the extracted feature was conducted by training a SVM. SVM is a supervised learning algorithm that is primarily used for classification, regression, and outlier detection tasks on both simple and complicated data [43]. It works by finding an optimum separating hyperplane (OSH) that acts as a boundary that separates classes of data; this is achieved by solving an optimization problem [44]. As a supervised learning algorithm, it learns to classify unseen data based on a set of labeled training data. Thus, the training phase is used to build a model; thereafter, new unseen data is mapped to one of those learned classes by computing the dot product of the model (weights) and image-data features [44].

Moreover, the evaluation was conducted via k-fold cross-validation (CV) techniques, where data is subsampled into k parts equally. CV is a statistical means of evaluating learning algorithms [45] that help to avoid sampling error. A special case of K-fold (where $k=10$) CV is used, where $k-1$ is used for training, and the remaining split is used for validation (testing). The iteration continues until all the subsamples are exhausted in such a way that each subsample is used for both training and testing, and finally, the average of all the tests gives the result as expressed mathematically in Equation (9.2) later

$$\text{Accuracy} = \frac{1}{n}\sum_{i=1}^{n} A_i \tag{9.2}$$

where n is the number of folds and A_i is the accuracy obtained in each iteration.

9.5 RESULTS AND DISCUSSION

The contingency table in Table 9.2 shows the summary of all the results obtained. The contingency table was used to display the performace output of the classifier, where the columns correspond to original classes fed into the algorithm, and the rows are the corresponding predicted outputs by the algorithm.

The labels (a–c) in Table 9.2 mark the individual class category. Healthy skin images were labeled as "a," superficial burn images were labeled as "b," and full-thickness burn images were labeled as "c". The diagonal value at cell position (a, a) gives the correct predicted samples of healthy skin (660 images), the value at cell position (b, b) gives the exact predicted samples of superficial burns (660), and the one at cell (c, c) is the correct predicted full-thickness images (659). Any other predictions other than those mentioned in the three cells are misclassified instances.

9.5.1 Terms Related to Contingency Table

To evaluate the result obtained in the contingency table as presented in Table 9.2, there are some related terms that need to be defined. These terms include true negatives (TN) that refer to actual negative samples (healthy images) of a class that are predicted as such (refer to cell [a, a] in Table 9.2). Negative images predicted as positives are referred to as false positives (FP). Those positive samples that were predicted as negatives are referred to as false negatives (FN). Burn images that are

TABLE 9.2
Classification Result

		Target Classes		
		a	b	c
Predicted classes	a	660	0	1
	b	0	660	0
	c	0	0	659

predicted as burns are referred to as TPs, and this can be obtained in cell [c, c]. Moreover, cell [b, b] also presents a TP class for superficial epidermal burns.

9.5.2 Classifier Performance

The performance evaluation of the classification algorithm in categorizing each instance class is presented in Table 9.3. The classifier mapped an instance from the three classes into one of the healthy, superficial burn, full-thickness burn class labels. The performance evaluation metrics such as precision, sensitivity (recall), and specificity are computed as well. Precision gives the proportion of relevant instances retrieved by the classifier (i.e., fraction of relevant instances among the total retrieved instances) [46] as represented in Equation (9.3). Recall is the number of accurately categorized positive samples divided by the number of positive samples contained in the data. Recall is a term sometimes referred to as sensitivity (accurate prediction of positive instances), as formulated in Equation (9.4). Specificity is the measure of negative sample recognition [47], as formulated in Equation (9.5). Moreover, the overall accuracy (OA) of the classification algorithm and the corresponding precision metric are shown in Table 9.4.

$$\text{Precision} = \frac{\text{TP}}{\text{TP} + \text{FP}} \tag{9.3}$$

$$\text{Sensitivity (Recall)} = \frac{\text{TP}}{\text{TP} + \text{FN}} \tag{9.4}$$

$$\text{Specificity} = \frac{\text{TN}}{\text{TN} + \text{FP}} \tag{9.5}$$

TABLE 9.3
Performance Metric of an Individual Class

Metrics	Healthy	Superficial	Full-Thickness
Precision (%)	99.848	100.000	100.000
Sensitivity (Recall) (%)	100.000	100.000	99.848
Specificity (%)	99.924	100.000	100.000

TABLE 9.4
OA and OP of the Classification Algorithm

OA (%)	OP (%)
99.949	99.924

The result, however, recorded a misclassified element from full-thickness burns, which is falsely predicted as healthy skin, as shown in Table 9.2 in cell position [a, c]. This misclassification could be as a result of waxy appearances of some of the full-thickness images as shown in Figure 9.8.

The full-thickness burns as shown in Figure 9.8 appear to be white, waxy, and shiny compared with the healthy skin. This shiny appearance of full-thickness seems like unaffected skin images; thus, this could be the reason for a single misclassification of full-thickness image, as obtained in the result.

OA of the classification can be obtained by summing the correct classifications (TP) in all classes and then divide it by the sum of all classifications.

$$\text{OA} = \frac{\text{TP(H)} + \text{TP(S)} + \text{TP(F)}}{\text{Total samples}} \tag{9.6}$$

where TP(S) is the TP for healthy skin class, TP(S) is the TP for superficial epidermal burns class, and TP(F) is the TP for the full-thickness burns class. The result obtained by Equation (9.6) is presented in Table 9.4.

The overall precision (OP) of the classification algorithms was obtained as expressed in Equation (9.7)

$$\text{OP} = \frac{\text{TP(S)} + \text{TP(F)}}{\text{TP(S)} + \text{TP(F)} + \text{FP}} \tag{9.7}$$

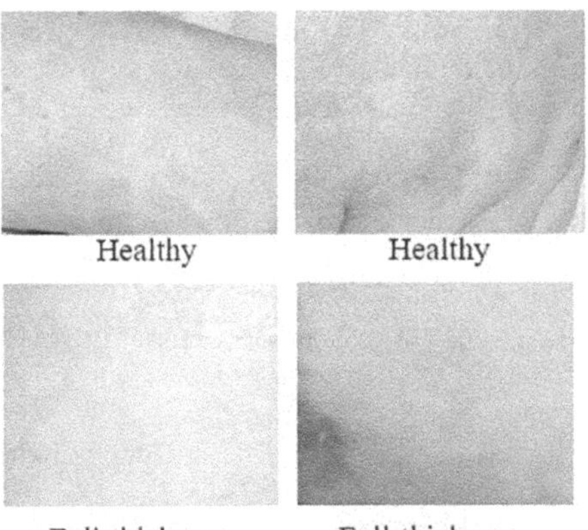

FIGURE 9.8 Healthy and full-thickness burn images.

The FP in Equation (9.8) stands for a negative instance that is misclassified as positive, and the result is presented in Table 9.4.

9.6 CONCLUSIONS

The conclusive objective of diagnosing burns is to provide immediate assessment, reduce length of hospital delays, avoid complications that are likely to be acquired due to the long hospitalization, and prevent loss of lives as a result of unavailability of medical facilities. Thus, ML algorithm is used to discriminate healthy skin, superficial burns, and full-thickness burns. The extraction of image features was conducted using a pretrained model (ResNet101) and a subsequently utilized SVM classifier for the classification task. The result shows a performance accuracy of up to 99. An interesting part of this investigation is that the future extraction model was trained on object recognition and classification task, but it seems good when applied to medical images for the feature extraction. Application of ML can be utilized to tackle the unavailability of medical personnel in some remote locations where access to health facility and experienced health practitioners is subjecting patients to difficult situations.

One of the limitations of the investigation in this work is noninclusion of intermediate burns. This was due to unavailability of such data at our disposal. Second, the investigation was conducted on Caucasian patients (people with white skin), as it is worth to include data from people of different ethnicities, such as burn images from skin of color ethnicities. Therefore, these provide the room for further investigation.

REFERENCES

1. C. Wearn, K. C. Lee, J. Hardwicke, A. Allouni, A. Bamford, P. Nightingale, and N. J. B. Moiemen, "Prospective comparative evaluation study of Laser Doppler Imaging and thermal imaging in the assessment of burn depth," *Burns*, vol. 44, no. 1, pp. 124–133, 2018.
2. A. D. Gilbert, E. Rajha, C. El Khuri, R. B. Chebl, A. Mailhac, M. Makki, and M. El Sayed, "Epidemiology of burn patients presenting to a tertiary hospital emergency department in Lebanon," *Burns*, vol. 44, no. 1, pp. 218–225, 2018.
3. P. Brassolatti, P. S. Bossini, H. W. Kido, M. C. D. Oliveira, L. Almeida-Lopes, L. M. Zanardi, M. A. Napolitano, L. R. d. S. de Avó, F. M. Araújo-Moreira, and N. A. Parizotto, "Photobiomodulation and bacterial cellulose membrane in the treatment of third-degree burns in rats," *Journal of Tissue Viability*, vol. 27, no. 4, pp. 249–256, 2018.
4. R. Kandiyali, J. Sarginson, L. Hollén, F. Spickett-Jones, and A. Young, "The management of small area burns and unexpected illness after burn in children under five years of age—A costing study in the English healthcare setting," *Burns*, vol. 44, no. 1, pp. 188–194, 2018.
5. S. Wall, N. Allorto, R. Weale, V. Kong, and D. Clarke, "Ethics of burn wound care in a low-middle income country," *AMA Journal of Ethics*, vol. 20, no. 6, p. 570, 2018.
6. D. McGill, K. Sørensen, I. MacKay, I. Taggart, and S. Watson, "Assessment of burn depth: a prospective, blinded comparison of laser Doppler imaging and videomicroscopy," *Burns*, vol. 33, no. 7, pp. 833–842, 2007.
7. A. D. Jaskille, J. C. Ramella-Roman, J. W. Shupp, M. H. Jordan, and J. C. Jeng, "Critical review of burn depth assessment techniques: part II. Review of laser Doppler technology," *Journal of Burn Care & Research*, vol. 31, no. 1, pp. 151–157, 2010.

8. S. Charuvila, M. Singh, D. Collins, and I. Jones, "A comparative evaluation of spectrophotometric intracutaneous analysis and laser doppler imaging in the assessment of adult and paediatric burn injuries," *Journal of Plastic, Reconstructive & Aesthetic Surgery*, Vol. 71(7): pp: 1015–1022, 2018.
9. M. M. Rybarczyk, J. M. Schafer, C. M. Elm, S. Sarvepalli, P. A. Vaswani, K. S. Balhara, L. C. Carlson, and G. A. Jacquet, "A systematic review of burn injuries in low-and middle-income countries: Epidemiology in the WHO-defined African Region," *African Journal of Emergency Medicine*, vol. 7, no. 1, pp. 30–37
10. C. Johnson, "Management of burns," *Surgery (Oxford)*, vol. 36, no. 8, pp. 435–440, 2018.
11. F. Pencle and M. Waseem, "First degree burn," StatPearls [Internet], NCBI, 2017, last accessed on July 31, 2019
12. R. A. Masood, Z. N. Wain, R. Tariq, and I. Bashir, "Burn cases, their management and complications: A review," *International Current Pharmaceutical Journal*, vol. 5, no. 12, pp. 103–105, 2016.
13. K. D. a. R. Smith, "One in six NHS patients 'misdiagnosed'," *The Telegraph*, 2009.
14. P. Gill, "The critical evaluation of laser Doppler imaging in determining burn depth," *International Journal of Burns Trauma*, vol. 3, no. 2, p. 72, 2013.
15. M. Scheffler, S. Koranyi, W. Meissner, B. Strauß, and J. Rosendahl, "Efficacy of non-pharmacological interventions for procedural pain relief in adults undergoing burn wound care: A systematic review and meta-analysis of randomized controlled trials," Burns, Vol: 44(7): pp: 1709–1720, 2017.
16. A. A. Kouadio, F. Jordana, P. Le Bars, and A. Soueidan, "The use of laser Doppler flowmetry to evaluate oral soft tissue blood flow in humans: A review," *Archives of Oral Biology*, vol. 86, pp. 58–71, 2018.
17. P. L. Ricketts, A. V. Mudaliar, B. E. Ellis, C. A. Pullins, L. A. Meyers, O. I. Lanz, E. P. Scott, and T. E. Diller, "Non-invasive blood perfusion measurements using a combined temperature and heat flux surface probe," *International Journal of Heat and Mass Transfer*, vol. 51, no. 23–24, pp. 5740–5748, 2008.
18. P. Gill, "The critical evaluation of laser Doppler imaging in determining burn depth," *International Journal of Burns and Trauma*, vol. 3, no. 2, p. 72, 2013.
19. Z. B. M. Niazi, T. J. H. Essex, R. Papini, D. Scott, N. R. McLean, and M. J. M. Black, "New laser Doppler scanner, a valuable adjunct in burn depth assessment," *Burns*, vol. 19, no. 6, pp. 485–489, 1993.
20. A. D. Jaskille, J. W. Shupp, M. H. Jordan, and J. C. Jeng, "Critical review of burn depth assessment techniques: Part I. Historical review," *Journal of Burn Care & Research*, vol. 30, no. 6, pp. 937–947, 2009.
21. A. J. A. Holland, H. C. O. Martin, and D. T. Cass, "Laser Doppler imaging prediction of burn wound outcome in children," *Burns*, vol. 28, no. 1, pp. 11–17, 2002.
22. S. Monstrey, H. Hoeksema, J. Verbelen, A. Pirayesh, and P. Blondeel, "Assessment of burn depth and burn wound healing potential," *Burns*, vol. 34, no. 6, pp. 761–769, 2008.
23. H. Hoeksema, K. Van de Sijpe, T. Tondu, M. Hamdi, K. Van Landuyt, P. Blondeel, and S. Monstrey, "Accuracy of early burn depth assessment by laser Doppler imaging on different days post burn," *Burns*, vol. 35, no. 1, pp. 36–45, 2009.
24. S. N. Jan, F. A. Khan, M. M. Bashir, M. Nasir, H. H. Ansari, H. B. Shami, U. Nazir, A. Hanif, and M. Sohail, "Comparison of Laser Doppler Imaging (LDI) and clinical assessment in differentiating between superficial and deep partial thickness burn wounds," *Burns*, Vol. 44(2): pp: 405–413, 2017.
25. K. Nguyen, D. Ward, L. Lam, and A. J. A. Holland, "Laser Doppler Imaging prediction of burn wound outcome in children: Is it possible before 48h?," *Burns*, vol. 36, no. 6, pp. 793–798, 2010.

26. U. Sarwar, M. Javed, and W. A. Dickson, "Diagnostic challenges of assessing the depth of burn injuries overlying intricate coloured tattoos," *Journal of Plastic, Reconstructive & Aesthetic Surgery*, vol. 67, no. 7, pp. e186–e187, 2014.
27. F. Musumeci, C. Rottondi, A. Nag, I. Macaluso, D. Zibar, M. Ruffini, and M. Tornatore, "An overview on application of machine learning techniques in optical networks," *IEEE Communications Surveys & Tutorials*, Vol. 21(2), pp: 1383–1408, 2018.
28. Z. Al-Kassim, and Memon, Q., "Designing a low-cost eyeball tracking keyboard for paralyzed people," *Computers & Electrical Engineering*, Vol. 58, pp. 20–29, 2017.
29. Memon, Q., "Smarter health-care collaborative network," Building Next-Generation Converged Networks: Theory and Practice, 2013, pp. 451–476.
30. J. Deng, W. Dong, R. Socher, L.-J. Li, K. Li, and L. Fei-Fei, "Imagenet: A large-scale hierarchical image database," in *Computer Vision and Pattern Recognition,* 2009. *CVPR* 2009. *IEEE Conference on*, 2009, pp. 248–255, Miami, USA
31. K. Simonyan and A. Zisserman, "Very deep convolutional networks for large-scale image recognition," *arXiv preprint arXiv:*1409.1556, 2014.
32. W. Nash, T. Drummond, and N. Birbilis, "A review of deep learning in the study of materials degradation," *NPJ Materials Degradation*, vol. 2, no. 1, p. 37, 2018.
33. R. Yamashita, M. Nishio, R. K. G. Do, and K. Togashi, "Convolutional neural networks: An overview and application in radiology," Insights into Imaging, vol. 9, pp. 1–19, 2018.
34. S. Sakib, N. Ahmed, A. J. Kabir, and H. Ahmed, "An Overview of Convolutional Neural Network: Its Architecture and Applications," 2018110546 (doi: 10.20944/preprints201811.0546.v1), 2018.
35. O. Russakovsky, J. Deng, H. Su, J. Krause, S. Satheesh, S. Ma, Z. Huang, A. Karpathy, A. Khosla, and M. Bernstein, "Imagenet large scale visual recognition challenge," *International Journal of Computer Vision*, vol. 115, no. 3, pp. 211–252, 2015.
36. W. Nash, T. Drummond, and N. Birbilis, "A review of deep learning in the study of materials degradation," *NPJ Materials Degradation*, vol. 2, no. 1, p. 37, 2018.
37. A. M. Bukar and H. Ugail, "Automatic age estimation from facial profile view," *IET Computer Vision*, vol. 11, no. 8, pp. 650–655, 2017.
38. J. Yosinski, J. Clune, Y. Bengio, and H. Lipson, "How transferable are features in deep neural networks?," in *Advances in Neural Information Processing Systems*, 27, pp: 3320–3328, 2014.
39. A. Krizhevsky, I. Sutskever, and G. E. Hinton, "Imagenet classification with deep convolutional neural networks," in *Advances in Neural Information Processing Systems*, 2012, pp. 1097–1105.
40. C. Szegedy, W. Liu, Y. Jia, P. Sermanet, S. Reed, D. Anguelov, D. Erhan, V. Vanhoucke, and A. Rabinovich, "Going deeper with convolutions," in *Proceedings of the IEEE Conference on Computer Vision and Pattern Recognition*, Boston, MA, 2015, pp. 1–9.
41. K. He, X. Zhang, S. Ren, and J. Sun, "Deep residual learning for image recognition," in *Proceedings of the IEEE Conference on Computer Vision and Pattern Recognition*, Las Vegas, NV, 2016, pp. 770–778.
42. A. S. Razavian, H. Azizpour, J. Sullivan, and S. Carlsson, "CNN features off-the-shelf: an astounding baseline for recognition," in *Computer Vision and Pattern Recognition Workshops (CVPRW), 2014 IEEE Conference on*, Columbus, OH, 2014, pp. 512–519.
43. H. Xue, Q. Yang, and S. Chen, *SVM: Support Vector Machines*: Chapman & Hall/CRC: London, UK, 2009.
44. V. Jakkula, "Tutorial on support vector machine (svm)," in *School of EECS, Washington State University*, vol. 37, 2006.
45. R. Kohavi, "A study of cross-validation and bootstrap for accuracy estimation and model selection," in *IJCAI*, vol. 14, pp. 1137–1145, 1995.

46. V. López, A. Fernández, S. García, V. Palade, and F. Herrera, "An insight into classification with imbalanced data: Empirical results and current trends on using data intrinsic characteristics," *Information Sciences*, vol. 250, pp. 113–141, 2013.
47. M. Sokolova and G. Lapalme, "A systematic analysis of performance measures for classification tasks," *Information Processing & Management*, vol. 45, no. 4, pp. 427–437, 2009.

10 A Study and Analysis of an Emotion Classification and State-Transition System in Brain Computer Interfacing

Subhadip Pal, Shailesh Shaw, Tarun Saurabh, Yashwant Kumar, and Sanjay Chakraborty
TechnoIndia

CONTENTS

10.1 INTRODUCTION

In the human world, emotion plays a vital role for successful communication. Ability to understand the emotional states of a person whom we are communicating with comes to humans naturally. But that is not the case with machines. Whenever we think about machine communication, we assume its artificial machine tone and lack of mutual rapport. However, considering the explosion of machines in our recent lives, there should be no hesitation to admit that any interface that disregards human affective states in the interaction will appear cold and socially inept to users. To approach the effective human–machine interaction, one of the most important prerequisites is a reliable emotion recognition system that stands tall in some parameters like recognition accuracy, robustness against any artifacts, and adaptability to practical applications. Emotion recognition is performed by classifying emotional features measured from the implicit emotion channels of human communication, such as speech, facial expression, gesture, pose, physiological response, etc. The first kind of approach revolves around the analysis of facial expression or speech [1–3]. However, these techniques can be made a victim of deception. Peripheral physiological signals provide the basis for another kind of approach. Changes in the autonomic nervous system in the periphery, such as electrocardiography (ECG), skin conductance (SC), respiration, and pulse can help us in detecting changes in the emotional state [4–6]. This is a more detailed and complex technique than audiovisual-based techniques.

The second kind of approach focuses on brain signals captured from the central nervous system (CNS), such as *electroencephalography (EEG),* where electrodes are placed at different regions on the skull, which measure how active the part of the brain is using voltage fluctuation (Figure 10.1), electrocorticography (ECoG), and functional magnetic resonance imaging (fMRI). Among these, EEG signals have proven to be more informative about emotional states.

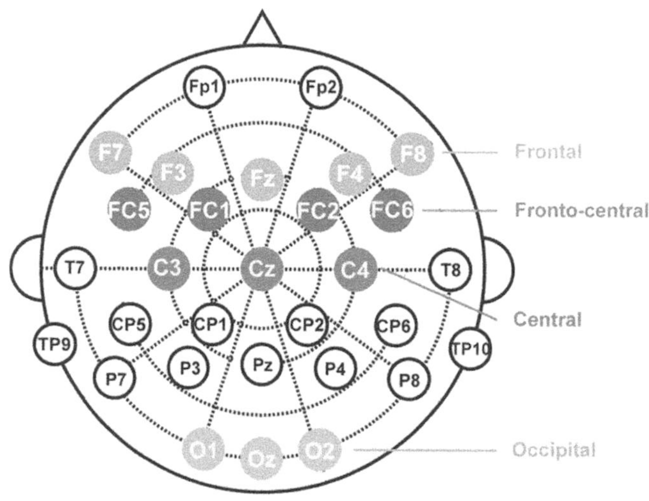

FIGURE 10.1 Electrode placement.

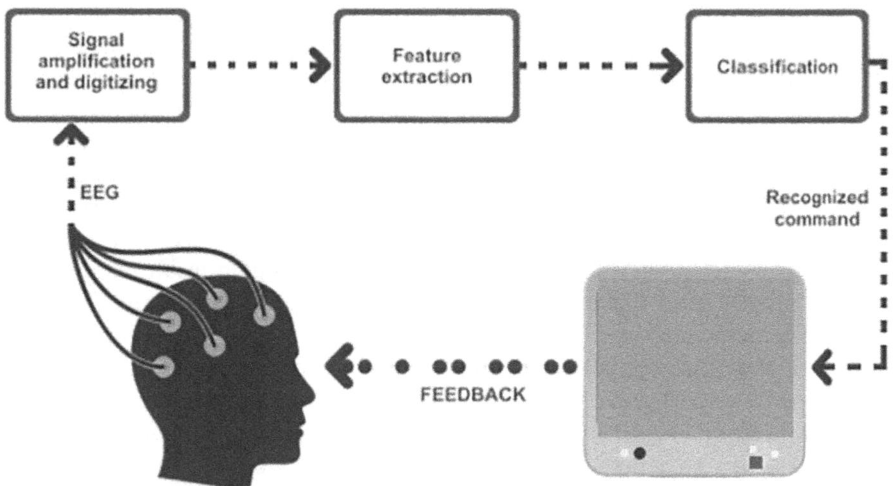

FIGURE 10.2 BCI setup.

The overall brain–computer interface (BCI) setup is shown in Figure 10.2. BCI is an interface between the brain and the outside world, without the direct intervention of muscular activities. BCI technology mainly consisted of four sections, such as signal acquisition, signal processing, feature extraction and classifications, and an application interface. The brain signal is captured using EEG signal acquisition technique and then the scalp voltage level is transferred into some physical movements. In paper [7,8], an efficient algorithm of cursor movement has been proposed to reach the desired target in minimum time. In our work, we have implemented an emotion classification technique, and we have also implemented a new model of emotional state-transition machine using a supervised learning approach. With the help of an emotional state-transition system, we can identify the change of emotional states at a fixed time interval. According to the two-dimensional model of emotion described by Davidson et al. [9], emotion is represented as a two-dimensional space (arousal and valence). There are multiple ways to explain emotion states, such as (i) visual (images/pictures), (ii) audiovisual (clips/video clips), (iii) audio (songs/sounds) [10], etc. In paper [2], the author has described emotion detection based on facial expression and speech. Physiological signal analysis is another popular approach to emotion recognition. Various studies show that peripheral physiological signals like ECG, skin conductive resistance (SCR), and blood volume pressure (BVP) can also change the emotions [11]. Davidson et al. [12] suggested that frontal brain activity is related to positive and negative emotions. In paper [13], the author has recorded EEG signals to analyze the oscillatory brain activity while the subjects were listening to music. Qing et al. [14] have implemented an integrated emotion recognition system using brian activity, facial expression, and gist to discover the emotions from the neutral scene. In paper [15], the author has developed a system that can recognize different emotional states by analyzing the speaker's voice/recordings. In our work, we have recognized four different types of emotion (positive, negative, depressed, and harmony). After classifying the emotions completely, we proceed for the transition of different emotional states from the neutral state at a fixed time interval using state-transition machine. In this chapter, we use support vector machine (SVM), *k*-nearest neighbors (KNN) and random forest for classification. We are trying to recognize the emotional state and predict the next state with some time interval. To recognize emotions defined by an accepted psychological model, in EEG, by the use of any classification algorithm (KNNs). This entails an investigation of the problems areas (emotions, EEG, and BCIs), i.e., commonly used computational methods in these areas; this should result in a method and an implementation that accomplish the goal of recognizing different emotions in EEG. The main motivation of this work is to classify emotions in such a way that helps in rational decision-making, perception, human interaction, and human intelligence processes. Hence, emotions are fundamental components of being human, as they motivate action and add meaning and richness to virtually all human experience. They hold a great potential for people who are paralyzed or otherwise unable to use their hands or who cannot talk.

This rest of the book chapter is organized as follows. A brief literature review has been done in Section 10.2. In Section 10.3, we have described our proposed work with a suitable flow chart diagram, which includes emotion classification and transition function for the state-transition system. Besides that, we have also analyzed a state-transition system based on the increased or decreased activity of specific cortex of the brain in Section 10.3. Then, we have done a detailed performance analysis of our approach in Section 10.4. We have compared our proposed work with previous studies related to different parameters of classification in Section 10.4, and finally, Section 10.5 describes the conclusion of this work.

10.2 LITERATURE REVIEW

The research activities on human emotions have been existing for a long time in psychology and psychophysiology. But in engineering, research in that field is relatively new. However, many efforts have been made to recognize human emotions using audiovisual channels of emotion expression like speech, facial expression, and gestures till now. These audiovisual techniques help in noncontact detection of emotions, and so the subject is always in comfort. These techniques are however deception prone. Some of the audiovisual recognition techniques are elaborately discussed in [1–3]. However, the attention is now gradually moving towards using physiological measures [4–7] and, most recently, towards EEG signal. Recent theories on emotion [16,17] explains that physiological activity is an important aspect of emotion, or rather, we can say that physiological activities help to construct a certain emotion. Several previous works show the associativity of increased activity of certain parts of our brain with basic emotions, which can be captured through an EEG signal. Yazdani et al. [18] proposed using a BCI based on P300 evoked potential to emotionally tag videos with basic emotions. This book chapter proposes an effective algorithm to classify EEG signals, explains the activities of different parts of the brain based on values measured with electrodes, and demonstrates how the transition of emotion can happen for a human. In Bhardwaj, Gupta, Jain,

Rani, and Yadav, 2015, they worked on six different emotional states and used linear discriminant analysis (LDA) and SVM classification to determine the different emotional states. Other papers have lower classification accuracy in comparison to our proposed algorithm, as in the papers (Anh et al., 2012; Jatupaiboon, Panngum, & Israsena, 2013; Jirayucharoensak et al., 2014; Wijeratne & Perera, 2012; Yoon & Chung, 2013). In this paper, a framework was proposed to optimize EEG-based emotion recognition systematically: (i) seeking emotion-specific EEG features and (ii) exploring the efficacy of the classifiers. They classified four emotional states (joy, anger, sadness, and pleasure) using SVM and obtained an averaged classification accuracy of 82.29%±3.06% across 26 subjects [19]. In paper [18], for implicit emotional tagging of multimedia content, the author proposed a BCI system based on P300 evoked potential. Their system performs implicit emotional tagging and naïve subjects. This system can be used efficiently on who have not participated in the training. They had given a subjective metric called as "emotional taggability." The recognition performance of the system was analyzed, and the degree of ambiguity was given, which exists in terms of emotional values that are associated with multimedia content [18]. BCI is able to access brain activity, which leads to understand the emotional state of any human. One very interesting work of EEG signal-based emotion classification using LDA classifier with "Correlation-Based Subset Selection" technique was introduced in [20].

In paper [21], they used this information in two manners: (i) The intention of the user is correctly interpreted in spite of signal deviations induced by the subject's emotional state, because the influence of the emotional state on brain activity patterns can allow the BCI to adapt its recognition algorithms. (ii) The user can use more natural ways of controlling BCI through affective modulation and can potentially lead to higher communication throughput by the ability to recognize emotions.

Inspired by the earlier works, we have used some popular supervised learning algorithms for classification (SVM, random forest, and KNNs) in our work. This study has been applied to various machine-learning algorithms to categorize EEG dynamics according to subject self-reported emotional states while watching videos. In our work, we have classified emotions into four states (negative emotion, positive emotion, depression, and harmony). Also, we have proposed a "State-Transition Algorithm" as well and tried to transit from one emotional state to another.

10.3 PROPOSED WORK

- Analysis of EEG data through signal processing

  Here we have used "kaggle NER" dataset [22], where the EEG signal is captured at a 1-min time gap, and the EEG signal is sampled at 600 Hz. Then, the EEG signal is preprocessed to check the presence of artifacts and which artifacts are removed using adaptive filtering. After filtering, we get useful features from the signal by applying a suitable feature subset selection method [20]. After some useful feature extraction [20], we proceed to emotion classification using SVM, KNN, random forest, and emotion transition. We have described the flow chart of the entire proposed

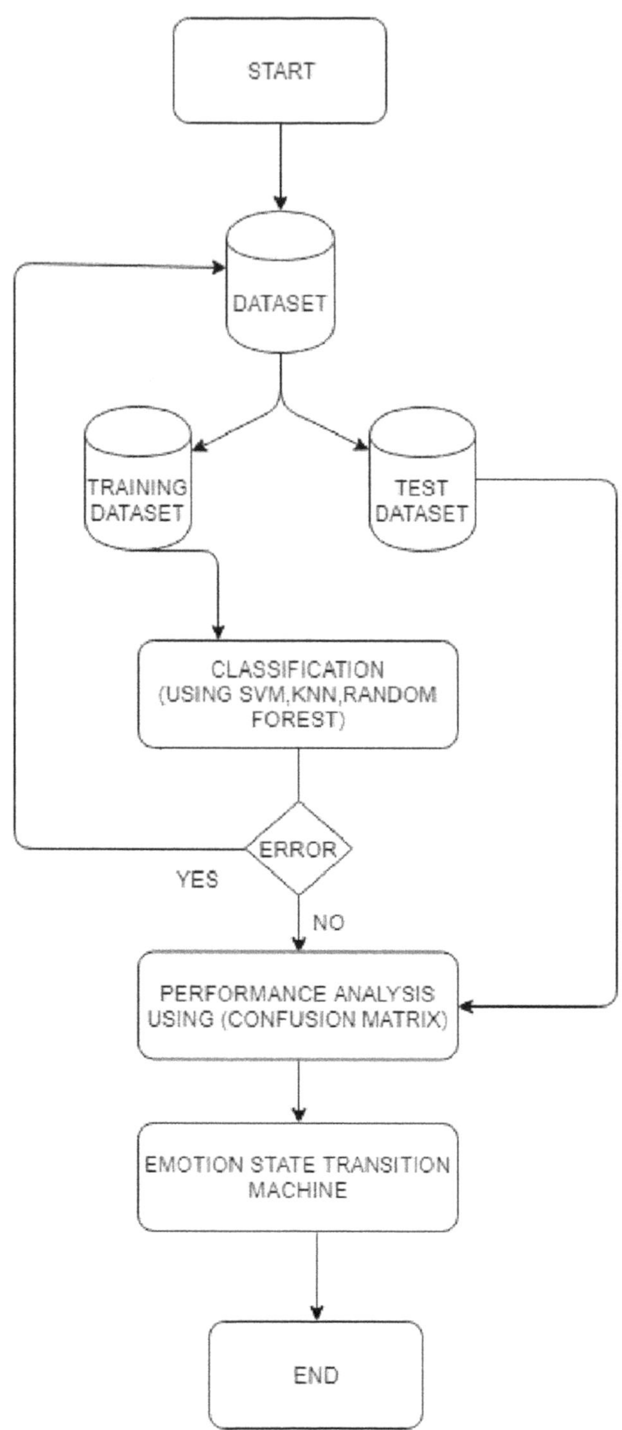

FIGURE 10.3 Flow chart of the overall proposed work.

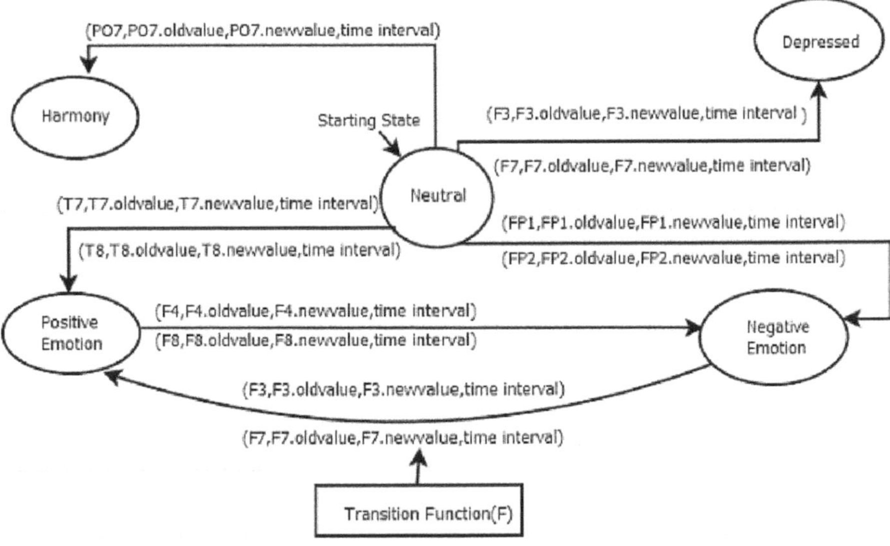

FIGURE 10.4 Model of an emotional state-transition machine.

work in Figure 10.3. We implement the algorithm for transition function of emotional state-transition machine and describe the model of emotional state-transition machine in Figure 10.4.

We have divided the proposed work into two parts:

1. Classifying the different emotional states of a human using SVM, KNN, and random forest to select an efficient model for emotion classification.
2. Implement a new state-transition machine for changing different emotional states of humans at some particular time intervals. To achieve our result, we have used a kaggle emotion dataset [22].

In Figure 10.3, we have described the flow chart of the entire proposed work that states the division of dataset into two parts: training and test data. The training set is used for training the classification models (SVM, KNN, random forest), and then the model is tested with respect to the test dataset. In Figure 10.3, we have shown the flow among the three processes: classification, performance analysis, and emotion state-transition machine. If classification models output any error, we go back to the start stage.

10.3.1 Classification Processes

10.3.1.1 SVM Classifier

SVM multiclass classification with "one against all" approach uses a set of binary classifiers that is trained to separate each class from all others. Then, each data object is classified to a particular class, for which the largest decision value was determined (Hsu and Lin, 2002) [23,24]. This method trains four SVMs (where four is the number of classes) and four decision functions. The basic idea is to reduce the multiclass into a set of binary SVM problems. Briefly, the approach is stated, as mentioned below:

```
{xi,yi}, i=1,2,..N, where N is the number of training samples.
For each c in (c1, c2, c3, c4)
yi∈ {-1, +1}, where N is the number of training samples, yi=+1
for class c and yi=-1 for class (c, rest of ci)
       Find hyperplane between two classes defined by vector w
with a scalar b: w.x + b=0.
       Find the distance between the hyperplane: 2 / |w|
       yi[wT. xi + b] ≥1; i =1,2...N.
Apply a Lagrangian formulation
Use polynomial kernel: K (xi, xj) = (γ xiTxj + r) d, γ >0; γ, r
and d are kernel parameters.
end
```

The training data are represented by $\{x_i, y_i\}$, $i = 1, 2, \ldots$, N and c_1, c_2, c_3, and c_4 are classes representing positive, negative, depressed, and harmony. For implementation purpose, we can use widely used SVM library in python [25] called scikit-learn [26].

10.3.1.2 KNN Classifier

KNN is also known as lazy learning classifier. Decision tree and rule-based classifiers are designed to learn a model that maps the input attributes to the class labels as soon as the training data becomes available, and thus, they are known as eager learning classifiers. Unlike eager learning classifier, KNN does not construct a classification model from data, and it performs classification by matching the test instance with *K* training examples and decides its class based on the similarity to KNNs. Briefly, the approach is stated, as mentioned below:

```
Let (Xi, Ci) where i = 1, 2..., n be data points. Xi denotes
feature values &Ci denotes labels for Xi for each i2.
Assuming the number of classes as 'c'.
ci ∈ {1, 2, 3, ......, c} for all values of i.
Let x be a point for which label is not known, and we would
like to find the label class using KNN algorithms.
For i = 1 to n do
```

$$d(x,x_i) = \sqrt{\sum_{1}^{i}(x - x_i)^2} \quad \text{(finding Euclidean distance)}$$

```
End for
Sort n Euclidean distances in non-decreasing order.
Let k be a positive integer, take the first k distances from
this sorted list.
Find those k-points corresponding to these k-distances.
Let ki denotes the number of points belonging to the ith class
among k points i.e. k ≥ 0.
If ki >kj ∀i ≠ j then put x in class i.
```

10.3.1.3 Random Forest Classifier

Briefly, the approach is stated, as mentioned below:

```
Precondition: A training set S: = (x1, y1), . . ., (xn, yn),
features F, and number of trees in forest B.
```

```
function RandomForest(S, F)
      H ← ∅
            for i∈ 1, . . ., B do
                  S (i) ← A bootstrap sample from S
                  hi ← RandomizedTreeLearn(S (i), F)
                  H ← H ∪ {hi}
            end for
      return H
end function
function RandomizedTreeLearn (S, F)
      At each node:
            f ← very small subset of F
            Split on best feature in f
      return The learned tree
end function
```

The algorithm works as follows: for each tree in the forest, we select a bootstrap sample from *S*, where *S*(*i*) denotes the *i*th bootstrap. We then learn a decision-tree using a modified decision-tree learning algorithm. The algorithm is modified as follows: at each node of the tree, instead of examining all possible feature splits, we randomly select some subset of features $f \subseteq F$, where *F* is the set of features. The node then splits on the best feature in *f* rather than *F*. In practice, *f* is much smaller than *F*. Deciding on which feature to split is oftentimes the most computationally expensive aspect of decision-tree learning. By narrowing the set of features, we drastically speed up the learning of the tree.

In KNN, the predicted output of the input feature vector is the average output value of all the "*K*" neighbors within the feature space. KNN is used to classify the spectrogram image using brainwave balancing application in BCI. It is also used to classify normal brain activity through EEG signal. This algorithm was used to classify chronic mental stress and performs satisfactory accuracy and sensitivity analysis. KNN algorithm is also used to classify the features of left- and right-hand motor imagery from EEG signals.

SVM uses optimal hyperplane to separate feature vectors between two classes. Different types of kernel functions are available for feature classification. The Gaussian kernel function in SVM has been applied in BCIs to classify P300 evoked potentials. SVM has implemented in multiclass channel analyzed EEG signal. In a similar work, a semisupervised SVM has significantly reduced the training effort of P300-based BCI speller. Often in data science, we have hundreds or even millions of features, and we want a way to create a model that only includes the most important features. This has three benefits. First, we make our model simpler to interpret. Second, we can reduce the variance of the model, and therefore overfitting. Finally, we can reduce the computational cost (and time) of training a model. The process of identifying only the most relevant features is called "feature selection." Random forests are often used for feature selection in a data science workflow. The reason is that the tree-based strategies used by random forests naturally ranks by how well they improve the purity of the node. This mean decreases in impurity over all trees. Nodes with the greatest decrease in impurity happen at the start of the trees, while notes with the least decrease in impurity occur at the end of trees. Thus, by pruning trees below a particular node, we can create a subset of most important features.

10.3.2 State-Transition Machine

The machine (M) can be described using six tuples that are mentioned as follows.

$$M = [Q, \Sigma, \Gamma_{old}, \Gamma_{new}, F, T]$$

where

Q = Finite number of emotional states: {Neutral, Harmony, Depressed, PositiveEmotion, NegativeEmotion, etc.}

Σ = Finite number of channels: {FP1, FP2, T7, T8, F3, F4, F7, F8, PO7}.

Γ_{old} = The old value of the channels.

Γ_{new} = The new value of the channels after a fixed ("T") time interval.

F = Transition function which is described as follows.

F: (q_1* Σ *Γ_{old} * Γ_{new} * T) -> $q_{2,}$ where q_1, $q_2 \subset Q$ and Presentstate = q_1 and Nextstate = q_2.

T = Fixed time interval for changing the states.

In the model of emotional state-transition machine, the "Neutral" is the starting state of emotion from where human emotion can be changed to different states based on alpha/beta/gamma/delta wave activation in the specific region of the hemisphere. The increased or decreased signal value of specific channels within the hemisphere are captured for a fixed time interval, and then transfer the current state, channel current value, and new value to the transition function. Then, transition function of the machine calls the proposed transition function algorithm, which will take the decision of state change with respect to the given inputs. A model of an emotional state-transition machine is illustrated in Figure 10.4.

where the emotions {Happy, Joy} $\subset$ PositiveEmotion
Emotions {Fear, Sad, Anxiety} $\subset$ NegativeEmotion
Emotions {Love, Serenity} $\subset$ Harmony

There is no predefined final state of this machine, as the emotion of people can be changed to any state (positive, negative, depressed, and harmony) from the neutral state after a fixed time interval and that state will be the final state of the machine at that particular time instance.

10.3.2.1 Proposed Algorithm of Emotional State Transition Based on Channel Value for a Fixed Time Interval

>Input: Present and new state, time interval, channel name, and their corresponding current and new values (after a fixed time interval).
>Output: The machine will move to the next state for correct inputs or remain in the same state for wrong inputs.

```
Set time interval= "t" minute
if (Presentstate == "Neutral" and Nextstate =="Depressed") then
        if ((Channel == ("F3" OR "F7")) AND
((F3.oldvalue> F3.Newvalue)
                    OR (F7.oldvalue> F7.Newvalue))) then
                    Moves to the next state.
              else
                     Remains in the same state.
        end
        if (Presentstate == "Neutral" and Nextstate ==
"PositiveEmotion") then
              if ((Channel == ("T7" OR "T8")) AND
((T7.oldvalue< T7.Newvalue)

                    OR (T8.oldvalue< T8.Newvalue))) then
                    Moves to the next state.
               else
                     Remains in the same state.
         end
        if (Presentstate == "Neutral" and Nextstate ==
"NegativeEmotion") then
              if ((Channel == ("FP1" OR "FP2")) AND ((FP1.
oldvalue <FP1.Newvalue)
                    OR (FP2.oldvalue < FP2.Newvalue))) then
              Moves to the next state.
               else
                    Remains in the same state.
      end
      if (Presentstate == "PositiveEmotion" and Nextstate ==
"NegativeEmotion") then
       if ((Channel == ("F4" OR "F8")) AND ((F4.oldvalue<
F4.Newvalue)
                    OR (F8.oldvalue< F8.Newvalue))) then
              Moves to the next state.
          else
               Remains in the same state.
       end
       if (Presentstate == "NegativeEmotion" and Nextstate ==
"PositiveEmotion") then
                    if ((Channel == ("F3" OR "F7")) AND ((F3.
oldvalue< F3.Newvalue)
                    OR (F7.oldvalue< F7.Newvalue))) then
                    Moves to the next state.
              else
               Remains in the same state.
                end
          if (Presentstate == "Neutral" and Nextstate ==
"Harmony") then
          if ((Channel == "PO7") AND ((PO7.oldvalue < PO7.
Newvalue) OR
```

```
                (PO7.oldvalue > PO7.Newvalue))) then
                Moves to the next state.
        else
                 Remains in the same state.
    end
```

10.4 RESULT ANALYSIS

10.4.1 Requirement

The proposed transition function algorithm and classification are performed on the following computing platform.

Application Environment: Python (version-3.0), and JFLAP (version-7.0) [23].
Hardware Environment: Operating system-Windows10 (64 bit), Processor-Intel Core(TM)i3, RAM-8GB, clock speed-2.26 GHZ. Graphics memory is 2 GB with AMD Radeon.

Initially, the signal is captured at a fixed time interval and then fast Fourier transform (FFT) is applied on that signal to remove the noise from the raw EEG signal. After filtering and smoothing processes, we classify the samples of the useful channels related to emotion detection. Once the classification process is over, then we proceed for implementation of the emotional state-transition machine.

10.4.2 Result Comparisons of SVM, KNN, and Random Forest Classifiers

In Figure 10.5 we have compared various algorithms like SVM, KNN, and random forest, taking the same parameters for each of them and found out that SVM performs better than other algorithms by a huge margin. From Figure 10.5 and Table 10.1, F-score of KNN was found to be least, which implies that it has the least accuracy based on precision and recall. SVM has the highest F-score and hence is best suited for our classification work. We have also compared different SVM kernels in Figure 10.6 and Table 10.2, and we found out that the polynomial kernel

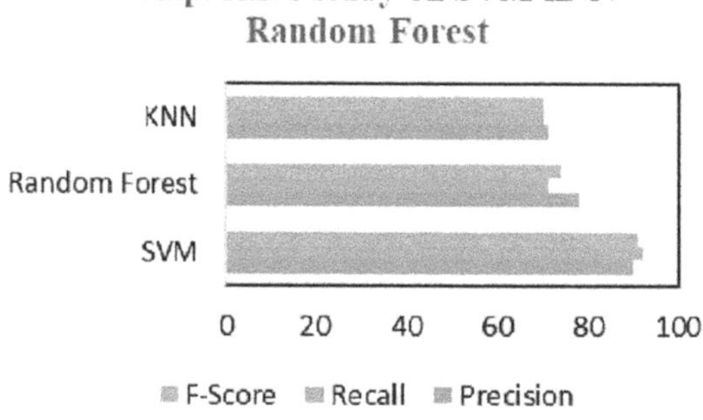

FIGURE 10.5 Algorithm comparison based on confusion matrix.

TABLE 10.1
Average Values of Precision, Recall, and f-Score

Algorithm	Precision (%)	Recall (%)	F-Score (%)
SVM	90	92	91
Random forest	78	71	74
KNN	71	70	70

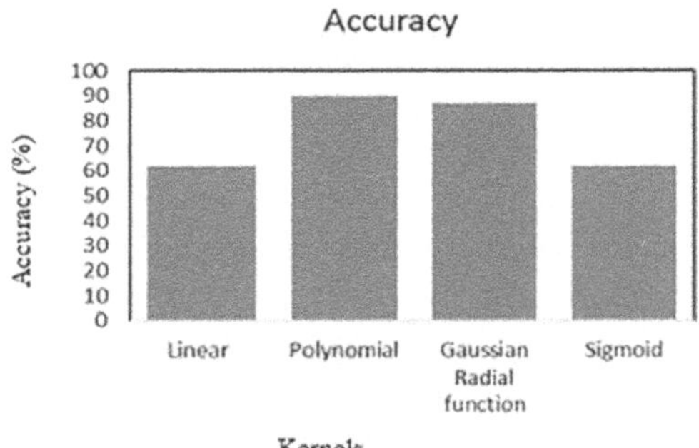

FIGURE 10.6 Kernel comparisons.

TABLE 10.2
Kernel Accuracy (SVM)

Kernels	Accuracy
Linear	62
Polynomial	90
Gaussian radial function	87
Sigmoid	62

(of degree 3$K(x_i, x_j) = (0.05x_i^T x_j + r)^3$*and regularization parameter,* C =100) gave the best accuracy of (90%) for our dataset. Also, the nonlinearity can be concluded from the linear kernel as well, as it gave the least accuracy of 62%. Hence, we select SVM polynomial kernel (degree 3) for our classification work.

10.4.3 SVM Polynomial Kernel Performance Analysis

In this work, we analyze the performance of the classification process using confusion matrix (Figure 10.7), which is also known as error matrix. The confusion matrix is a table that is used to visualize the performance of a classification model on a set of test data for which true values are already known, and hence is typically used for supervised learning (in case of unsupervised, it is called matching matrix). For each class, the confusion matrix gives a summarized form of the number of correct and incorrect predictions. With the help of the confusion matrix, we can simply observe

$$\begin{pmatrix} 25849 & 774 & 244 & 35 \\ 507 & 25530 & 760 & 47 \\ 305 & 873 & 25419 & 315 \\ 79 & 143 & 449 & 26218 \end{pmatrix}$$

FIGURE 10.7 Confusion matrix.

how much our classification model gets "confused" during predictions. It gives us insight not only on the errors being made by a classifier but also, more importantly, on the types of errors that are being made. Each row of the table represents the instances in a predicted class versus the instances in an actual class in the column. The name is taken from the fact that it makes it easy to see whether the system is confusing two classes (i.e., commonly mislabeling one as another).

Here, we have evaluated the classification accuracy of our proposed work using the confusion matrix. *Sensitivity/recall* refers to the true prediction of each class (emotional state) when it is actually true. *Specificity* refers false prediction of each class (emotional state) when it is actually false. *Precision* means how many of the truly classified samples are relevant in each class. *F-measure* computes the accuracy of the multiclass problem, and it is calculated using precision and recall values.

From Table 10.3 and Figure 10.8, we can conclude that emotional state "Harmony" (class 4) has the maximum sensitivity, so the classifier shows the maximum true prediction rate for that class, whereas emotional states "Positive emotion" (class 1), "Angry" (class 3), and "Harmony" (class 4) have specificity in the same range with least specificity for state "Negative emotion" (class 2). "Negative emotion" (class 2) also has the least precision value, indicating the minimum number of correctly classified samples that belong to a given class. Class 4 has the highest accuracy based on maximum F-measure, with class 2 giving us the least. We also observe that class 3 also has the least sensitivity and hence produces the least positive outcome when it is actually positive(True).

10.4.4 Analysis of the State-Transition Machine

After effective classification of different emotional states, we proceed to the second phase of our proposed work, implementing a model of state-transition machine. In

TABLE 10.3
Predefined Classwise Statistical Feature Values

	Performance Analysis of Each Class Using Statistical Parameters			
Parameters (%)	**Positive Emotion**	**Negative Emotion**	**Angry**	**Harmony**
Sensitivity	90	92	86	96
Precision	94	80	92	97
Specificity	96	92	97	98
F-measure	91	86	89	97

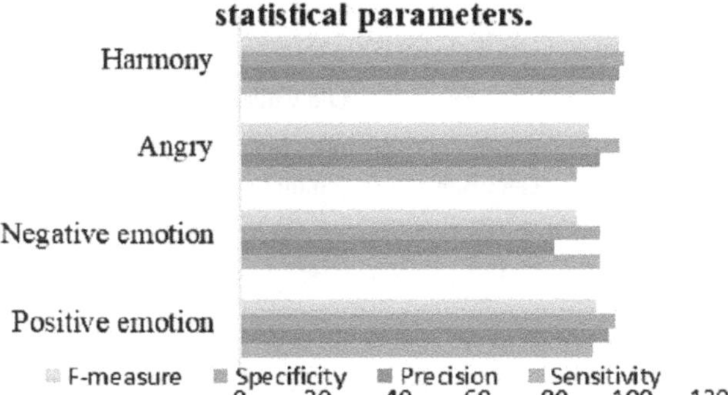

FIGURE 10.8 Classwise statistical feature analysis.

this model, we illustrate how the emotion of a subject has changed from the neutral state to a different emotional state after a fixed time interval based on alpha/beta/gamma/delta activity of EEG signal on different cortices (frontal, frontopolar, temporal, etc.). The increased or decreased activity on the left or right side of the different cortices are measured by the proper channel (frontal-F3/F4, frontopolar-FP1/FP2, temporal-T7/T8) placed over those cortex regions. The emotional state of a subject can be changed using our proposed transition function algorithm based on the increased or decreased value of each channel. Initially, from the neutral or starting state, if the transition algorithm finds suitable parameters, then the machine will move to the next state; otherwise, the machine will remain in the same state. Different emotional states and corresponding channel value after a fixed time interval mapping is mentioned in Table 10.4. The state-transition activity based on the old and new value of each channel is shown (Figures 10.9–10.13). There is no predefined final state of this machine as the emotion of people can be changed to any states (positive, negative, depressed, and harmony) from the neutral state after a fixed time interval and that state will be the final state of the machine at that particular time instance.

The emotional state-transition machine describes the transition from one emotional state to another using a modified value of different channels in a fixed time interval. Here we assume that neutral is the initial or starting emotional state of the machine. Electrodes placed over the frontal section of the brain are responsible for the transition from a neutral state to depressed emotion and negative to positive emotion. Electrodes used in the temporal section are useful for the transition from neutral to positive emotion. The machine moves from one state to another using the proposed transition function algorithm. The transition function algorithm was described in the "proposed work" section. For implementing the state-transition process, we have noticed the initial and final signal values of different channels for each subject, which is described in the state-channel table (Table 10.4). The table contains different emotional states and the initial, new value of all the required channels at a specific instance of time for all combination of transition possible on our state-transition diagram. After some fixed time interval (1-min), we check the value of those required channels with respect to alpha/beta/gamma activity of EEG and

TABLE 10.4
State-Channel Table

Sl. No.	Current State	Electrode	Old Value Time (min): 0.005	New Value Time (min): 1.005	Next State
1	Neutral	T7	–861	–706	Positive
2	Neutral	FP1	–840	–657	Negative
3	Neutral	F3	–628	–851	Depressed
4	Neutral	PO7	193	202	Harmony
5	Negative	F3	147	319	Positive
6	Positive	F4	246	412	Negative
7	Neutral	F7	298	197	Depressed
8	Neutral	PO7	202	193	Harmony
9	Neutral	T8	378	653	Positive
10	Neutral	FP2	103	263	Negative
11	Negative	F7	–177	–113	Positive
12	Positive	F8	600	717	Negative
13	Positive	(F4, F8)	{511, 471}	{356, 128}	Positive
14	Negative	(F3, F7)	{319, 424}	{105, 295}	Negative
15	Neutral	T7	175	102	Neutral

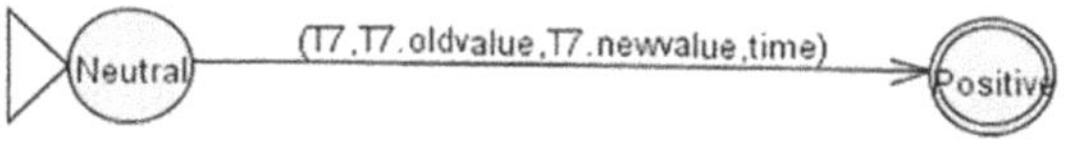

Neutral —(T7,T7.oldvalue,T7.newvalue,time)→ Positive

$\sum$ = T7

Γ_{old} = T7.oldvalue(-861.501)

Γ_{new} = T7.newvalue(-706.549)

T = time(1 min)

Transition Function(F) = F($q_1 * \sum * \Gamma_{old} * \Gamma_{new} * T$) => q_2

Present State(q_1) = "Neutral" and Final State (q_2) = "Positive"

FIGURE 10.9 State transition from neutral (starting) state to positive (final) state with correct inputs.

Neutral —(FP1,FP1.oldvalue,FP1.newvalue,time)→ Negative

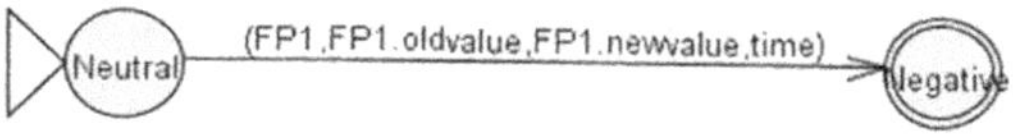

$\sum$ = FP1

Γ_{old} = FP1.oldvalue(-840.365)

Γ_{new} = FP1.newvalue(-657.527)

T = time(1 min)

Transition Function(F) = F($q_1 * \sum * \Gamma_{old} * \Gamma_{new} * T$) => q_2

Present State(q_1) = "Neutral" and Final State (q_2) = "Negative"

FIGURE 10.10 State transition from neutral (starting) state to negative (final) state with correct inputs.

Neutral —(F3,F3.oldvalue,F3.newvalue,time)→ Depressed

$\sum = F3$

$\Gamma_{old} = $ F3.oldvalue(-628.257)

$\Gamma_{new} = $ F3.newvalue(-851.477)

T = time(1 min)

Transition Function(F) = $F(q_1 * \sum * \Gamma_{old} * \Gamma_{new} * T) \Rightarrow q_2$

Present State(q_1) = "Neutral" and Final State (q_2) = "Depressed"

FIGURE 10.11 State transition from neutral (starting) state to depressed (final) state with correct inputs.

Neutral —(PO7,PO7.oldvalue,PO7.newvalue,time)→ Harmony

$\sum = PO7$

$\Gamma_{old} = $ PO7.oldvalue(-861.501)

$\Gamma_{new} = $ PO7.newvalue(-706.549)

T = time(1 min)

Transition Function(F) = $F(q_1 * \sum * \Gamma_{old} * \Gamma_{new} * T) \Rightarrow q_2$

Present State(q_1) = "Neutral" and Final State (q_2) = "Harmony"

FIGURE 10.12 State transition from neutral (starting) state to harmony (final) state with correct inputs.

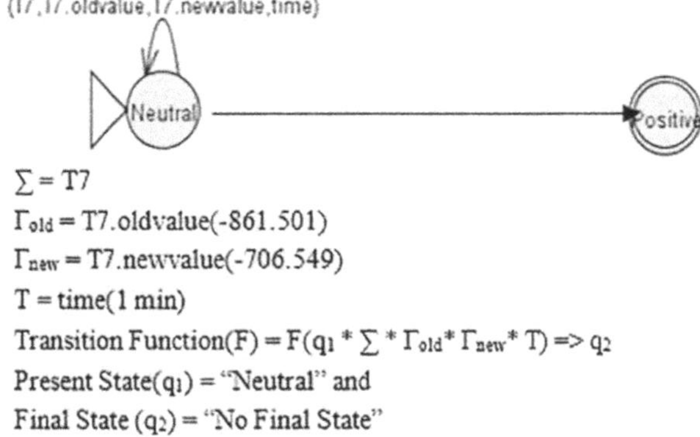

FIGURE 10.13 State transition from neutral (starting) state to the same state with wrong inputs.

execute our transition function algorithm. The next state form of the finite machine is verified using the prediction of the classification algorithm. We describe the state channels table using both initial and final values of the machine in Table 10.4 and illustrate the process of state transition in Figures 10.9–10.13. The state-transition machine (M) is described using six tuples mentioned in the proposed work section.

$$M = [Q, \Sigma, \Gamma_{old}, \Gamma_{new}, F, T]$$

In Figure 10.13, we have shown how the different emotional states are classified based on different electrode values discussed in Table 10.4. From Table 10.4, we can conclude that increased value of channel (PO7) placed over parietooccipital area leads to harmony emotions from the neutral state. Harmony emotions have a high correlation in alpha band power over the parietooccipital area [27]. A depressed subject has less activation in the left frontal area than the normal subject [28]. An increase in the value of the frontopolar area leads to negative emotions (anger, fear, sad, etc.). The EEG activity in the frontal and frontopolar cortex is strongly correlated to adjust emotion regulation [29]. Hence, negative emotions and depressed states are highly correlated. An increase in the value of channels placed over a temporal area leads to positive emotions. In paper [30], the author describes that stable patterns of positive emotions exhibited due to an increased value of lateral temporal area for beta and gamma bands and stable pattern of higher gamma response in frontopolar cortex leads to negative emotions. The state-transition diagram has been drawn using "JFLAP" software [31]. From Figures 10.9, 10.10, and 10.12, we have observed that the subject's emotion has changed from neutral to positive, negative and harmony states due to the increased value of specific channels, but the emotion of the subject has changed to a depressed state due to the decreased value of channel, as shown in Figure 10.11. All the inputs are taken from Table 10.4. Here we assume that, initially, all the subjects are in "Neutral" state, so we consider "Neutral" state as a starting state in our machine. Though we have captured the brain signal through various channels, the machine moves to only one destination state at a fixed time interval after taking the user inputs, so we only consider one specific channel value (T7/FP1/F3/PO7, etc.) as input at a fixed time interval. For wrong inputs (T7. oldvalue and T7.newvalue), machine does not move to the final emotional state (positive), but it only loops within the same starting state (Figure 10.1). In Figure 10.14, we have plotted the state transition based on the changed value of different electrodes, and increased channel value of FP1, T8, and PO7 indicate negative, positive, and harmony emotions, respectively, whereas decreased channel value of F3 indicates a depressed state. Neutral state (starting state in emotional state-transition machine) of the machine shows no change in the channel's value. Figure 10.15 (3D plot) describes

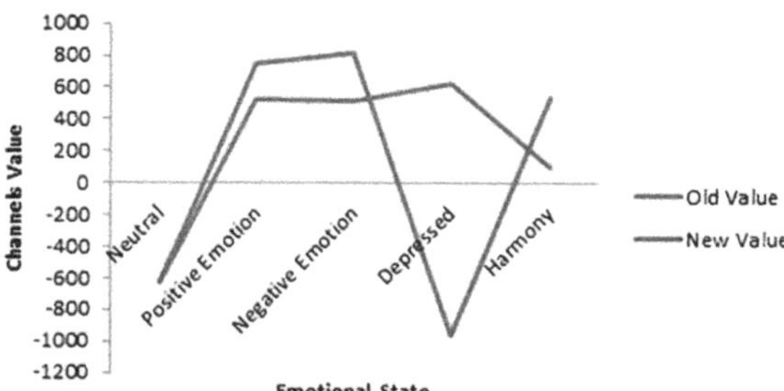

FIGURE 10.14 Emotional state transition.

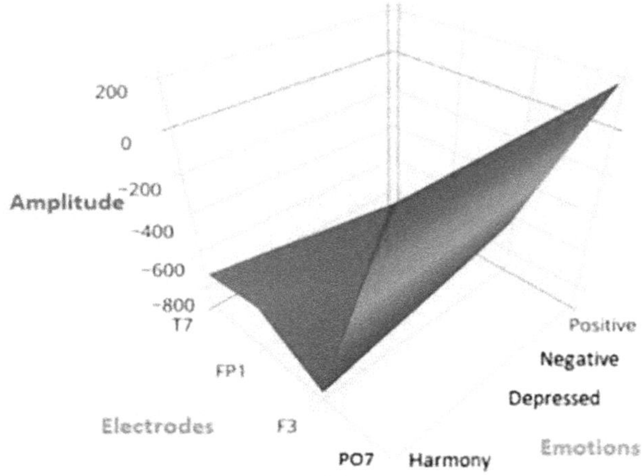

FIGURE 10.15 3D plotting of emotions and corresponding electrodes with the amplitude value of those electrodes.

the different emotional states based on the corresponding electrodes with their amplitude/channel value.

10.4.5 Comparison with Previous Works

In Figure 10.16, we have made a comparison of our proposed work with other models (in Table 10.5) based on the accuracy factor and showed that our work stands apart from all other previous works. The model referred by [20] has an accuracy of 82% for

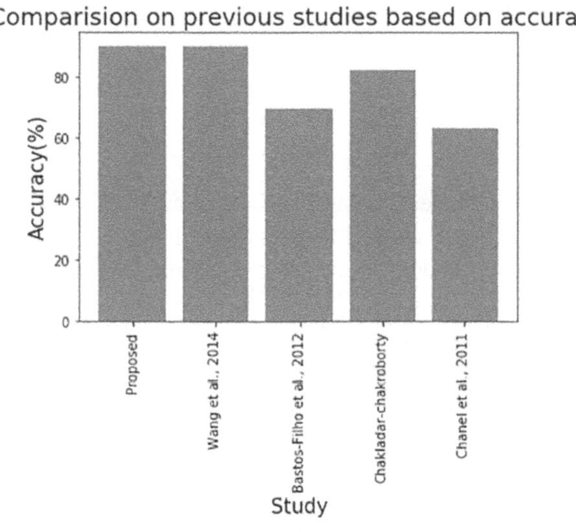

FIGURE 10.16 Accuracy comparison of the proposed approach with some popular approaches.

TABLE 10.5
Comparison of Proposed Model with Other Popular Models

Study	Number of Electrodes	Sampling Frequency (Hz)	Property of Study			
			Classifier	Number of Emotions	Stimuli	Accuracy
Proposed	13	512	SVM	4	Video	90
Wang et al., 2014	128	200	SVM	3	Music, video	90
Bastos-Filho et al., 2012	32	256	kNN	2	Video	69.5
D. Chakladar, S. Chakrobarty	4	512	LDA	4	Video	82
Chanel et al., 2011	19	256	LDA	3	Video, games	63

four emotions than the model referred by [27], but the number of emotions recognized by our model is four, which is greater thanthat of [27]. So, considering all the parameters, we can conclude that our proposed classification work stands apart from the other models of emotion detection.

10.4.6 Computational Complexity

In case of classification, training of dataset using SVM will execute with time complexity, which depends on both training set and support vector, and can be bounded by function n^3 approximately, hence, $O(n^3)$. For state-transition process, the algorithm will run with linear time complexity $O(n)$, for n input size, each input takes a constant time for execution.

10.5 CONCLUSION

In this chapter, we have done a comprehensive survey of various classification methods used in BCI research as well as investigated different emotional states of humans. SVM, KNN, and random forest are the techniques that we have been specifically interested in and worked on. The classification accuracy has been computed using the confusion matrix and different statistical parameters (sensitivity, specificity, prevalence, and precision). After comparing with other well-known models, we can conclude that our proposed model outperforms others. The classification process for a large number of nonlinear data using different SVM kernels and figures shows that the efficient kernel for emotion classification is the polynomial kernel. After achieving 90% accuracy in classification, we use the state-transition model for different emotional states. For performance analysis of the classifier, we have used a confusion matrix, and based on different parameters (sensitivity, specificity, prevalence, precision), we measure the accuracy and prediction of the classification process. Both cases of performance analysis show quite optimistic results. The state-transition process will be useful for emotional state prediction method in the near future. We also try to implement the state-transition machine for paralyzed people in the near future, and this will boost the performance of BCI system.

ACKNOWLEDGMENT

No research funding has been received for this work.

REFERENCES

1. Petrushin, V. (1999, November). Emotion in speech: Recognition and application to call centers., In Engr, St. Louis, MO, pp: 7–10, 1999.
2. Black, M. J., & Yacoob, Y. (1997). Recognizing facial expressions in image sequences using local parameterized models of image motion. *International Journal of Computer Vision*, 25(1), 23–48.
3. Anderson, K., & McOwan, P. W. (2006). A real-time automated system for the recognition of human facial expressions. *IEEE Transactions on Systems, Man, and Cybernetics, Part B (Cybernetics)*, 36(1), 96–105.
4. Wagner, J., Kim, J., & André, E. (2005, July). From physiological signals to emotions: Implementing and comparing selected methods for feature extraction and classification. In *Multimedia and Expo, 2005. ICME 2005. IEEE International Conference on* (pp. 940–943). IEEE., Amsterdam, Netherlands
5. Kim, K. H., Bang, S. W., & Kim, S. R. (2004). Emotion recognition system using short-term monitoring of physiological signals. *Medical and Biological Engineering and Computing*, 42(3), 419–427.
6. Brosschot, J. F., & Thayer, J. F. (2003). Heart rate response is longer after negative emotions than after positive emotions. *International Journal of Psychophysiology*, 50(3), 181–187.
7. Memon, Q., & Mustafa, A. (2015). Exploring mobile health in a private online social network. *International Journal of Electronic Healthcare*, 8(1), 51–75.
8. Murugappan, M., Ramachandran, N., & Sazali, Y. (2010). Classification of human emotion from EEG using discrete wavelet transform. *Journal of Biomedical Science and Engineering*, 3(04), 390.
9. Chakladar, D. D., & Chakraborty, S. (2017, March). Study and analysis of a fast moving cursor control in a multithreaded way in brain computer interface. In *International Conference on Computational Intelligence, Communications, and Business Analytics* (pp. 44–56). Springer, Singapore.
10. Davidson, R. (1979). Frontal versus perietal EEG asymmetry during positive and negative affect. *Psychophysiology*, 16(2), 202–203.
11. Chakladar, D. D., & Chakraborty, S. (2018). Multi-target way of cursor movement in brain computer interface using unsupervised learning. *Biologically Inspired Cognitive Architectures*, Elsevier, 25, 88–100.
12. Picard, R. W. (2000). Toward computers that recognize and respond to user emotion. *IBM Systems Journal*, 39(3.4), 705–719.
13. Davidson, R. J., & Fox, N. A. (1982). Asymmetrical brain activity discriminates between positive and negative affective stimuli in human infants. *Science*, 218(4578), 1235–1237.
14. Baumgartner, T., Esslen, M., & Jancke, L. (2006). From emotion perception to emotion experience: Emotions evoked by pictures and classical music. *International Journal of Psychophysiology*, 60(1), 34–43.

15. Zhang, Q., & Lee, M. (2010). A hierarchical positive and negative emotion understanding system based on integrated analysis of visual and brain signals. *Neurocomputing*, 73(16), 3264–3272.
16. Cornelius, R. R. (1996). *The Science of Emotion: Research and Tradition in the Psychology of Emotions*. Prentice-Hall, Inc.
17. Sander, D., Grandjean, D., & Scherer, K. R. (2005). A systems approach to appraisal mechanisms in emotion. *Neural Networks*, 18(4), pp. 317–352.
18. Yazdani, A., Lee, J. S., & Ebrahimi, T. (2009, October). Implicit emotional tagging of multimedia using EEG signals and brain computer interface. In *Proceedings of the First SIGMM Workshop on Social Media* (pp. 81–88). ACM., Beijing, China
19. Lin, Y. P., Wang, C. H., Jung, T. P., Wu, T. L., Jeng, S. K., Duann, J. R., & Chen, J. H. (2010). EEG-based emotion recognition in music listening. *IEEE Transactions on Biomedical Engineering*, 57(7), 1798–1806.
20. Chakladar, D. D., & Chakraborty, S. (2018). EEG based emotion classification using correlation based subset selection. *Biologically Inspired Cognitive Architectures*, Elsevier, 24, 98–106.
21. Molina, G. G., Tsoneva, T., & Nijholt, A. (2009, September). Emotional brain-computer interfaces. In *2009 3rd International Conference on Affective Computing and Intelligent Interaction and Workshops* (pp. 1–9). IEEE., Amsterdam, Netherlands.
22. Kaggle dataset on eeg based emotion detection. www.kaggle.com/c/inria-bci-challenge.
23. Hsu, C.-W., Chang, C.-C., & Lin, C.-J. (2003). A Practical Guide to Support Vector Classification. Dept. of Computer Sci. National Taiwan Uni, Taipei, 106, Taiwan.
24. Wang, Z., & Xue, X. (2014). Multi-class support vector machine. In *Support Vector Machines Applications,* Editors: Yunqian Ma, Guodong Guo, (pp. 23–48). Springer, Cham.
25. Python Software Foundation. Python Language Reference, version 2.7.
26. Buitinck, L., Louppe, G., Blondel, M., Pedregosa, F., Mueller, A., Grisel, O., & Layton, R. (2013). API design for machine learning software: experiences from the scikit-learn project. arXiv preprint arXiv:1309.0238.
27. Hu, X., Yu, J., Song, M., Yu, C., Wang, F., Sun, P., & Zhang, D. (2017). EEG correlates of ten positive emotions. *Frontiers in Human Neuroscience*, 11, 26.
28. Henriques, J. B., & Davidson, R. J. (1991). Left frontal hypoactivation in depression. *Journal of Abnormal Psychology*, 100(4), 535.
29. Dennis, T. A., & Solomon, B. (2010). Frontal EEG and emotion regulation: Electrocortical activity in response to emotional film clips is associated with reduced mood induction and attention interference effects. *Biological Psychology*, 85(3),456–464.
30. Zheng, W. L., Zhu, J. Y., & Lu, B. L. (2017). Identifying stable patterns over time for emotion recognition from EEG. Preprint. *IEEE Transactions on Affective Computing.*
31. www.jflap.org/.

Part III

Applications and New Trends in Data Science

11 Comparison of Gradient and Textural Features for Writer Retrieval in Handwritten Documents

Mohamed Lamine Bouibed, Hassiba Nemmour, and Youcef Chibani
University of Sciences and Technology
Houari Boumediene

CONTENTS

11.1 INTRODUCTION

In the past years, numerous handwriting analysis systems were developed for real-life purposes, such as identity verification and bank check reading. Such systems are commonly based on the recognition of handwritten signatures, words, or digits. More recently, new applications have emerged in this field. For instance, the requirement of navigation and indexing tools dealing with the huge amount of digitized handwritten documents promoted the use of word spotting and Writer Retrieval. Introduced by Atanasiu et al. in 2011 [1], the Writer Retrieval aims to find all documents belonging to the same writer, despite of being written at different moments, in different languages, and so, having different textual contents. The retrieval is achieved without the need of any prior knowledge about the writer's identity. In fact, the system tries to select all documents sharing the same writing traits with the query. Conventionally, Writer Retrieval systems are composed of feature generation and matching steps (see Figure 11.1).

To generate features, all descriptors employed for handwriting recognition, such as gradient features, textural features, and topological features can be used [1]. The matching step is performed through simple dissimilarity measures. Thereby, to improve retrieval scores, researchers focused on developing robust features.

Presently, we investigate the performance of different kinds of feature generation schemes for solving Writer Retrieval. Precisely, local binary patterns (LBP) and the rotation invariance LBP are used for texture characterization. As gradient feature, we propose the use of histogram of oriented gradients (HOG) and gradient LBPs (GLBP). Finally, pixel density and run length feature (RLF) are used as topological features. To achieve the retrieval stage, various similarity and dissimilarity measures are used. Experiments are conducted on the two versions of ICDAR (International Conference on Document Analysis and Recognition)-2011 writer identification dataset.

The remaining chapter is organized as follows: Section 11.2 reviews the state of the art. Section 11.3 describes the methods adopted for feature generation, while Section 11.4 presents the similarity measures employed in the retrieval step. Experiments and discussion are reported in Sections 11.5 and 11.6, respectively. The last section gives the conclusion of this work.

11.2 LITERATURE REVIEW

Writer Retrieval is a new topic in the handwriting recognition field introduced in 2011 [1]. For this reason, there are few research works dealing with this topic that is commonly overlapped with writer identification. Since the retrieval step is carried out through distance-based matching, the performance is controlled using the feature generation step. Therefore, the state of the art reveals the research efforts to find robust features for characterizing handwritten documents. Roughly, various kinds of features such as statistical, topological, as well as trainable features have

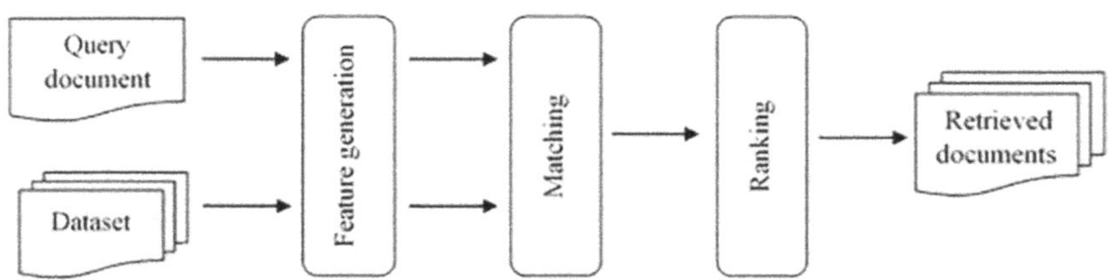

FIGURE 11.1 Writer Retrieval system.

been employed for Writer Retrieval. In [1], Atanasiu et al. employed the local orientation of handwriting contours to generate discriminative handwriting features. The retrieval step was based on Euclidean distance. Experiments were conducted using samples of 301 writers selected from the Institute of Applied Mathematics (IAM) dataset [2]. Each writer is represented by five samples. Results reveal that the principal mode of the orientation pdf provides the best performance. On the other hand, Shirdhonkar et al. [3] proposed the use of contourlet transform associated with Canberra distance. Experiments were carried out on a private dataset of 15 writers represented by 16 handwritten texts. The contourlet was compared with a curvelet transform. Using soft retrieval ranking, both methods derived precisions lower than 60%, which means that precisions will be substantially worse when using a hard retrieval score. This outcome reveals that global transforms do not provide a good characterization. Therefore, Fiel and Sablatnig [4] introduced the use of scale-invariant feature transform (SIFT) as local document features. The Chi-square distance was adopted for the matching step. The retrieval accuracy obtained for the IAM dataset reached 97.2% by considering the percentage of correct ranking among the two TOP similar documents (TOP-2). In [5], retrieval tests were performed on the Computer and Vision Laboratory (CVL) dataset. Several features, such as contour orientation, edge orientation, run length, and local microstructure features, were compared using Euclidian distance. The best retrieval score that is about 96.8% at TOP-2 is obtained by using microstructure features. Thereafter, this score has been improved to 97.1% by using local SIFT features, for which visual words were represented using Gaussian mixture models [6]. However, the results obtained were not replicable for other datasets. For instance, a substantially lower performance was obtained when applying the same system on ICDAR-2011 dataset, which contains images of two-line texts. For this dataset, the retrieval score did not exceed 87.7%. Furthermore, in [7], authors employed a histogram of templates (HOT) for feature generation associated with reinforced retrieval step that uses support vector machine (SVM). The result obtained on CVL dataset is about 70% at TOP-2. After that, the same research team replicated the system using gradient features. Experiments conducted on ICDAR-2011 dataset shows a high accuracy [8]. Finally, in [9], Fiel and Sablatnig introduced the use of convolutional neural networks (CNN) as feature generators, where CVL score was improved to 98.3% at TOP-2.

The common observation on all these research works evince that the retrieval performance is mainly related to the feature's robustness, since the retrieval step is carried out through distance measures. Moreover, experiments reveal that there is no feature that can give an optimal performance whatever the dataset. In this respect, this chapter addresses the comparison of different descriptors and analyses their behavior with different similarity measures. Precisely, new features such as LBPs and gradient features are proposed and evaluated with respect to classical features.

11.3 ADOPTED FEATURES

To develop the Writer Retrieval system, a set of textural, gradient, and topological features are investigated. These features are presented in the following sections.

11.3.1 Local Binary Pattern

The term LBP was introduced by Ojala et al. [10], which helps to determine the gray level distribution in the neighborhood of each pixel. This is carried out by comparing the gray level value of the central pixel with the neighboring gray levels. Precisely, the value of each neighboring pixel is substituted by 1 or 0, as expressed in Equations (11.1) and (11.2).

$$\mathrm{LBP}_{(P,R)}(x,y)=\sum_{p=0}^{P}\left(s\left(g_p-g_c\right)\times 2^p\right) \tag{11.1}$$

With

$$s(l)=\left\{\begin{array}{cc} 1 & l\geq 0 \\ 0 & l<0 \end{array}\right\} \tag{11.2}$$

g_c: Gray level value of the central pixel $P(x, y)$.
g_p: Gray value of the pth neighbor.

The LBP code is obtained by multiplying the thresholded values with weights given by the corresponding pixels and summing up the result, which replaces the central pixel value, as shown in Figure 11.2.

A major drawback of LBP is related to its variability towards small rotation or illumination changes. To get rotation invariance within LBP, the rotation invariant uniform LBP $\mathrm{LBP}^{\mathrm{riu}}$ was introduced in [11,12]. It is calculated according to the following equation:

$$\mathrm{LBP}^{\mathrm{riu}}(x,y)=\left\{\begin{array}{cc} \sum_{p=0}^{P} s\left(g_p-g_c\right), & U(P)>2 \\ P+1, & \text{otherwise} \end{array}\right\} \tag{11.3}$$

$U(P)$ is a uniformity measure calculated as

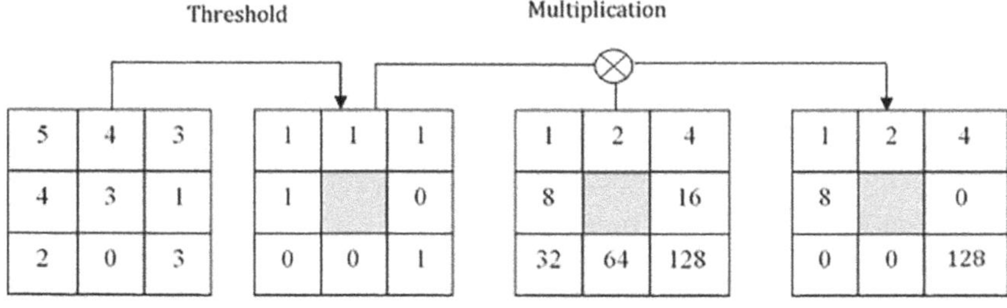

FIGURE 11.2 LBP calculation in a 3 × 3 neighborhood.

$$U(P) = \left|s(g_{p-1} - g_c) - s(g_0 - g_c)\right| + \sum_{p=1}^{P-1} \left|s(g_p - g_c) - s(g_{p-1} - g_c)\right| \tag{11.4}$$

With $g_0 = g_p$

11.3.2 Histogram of Oriented Gradients

This descriptor was introduced by Dalal and Triggs for human detection [13]. It aims to describe local shape orientation through the distribution of gradient intensities. HOG was successfully employed in several handwritten analysis applications, such as signature verification [14], writer soft-biometrics prediction [15], and word spotting [16]. Basically, HOG is calculated by applying a dense grid over images where a local histogram is computed for each region. Obtained histograms are concatenated to form a HOG descriptor [17]. More specifically, the HOG calculation in a given cell is summarized in Algorithm 1.

Algorithm 1. HOG Calculation

1. For each pixel, calculate horizontal and vertical gradients G_x and G_y.

$$G_x(x,y) = I(x+1,y) - I(x-1,y) \tag{11.5}$$

$$G_y(x,y) = I(x,y+1) - I(x,y-1) \tag{11.6}$$

2. Calculate the gradient magnitude and phase by using Equations (11.7) and (11.8), respectively:

$$\|G(x,y)\| = \sqrt{G_x{}^2(x,y) + G_y{}^2(x,y)} \tag{11.7}$$

$$\theta(x,y) = \tan^{-1}\left(\frac{G_y(x,y)}{G_x(x,y)}\right) \tag{11.8}$$

3. Histogram calculation: Accumulate magnitude values for each range of orientations that are taken in the interval [0°, 360°] (see Figure 11.3), according to [13].

11.3.3 Gradient Local Binary Pattern

GLBP was introduced for human detection [17]. Recently, it has been successfully employed for handwritten signature verification and soft biometrics prediction [8,16,18,19]. GLBP consists of computing the gradient information at the edge position. Precisely, the edge corresponds to transitions from 1 to 0 or from 0 to 1 in the LBP code. The GLBP calculation on a document image is summarized as follows:

For each pixel in the document image, GLBP feature is obtained as

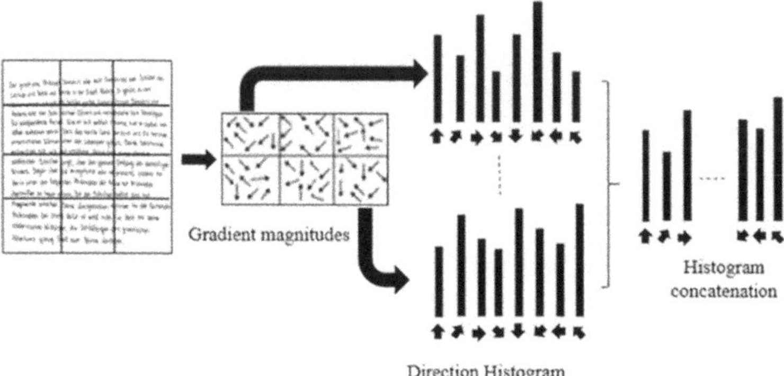

FIGURE 11.3 Summary of the HOG calculation.

1. Compute LBP code.
2. Compute width and angle values from the uniform patterns as follows:
   - The width value corresponds to the number of "1" in LBP code.
   - The angle value corresponds to the Freeman direction of the middle pixel within the "1" area in LBP code [20].
3. Compute the gradient at the 1 to 0 (or 0 to 1) transitions in uniform patterns.
4. Width and angle values define the position within the GLBP matrix, which is filled by gradient information.

The size of the GLBP matrix is defined by all possible angle and width values. Specifically, there are eight possible Freeman directions for angle values, while the number of "1" in uniform patterns can go from 1 to 7 (see Figure 11.4). This yields a 7 × 8 GLBP matrix, in which gradient features are accumulated. Finally, the L2 normalization is applied to scale features in the range [0, 1].

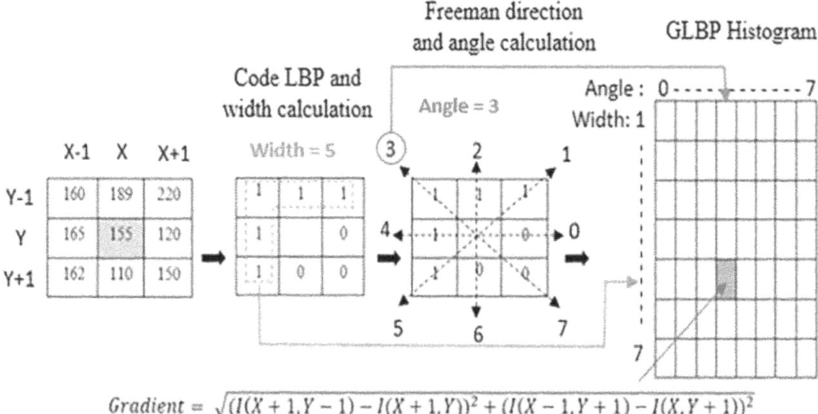

FIGURE 11.4 Summary of the GLBP calculation for a central pixel from [19].

11.3.4 Pixel Density

Density features are calculated by applying a uniform grid over document images (see Figure 11.5). This consists of dividing the document images into a predefined number of cells having the same size as much as possible. For each cell, the pixel density corresponds to the ratio between the number of text pixels and the cell size.

$$\text{Density} = \frac{\text{Number of text pixels}}{\text{Number of pixels in the cell}} \tag{11.9}$$

In this respect, each cell is substituted by the density value. Then, the size of the density feature vector is equal to the number of cells.

11.3.5 Run Length Feature

RLF has been introduced for writer identification to approximate the shape of handwritten text [21]. Basically, RLF seeks text and background pixel sequences along the horizontal, vertical, and diagonal directions. For each direction, a $(N \times M)$ matrix is calculated, where N is the number of pixel values and M is the maximum possible run length in the corresponding image. Presently, all document images undergo a binarization step using Otsu method [22]. Thereby, $P\ (N \times M)$ is a two-line matrix. Figure 11.6 shows an example of an RLF calculation.

To get the final feature vector, matrices are reshaped into normalized vectors that are subsequently concatenated to form the RLF vector.

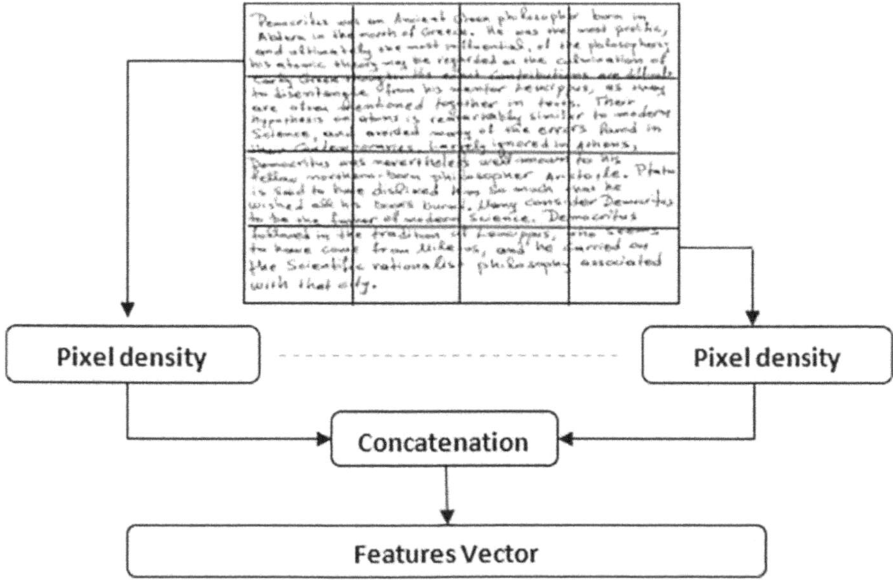

FIGURE 11.5 Pixel density features calculation using a uniform grid.

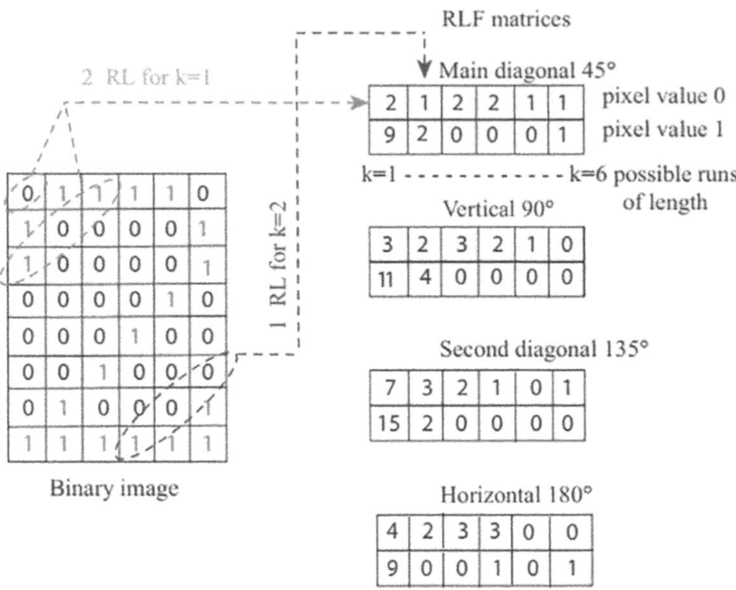

FIGURE 11.6 RLF feature calculation.

11.4 MATCHING STEP

The matching step consists in evaluating the similarity between the feature vector of the query document with feature vectors of all documents of a given database. In this stage, dissimilarity or similarity measures can be used. To calculate retrieval accuracies, similarities obtained for all referential documents are sorted from the largest to the smallest value, and in case of dissimilarities, from the smallest to the largest value. Table 11.1 reports the similarity and dissimilarity measures adopted in this work

The matching process is carried out by calculating the dissimilarity between the query and all documents available in the reference dataset. Then, documents expressing highest similarities with the query are supposed to belong to the query writer.

11.5 EXPERIMENTAL EVALUATION

Experiments were conducted on a public dataset, which was proposed for writer identification contest organized in International Conference on Document Analysis and Recognition (ICDR 2011). The original version of this dataset consists of 208 documents written by 26 writers. Each writer contributed eight documents containing text in four languages (English, French, German, and Greek).[1] A cropped corpus has been extracted by considering the first two lines

[1] www.cvc.uab.es/icdar2011competition.

TABLE 11.1
Adopted Similarity and Dissimilarity Measures

Dissimilarity/Similarity Measure	Equation
Euclidean distance	$D_{\text{Euclidean}}(a,b)=\left(\sum_{i=1}^{n}\lvert a_i-b_i\rvert^2\right)^{\frac{1}{2}}$
Manhattan distance	$D_{\text{Manhattan}}(a,b)=\sum_{i=1}^{n}\lvert a_i-b_i\rvert$
Canberra distance	$D_{\text{Canberra}}(a,b)=\sum_{i=1}^{n}\frac{\lvert a_i-b_i\rvert}{\lvert a_i+b_i\rvert}$
Chi-square distance	$D_{\chi 2}(a,b)=\sum_{i=1}^{n}\frac{(a_i-b_i)^2}{b_i}$
Cosine similarity	$D_{\text{Cosine}}(a,b)=\frac{a\cdot b}{\lVert a\rVert\cdot\lVert b\rVert}$
Jaccard distance	$D_{\text{Jaccard}}(a,b)=\frac{\sum_{i=1}^{n}(a_i*b_i)}{\sum_{i=1}^{n}a_i+\sum_{i=1}^{n}b_i-\sum_{i=1}^{n}(a_i*b_i)}$

a and b are two feature vectors, and n is the size of these vectors.

of each document [23]. Figure 11.7 shows the samples from the original and cropped corpuses.

11.5.1 Evaluation Criteria

Two typical performance measures for information retrieval systems are used. The first is the TOP-N precision, which is the most commonly used criterion [3,5,6]. It corresponds to the percentage of correct documents among the N most similar documents to the query. Since each writer is represented by eight documents, we consider the retrieval from TOP-2 to TOP-7.

The second criterion is the mean average precision (MAP). It allows a global evaluation of the Writer Retrieval system, since it considers the mean of the average precision (AP), which is calculated for all query documents. Considering only the ranks where documents are relevant, the AP is calculated as follows:

$$\text{AP}=\frac{\sum_{k=1}^{M}[P@k]\times\text{rel}(k)}{M} \tag{11.10}$$

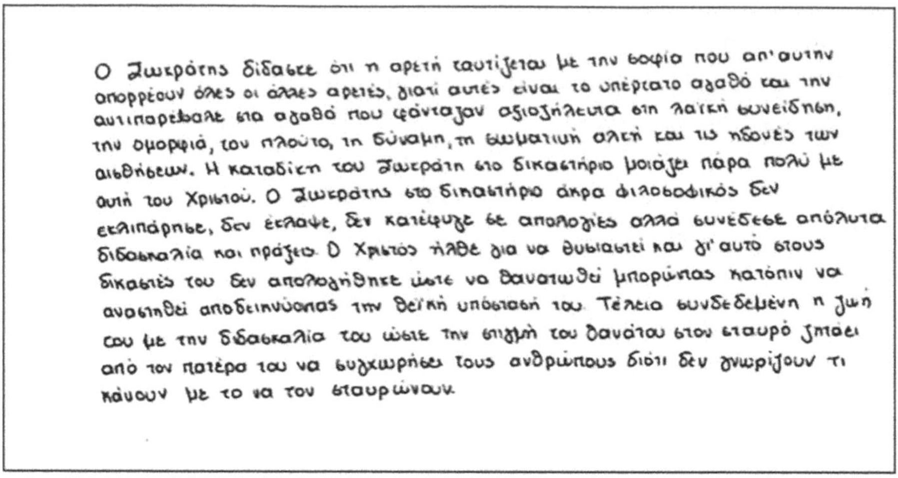

Ο Σωκράτης δίδασκε ότι η αρετή ταυτίζεται με την σοφία που απ' αυτήν
απορρέουν όλες οι άλλες αρετές, διότι αυτές είναι το υπέρτατο αγαθό και την
αντιπαρέβαλε στα αγαθά που φάνταζαν αξιοζήλευτα στη λαϊκή συνείδηση,
την ομορφιά, τον πλούτο, τη δύναμη, τη σωματική αλκή και τις ηδονές των
αισθήσεων. Η καταδίκη του Σωκράτη στο δικαστήριο μοιάζει πάρα πολύ με
αυτή του Χριστού. Ο Σωκράτης στο δικαστήριο άκρα φιλοσοφικός δεν
εκλιπάρησε, δεν έκλαψε, δεν κατέφυγε σε απολογίες αλλά συνέδεσε απόλυτα
διδασκαλία και πράξη. Ο Χριστός ήλθε για να θυσιαστεί και γι' αυτό στους
δικαστές του δεν απολογήθηκε ώστε να θανατωθεί μπορώντας κατόπιν να
αναστηθεί αποδεικνύοντας την θεϊκή υπόστασή του. Τέλεια συνδεδεμένη η ζωή
του με την διδασκαλία του ώστε την στιγμή του θανάτου στον σταυρό ζητάει
από τον πατέρα του να συγχωρήσει τους ανθρώπους διότι δεν γνωρίζουν τι
κάνουν με το να τον σταυρώνουν.

Original

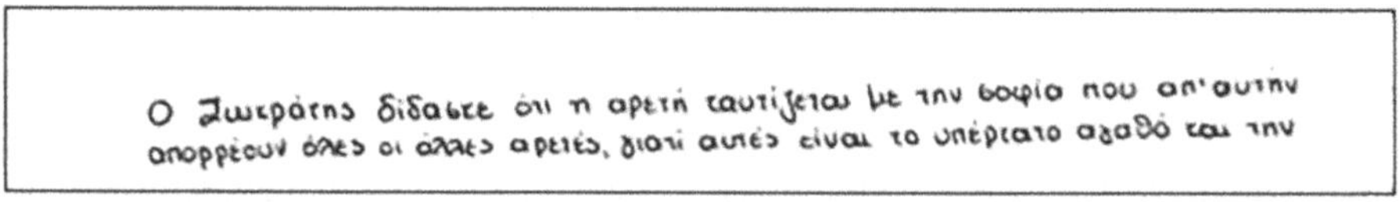

Ο Σωκράτης δίδασκε ότι η αρετή ταυτίζεται με την σοφία που απ' αυτήν
απορρέουν όλες οι άλλες αρετές, διότι αυτές είναι το υπέρτατο αγαθό και την

Cropped

FIGURE 11.7 Handwritten samples from ICDAR-2011 dataset.

With

$$\text{rel}(k) = \begin{cases} 1 \text{ if document at rank } k \text{ is relevant} \\ 0 \text{ if document at rank } k \text{ is not relevant} \end{cases} \tag{11.11}$$

P: Precision at rank *k*.

M: The number of relevant documents in the dataset (seven for ICDAR-2011).

11.5.2 Experimental Setup

Some of the adopted features such as LBP and HOG require an experimental tuning of their setup parameters. Therefore, a set of experiments is performed for all possible configurations to find the best retrieval performance. For this step, we consider the original ICDAR. In the case of LBP, experiments were conducted over the whole images by varying the number of considered neighbors P as well as the radius of the neighborhood R. Results obtained are shown in Figure 11.8, where the $LBP_{16\times2}$ gives the best MAP, which is about 78.34%.

Pixel density and HOG are locally calculated by applying a uniform grid over the document images. Consequently, the suitable number of cells needs an experimental

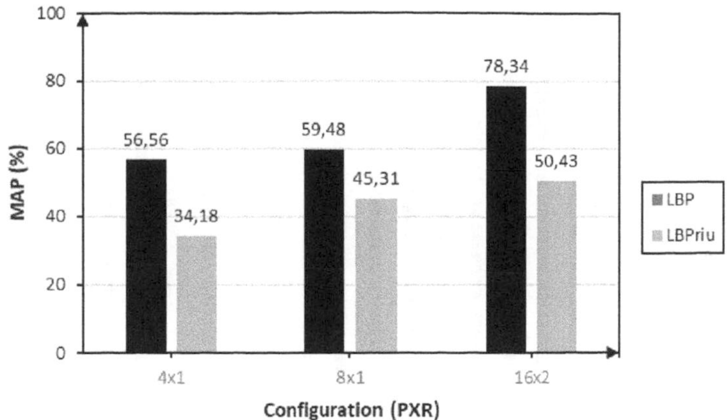

FIGURE 11.8 Variation of MAP (%) according to different LBP configurations.

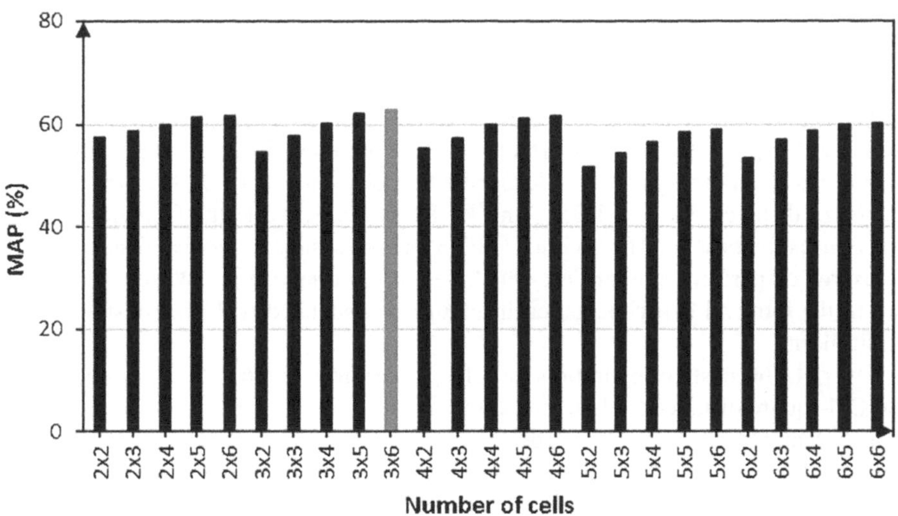

FIGURE 11.9 Variations of MAP according to the number of cells used to calculate HOG. (The best score appears in light gray in pdf version.)

selection. In this respect, we varied this parameter from 4 to 36 cells for HOG and from 4 to 100 for pixel density. Figures 11.9 and 11.10 depict the MAP scores. Roughly, the MAP fluctuates in the range of 50%–70% for HOG features, while it varies from 27% to 43% when using pixel density. This weak performance is due to the fact that we substitute each cell by a single value that is pixel density. On the contrary, HOG provides 81 features for each cell. Thereby, its best MAP, which is about 62.67%, is obtained for 3×6 cells. Pixel density exhibits a lower score with 41.87% for 8×6 cells.

In a second experiment, features were tested with all dissimilarity and similarity measures to select the suitable measure for each descriptor. Table 11.2 presents the

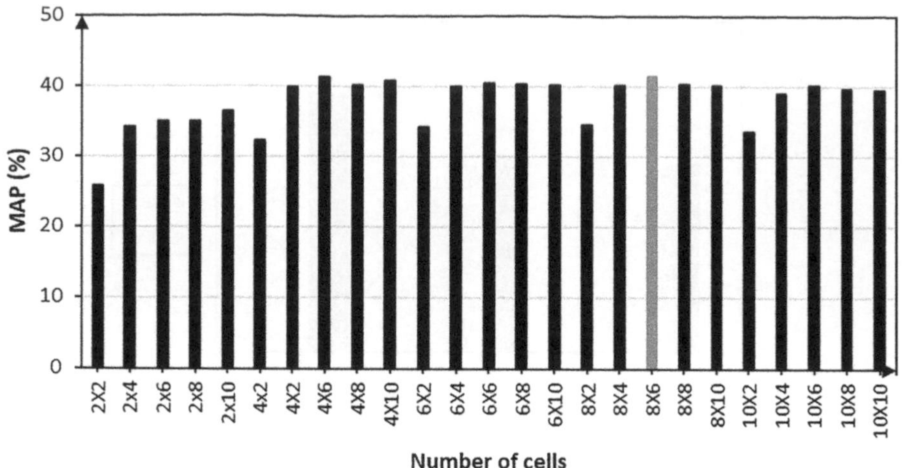

FIGURE 11.10 Variation of MAP according to the number of cells used to calculate pixel density. (The best score appears in light gray.)

MAP scores obtained for original ICDAR-2011. It is clear that there is no similarity measure that gives the best performance for all features. We note that Manhattan distance is the most suitable for HOG, pixel density, and RLF. On the other hand, Canberra distance fits with LBP_{riu} and GLBP features, while the Chi-square distance gives the best performance for LBP. Moreover, one can see that dissimilarity measures outperform similarities, which seem less effective for Writer Retrieval. So, in the retrieval experiments, each feature is associated with the most suitable dissimilarity.

Overall, the best configuration for LBP descriptors is (16×2) associated with the Chi-square distance for LBP and Canberra distance for LBP^{riu}. HOG and pixel density features reach their best performance using uniform grids with (3×6) and (8×6) cells, respectively. On the other hand, Manhattan distance gives the best performance for pixel density, HOG, and RLF. Finally, GLBP is more efficient

TABLE 11.2
MAP (%) Obtained Using different Similarity and Dissimilarity Measures

Descriptor	Manhattan	Euclidean	Canberra	Chi-square	Cosine	Jaccard
LBP_{16x2}	54.74	27.97	63.10	**78.34**	60.96	6.34
$LBP^{riu}_{16\times2}$	27.17	22.51	**50.43**	42.91	33.63	6.34
HOG	**62.67**	53.45	54.33	49.11	53.29	50.99
GLBP	49.25	46.77	**79.01**	58.55	61.55	9.62
Pixel density	**41.87**	41.38	41.63	41.33	31.89	7.92
RLF	**92.22**	85.15	87.84	90.88	84.87	50.48

when it is associated with Canberra distance. Once each feature has been associated to its suitable dissimilarity measure, we performed the Writer Retrieval experiments.

11.5.3 Retrieval Results

Writer Retrieval test is carried out by considering each document from the dataset as a query. Specifically, a vector containing dissimilarities between the query and all existing documents in the dataset is calculated and ranked to select the most similar documents. Tables 11.3 and 11.4 jointly report the retrieval results obtained at different ranks and MAP scores using all descriptors with their suitable dissimilarities. For both datasets, RLF-based system outperforms all other systems. In fact, the precision reaches 99.03% at TOP-2 for the original ICDAR-2011 and 96.15% for the cropped version. The MAP scores are about 92.22% and 83.16%, respectively. On the other hand, GLBP and $LBP_{16\times2}$ show competitive performance. However, the accuracy drops significantly for farther ranks. For instance, it falls from 96.39% at TOP-2 to 80.15% at TOP-7 when using $LBP_{16\times2}$, and from 97.59% at TOP-2 to 79.87% at TOP-7 for GLBP features.

TABLE 11.3
Retrieval Results (%) for Original ICDAR-2011

	TOP-2	TOP-3	TOP-4	TOP-5	TOP-6	TOP-7	MAP
$LBP_{16\times2}$ + Chi-square	96.39	93.31	90.62	87.78	84.37	80.15	78,34
$LBP^{riu}_{16\times2}$ + Canberra	88.70	78.36	70.31	65.48	61.21	56.79	50,43
HOG + Manhattan	88.70	81.50	77.16	72.30	69.15	65.65	62.67
GLBP + Canberra	97.59	95.51	92.30	88.36	83.73	79.87	79,01
Pixel density + Manhattan	75.00	65.54	59.02	54.51	50.32	46.91	41,87
RLF + Manhattan	**99.03**	**98.55**	**97.43**	**96.27**	**94.61**	**92.14**	**92,22**

TABLE 11.4
Retrieval Results (%) for Cropped ICDAR-2011

	TOP-2	TOP-3	TOP-4	TOP-5	TOP-6	TOP-7	MAP
$LBP_{16\times2}$ + Chi-square	76.68	65.54	58.53	53.55	50.00	47.18	40.93
$LBP^{riu}_{16\times2}$ + Canberra	64.90	51.12	43.99	37.98	33.01	30.35	21.71
HOG + Manhattan	73.07	60.89	52.76	47.78	43.58	40.93	32.83
GLBP + Canberra	89.18	82.37	78.24	73.84	70.59	66.27	62.89
Pixel density + Manhattan	69.47	57.37	49.87	45.19	41.26	37.98	28.50
RLF + Manhattan	**96.15**	**93.50**	**92.30**	**90.14**	**86.92**	**82.45**	**83.16**

11.6 DISCUSSION AND COMPARISON

This chapter focuses on evaluating different descriptors for Writer Retrieval. Several textural (LBP), gradient (HOG and GLBP), and topological (RLF and pixel density) features were proposed. Besides, a set of dissimilarity and similarity measures were tested to find the suitable measure for each feature. Referring to the literature, there is only one research work that uses ICDAR-2011 dataset. Tables 11.5 and 11.6 compare the results obtained herein with those obtained in [3]. It is important to note that in [3] authors used trainable features by associating SIFT features with Gaussian mixture models. Nevertheless, the (RLF-Manhattan) based system gives comparable result while being much simpler.

From all experiments, we can note

TABLE 11.5
Comparison with the State of the Art for Original ICDAR-2011

References	Feature	Matching	TOP-2	TOP-7
[6]	SIFT and GMM	Cosine similarity	99.3%	91.2%
[8]	HOG	Cosine Similarity	86.5%	53.3%
	GLBP	Cosine Similarity	88.5%	65.9%
Proposed methods	$LBP_{16\times2}$	Chi-square distance	96.4%	80.2%
	$LBP^{riu}_{16\times2}$	Canberra distance	88.7%	56.9%
	HOG	Manhattan distance	88.7%	65.7%
	GLBP	Canberra distance	97.6%	79.9%
	Pixel density	Manhattan distance	75.0%	46.9%
	RLF	**Manhattan distance**	**99.0%**	**92.1%**

TABLE 11.6
Comparison with State of the Art for Cropped ICDAR-2011

References	Feature	Matching	TOP-2	TOP-7
[6]	SIFT and GMM	Cosine similarity	87.0%	69.2%
[8]	HOG	Cosine similarity	69.2%	41.8%
	GLBP	Cosine similarity	80.8%	48.9%
Proposed methods	LBP_{16x2}	Cosine similarity	76.7%	47.2%
	$LBP^{riu}_{16\times2}$	Chi-square distance	64.9%	30.4%
	HOG	Canberra distance	73.1%	40.9%
	GLBP	Manhattan distance	89.2%	66.3%
	Pixel density	Canberra distance	69.5%	37.9%
	RLF	Manhattan distance	96.2%	82.5%

- Topological features are the most adequate for Writer Retrieval in handwritten documents. RLF descriptor shows the best accuracy for the two datasets. Actually, even with two lines of text, the system can achieve a high precision with small loss at farther ranks. In fact, when testing cropped ICDAR-2011, the MAP reaches 83.16%, and the loss of accuracy from TOP-2 to TOP-7 is about 13.7%, which is small compared with all other systems.
- The pixel density descriptor provides insufficient information, since it replaces each cell by a single density measure. Such characterization cannot reflect the writing style of each individual, the reason for which retrieval scores were substantially weak. Compared with RLF, there is a loss of accuracy of about 24% at TOP-2 and 45% at TOP-7.
- Furthermore, classical LBP outperforms LBP^{riu} with a gain of 7.69% and 12% at TOP-2, with original and cropped sets, respectively. However, this descriptor seams to need more than two lines of text, since its performance drops significantly with the cropped dataset. Actually, the MAP loses 37.41%.
- Gradient descriptors show competitive performance, especially GLBP, for which the precision reaches 97.6% for the original set and 89.2% for the cropped set. HOG feature requires a local calculation using uniform grid to enhance retrieval accuracies. However, it scores still lower than those collected using GLBP, RLF, and $LBP_{16\times2}$.

11.7 CONCLUSION

Currently, libraries contain a huge number of digitized documents, including historical manuscripts and other archives. Since the manual handling of these data is not trivial, researchers tried to develop automatic navigation tools such as spotting and retrieval techniques. In this respect, Writer Retrieval systems are developed to find all documents belonging to the same person. Specifically, such systems should recognize the handwriting of the same individual, whatever the language and textual content. In return, they should differentiate handwritings of different individuals. This can be done by using features that reduce the intrawriter variability while increasing the interwriter variability. To find suitable features for the Writer Retrieval task, this work evaluates various kinds of features. Specifically, we used LBPs and its uniform versions LBP^{riu}, HOGs, and GLBP as gradient descriptors. Besides, two topological features such as RLF and pixel density are used. To make a straightforward evaluation, several similarity and dissimilarity measures, including Manhattan, Euclidean, Canberra, Chi-square, Jaccard distance, and Cosine similarity are employed in experiments. Experiments conducted on the two ICDAR-2011 datasets showed that RLF descriptor outperforms all other descriptors as well as the state of the art. These findings reveal that pixel run length provides a pertinent characterization of the writing style. On the other hand, textural and gradient features, such as $LBP_{16\times2}$, GLBP, and HOG provide a satisfactory retrieval, especially when associated with a suitable dissimilarity measure.

REFERENCES

1. V. Atanasiu, L. Likforman-sulem, N. Vincent, Writer retrieval exploration of a novel biometric scenario using perceptual features derived from script orientation, in: *11th International Conference on Document Analysis and Recognition*, 18–21 September, Beijing, 2011, pp. 628–632.
2. C. Djeddi, I. Siddiqi, L. Souici-Meslati, A. Ennaji, Text-independent writer recognition using multi-script handwritten texts, *Pattern Recognition Letters* 34 (2013) 1196–1202.
3. M.S. Shirdhonkar, M.B. Kokare, Writer based handwritten document image retrieval using contourlet transform, Nagamalai D., Renault E., Dhanuskodi M. (eds.) Advances in Digital Image Processing and Information Technology. Communications in Computer and Information Science 205 (2011) 108–117.
4. S. Fiel, R. Sablatnig, Writer retrieval and writer identification using local features, in: *10th IAPR International Workshop on Document Analysis Systems*, 27–29 March, Queensland, 2012, pp. 145–149.
5. F. Kleber, S. Fiel, M. Diem, R. Sablatnig, CVL-database: An off-line database for writer retrieval, writer identification and word spotting, in: *International Conference on Document Analysis and Recognition*, 28–21 September, Beijing, 2013, pp. 560–564.
6. S. Fiel, R. Sablatnig, Writer identification and writer retrieval using the fisher vector on visual vocabularies, in: *International Conference on Document Analysis and Recognition*, 25–28 August, Washington, 2013, pp. 545–549.
7. M.L. Bouibed, H. Nemmour, Y. Chibani, Writer retrieval using histogram of templates features and SVM, in: *3rd International Conference on Electrical Engineering and Control Applications*, 25–26 November, Constantine, 2017, pp. 537–544.
8. M.L. Bouibed, H. Nemmour, Y. Chibani, Evaluation of gradient descriptors and dissimilarity learning for writer retrieval, in: *8th International Conference on Information Science and Technology*, 2–6 June, Cordoba, 2018, pp. 252–256.
9. S. Fiel, R. Sablatnig, Writer identification and retrieval using a convolutional neural network, in: *International Conference on Computer Analysis of Images and Patterns*, 2–4 September, Valetta, 2015, pp. 26–37.
10. T. Ojala, M. Pietikäinen, D. Harwood, A comparative study of texture measures with classification based on featured distributions, *Pattern Recognition* 29 (1996) 51–59.
11. J.F. Vargas, M.A. Ferrer, C.M. Travieso, J.B. Alonso, Off-line signature verification based on grey level information using texture features, *Pattern Recognition* 44 (2) (2011) 375–385.
12. M. Pietikäinen, A. Hadid, G. Zhao, T. Ahonen. *Computer Vision Using Local Binary Patterns*. Springer-Verlag, London, 2011.
13. N. Dalal, B. Triggs, Finding people in images and videos, PhD thesis, French National Institute for Research in Computer Science and Control (INRIA), July 2006.
14. M.B. Yilmaz, B. Yanikoglu, C. Tirkaz, A. Kholmatov, Offline signature verification using classifier combination of HOG and LBP features, In: *International Joint Conference on Biometrics*, 11–13 October, Washington DC, 2011, pp. 1–7.
15. N. Bouadjenek, H. Nemmour, Y. Chibani, Histogram of Oriented Gradients for writer's gender, handedness and age prediction, in: *International Symposium on Innovations in Intelligent Systems and Applications*, 2–4 September, Madrid, 2015, pp. 1–5.
16. M.L. Bouibed, H. Nemmour, Y. Chibani, New gradient descriptors for keyword spotting in handwritten documents, in: *3rd International Conference on Advanced Technologies for Signal and Image Processing*, 22–24 May, Fez, 2017, pp. 1–5.
17. N. Jiang, J. Xu, W. Yu, S. Goto, Gradient local binary patterns for human detection, in: *International Symposium on Circuits and Systems*, 19–23 May, Beijing, 2013, pp. 978–981.

18. N. Bouadjenek, H. Nemmour, Y. Chibani, Age, gender and handedness prediction from handwriting using gradient features, in: *13th International Conference on Document Analysis and Recognition*, 23–26 August, Nancy, 2015, pp. 1116–1120.
19. Y. Serdouk, H. Nemmour, Y. Chibani. New off-line handwritten signature verification method based on artificial immune recognition system. *Expert Systems with Applications*, 51 (2016) 186–194.
20. H. Freeman. On the encoding of arbitrary geometric configurations, *IRE Transactions on Electronic Computers* EC-10 (1961) 260–268.
21. C. Djeddi, I. Siddiqi, L. Souici-Meslati, A. Ennaji, Text-independent writer recognition using multi-script handwritten texts, *Pattern Recognition Letters* 34 (2013) 1196–1202.
22. J. Liu, W. Li, Y. Tian. Automatic thresholding of gray-level pictures using two-dimension Otsu method. In *International Conference on Circuits and Systems*, 15–16 June, Shenzhen, 1991, pp. 325–327.
23. G. Louloudis, N. Stamatopoulos, B. Gatos, ICDAR 2011 writer identification contest, in: *11th International Conference on Document Analysis and Recognition*, 18–21 September, Beijing, 2011, pp. 1475–1479.

12 A Supervised Guest Satisfaction Classification with Review Text and Ratings

Himanshu Sharma, Aakash, and Anu G. Aggarwal
University of Delhi

CONTENTS

12.1 INTRODUCTION

The digital revolution has resulted in the transformation of many businesses to online mode. Digitalization has modified the operational strategy of online firms due to advantages such as cost-effectiveness, 24 × 7 availability, no geographical limitations, and low entry and exit barriers, to name a few [1–2]. The upward trend in online marketing is especially noticed in the service sector. Few popular categories under the service sector are retails, banks, hotels, airlines, healthcare, and education, to name a few. Automation in conducting business practices has forced market practitioners to realign their strategies to online domain to streamline their daily functioning [3]. Electronic service (e-service) is defined as the process of providing services through a digital medium. These services may be commercial, such as booking movie tickets, hotel rooms, and airline tickets, or they may come under the noncommercial category, such as the service provided by government [4–5]. The worldwide e-service revenue was $165.3 billion in 2017 and is expected to reach $224 billion by 2022 [6]. Most popular activities under the service industry come under the hospitality sector. It has been asserted that most of the customers approach the requirements through online travel agencies (OTAs), which are third-party sites that allow customers to book hotel services through a single platform [7]. Customers prefer these web sites, as they provide various incentives, discounts, gift vouchers, and economical package, which may not be available on the direct hotelier's web site. Flourishing awareness of these e-services may be attributed to the active customer–retailer interaction, while providing the service and after-service delivery, providing solution to the problems of customers, supporting secure and efficient transaction process, and thereby improving the satisfaction of consumers.

Customer satisfaction in corporate terms is defined as "a business philosophy that highlights the importance of creating value for customers, anticipating and managing their expectations, and demonstrating the ability and responsibility to satisfy their needs" [8]. Instilling purchasers and consumers with a feeling of contentment is a very difficult task for marketers in service industry. In the hospitality sector, an evaluating measure of hotel's performance is obtained by judging the satisfaction level of the guests [9]. In order to gain competitive advantage, the hoteliers must emphasize on better handling of factors that ultimately result in the satisfaction of a guest. Researchers argue that satisfied perceptions lead to behavioral loyalty of online customers, i.e., an intention to repurchase and suggest it to others [10,11]. Previous studies in the tourism sector adopted various methodologies and developed various conceptual frameworks to study the satisfaction level of customers (guests) [12]. However, most of them provided empirical studies using measurable and nonmeasurable data.

Technological advances have enabled the users to share feedbacks related to a hotel and its several aspects or subattributes, such as cleanliness, value, location, and many more, and also their recommendations. These feedbacks are termed as electronic word-of-mouth (EWOM) or user-generated content (UGC). Potential customers look out for reviews posted by previous users, especially from those who are close to them, and also their opinions regarding the service [13]. These EWOM highly influences the purchasing decision of customers of whether to adopt the services of the firm or not. According to a report, 95% people read online reviews before making a purchase, and displaying of online feedbacks by a retailer increases the conversion rate by 720% [14]. Also, these reviews help the retailers in providing the services demanded by customers and understanding their preferences, which in turn helps them in enhancing their sale and financial performance [15]. Therefore, EWOM can be described as a form of casual information exchange with other consumers through online technological platforms talking about usage and characteristics of a product or service or their seller [16,17]. Since the OTA web sites allow users in booking hotel and other services, they encounter an ever-increasing amount of online textual data in the form of feedbacks and online reviews.

The hoteliers are recognizing the impact of guest satisfaction on their present and future business prospects, and are thus being highlighted in recent studies. Guest satisfaction evaluates how much of the customers' expectation parameters are fulfilled by the service provider [18]. Since the content available on hotel web sites plays a part in creating awareness for potential travelers, e-service is efficiently utilizing EWOM as a promotion tool by encouraging users to provide their feedbacks on its platform [19]. OTA web sites enable its users to rate the hotel and share their experience in textual form. These overall ratings and textual reviews empirically represent guest experience, which is observed by people from travel community as well as hoteliers to make informed decisions, in the customer's welfare [20]. Satisfaction of the guests affect the sales, revisit intention, positive EWOM probability, and market reputation of the firm [21]. Since the nature of hospitality industry matches with that of experienced goods, researchers suggest that EWOM holds particular importance for experienced goods as their quality level is obscured before consumption [22]. Online reviews have a notable effect in hotel sector when compared with other tourism segments, since they are referred by most of the users for making stay decisions. Therefore, guest experience holds the ability to affect all the aspects of hotel business.

There exist some latent dimensional variables that are a representation of large number of attributes, but the consumers might not explicitly mention. There occurs a need to introduce techniques to evaluate these collections (termed as documents) by sorting, probing, tagging, and searching by using computers. With the help of machine learning (ML), extant researchers have successfully proposed a model that finds a pattern of words in these documents under hierarchical probabilistic models. This is referred to as topic modeling. The rationale behind topic modeling is to identify the word-use pattern and how the documents portraying similar pattern should be connected. Under text analytics, the model makes use of bag-of-words concept and ignores the word ordering [23]. The topic modeling generally depends on the four methods, namely latent semantic analysis (LSA), probabilistic LSA (PLSA), latent Dirichlet allocation (LDA), and correlated topic model (CTM). The LSA, earlier known as latent semantic indexing (LSI), creates vector-based representation of texts to make semantic content, by making use of a predefined dictionary [24]. PLSA automates document indexing based on a statistical model for factor analysis of count data, without referring to a predefined dictionary [25]. LDA is a Bayesian-based unsupervised technique for topic discovery in abundant documents, without considering any parental distribution [24]. CTM helps in discovering the topics in a group of documents, underlined by a logistic normal distribution [26].

Another term that has gained popularity with the widespread use of internet technology is ML. ML is an algorithm that enables the systems to learn and predict automatically from previous experiments, without using any programming [27]. It is divided into two categories, namely supervised and unsupervised. Under supervised learning, the future events can be predicted with the help of labeled past data, whereas unsupervised learning does not require a labeled dataset [28]. However, if both types of data are available, then semisupervised learning is a good option. A key characteristic of ML is classification algorithms. Few popular classifiers adopted in this study are naïve Bayes (NB), decision tree (DT), random forest (RF), support vector machine (SVM), and artificial neural networks (ANN). NB is a set of algorithms using Bayes' theorem at grass-root level. It considers all the features to be independent of each other. DT organizes the testing conditions in a tree structure, where root and internal nodes represent different test conditions and their characteristics. RF is an ensemble model that considers many DTs at one time. The result from this model is usually better than the result from one of the individual models. SVM makes use of hyperplanes in a multidimensional space that divides cases of diverse class labels. It consists of regression and classification by making use of continuous and categorical variables. ANN comprises of layers of neurons, which convert an input vector into some output. Each unit takes an input, applies a nonlinear function to it, and then forwards the output to the succeeding layer.

This chapter aims to evaluate the satisfaction level of guests from a hotel review dataset obtained from Tripadvisor. We make use of overall star ratings as a proxy variable for guest satisfaction level of labeled data. First, topic modeling is applied with the help of LSA, which results in topics that represent the whole useful review information. Then, for measuring the satisfaction level, we make use of classifiers such as NB, DT, RF, SVM, and ANN, to check their accuracy using performance measures. Thus, we aim to solve the quest for the following research questions:

a. What are the vital determinants of hotel guest satisfaction as articulated by online reviews?
b. What are the most important features influencing guest satisfaction in EWOM based on ML techniques with the help of a large dataset?
c. How do these features vary across hotel guest satisfaction (or overall ratings)?

The remaining chapters are systematically confined as follows: A detailed overview of previous works will be discussed in Section 12.2. Section 12.3 presents the adopted methodology and the data description. The results of the applied techniques are provided in Section 12.4. Section 12.5 presents the discussions and conclusion based on the findings. Implications for researchers and practitioners are given in Section 12.6. Finally, limitations and future scope relevant to the present study are provided.

12.2 RELATED LITERATURE

12.2.1 Guest Satisfaction and Online Reviews

EWOM is defined as "any positive or negative statement made by potential, actual, or former customer about a product or company, made available to multitude of people and institutions via internet" [29]. Due to the innate uncertainty associated with an untried product/service, consumers often depend upon word-of-mouth for making purchase decisions. However, in this new digital age where communication of individual's opinions knows no bounds, the UGCs have dethroned the hoteliers from the role of travel opinion leaders [13]. Purchase involvement and service experience have also been identified as vital antecedents of review, providing motivation. A travel trip whether for recreational or work purpose is always followed by an experience felt or encountered by the traveler. These guest experiences are the major source of information for the service provider [30]. It can provide a snapshot of hotel performance, which can be evaluated and analyzed for various managerial purposes. Understanding of customer experience involves various intricate factors majorly, because of its personal nature. Due to the competitive nature of hospitality industry, it is vital for the hoteliers to understand guest experience and improve their satisfaction level.

A study consisting of 60,000 reviews made by travelers on a distributional web site was conducted to determine the factors impacting the satisfaction of guests [19]. They found the importance of qualitative and quantitative aspects of reviews for satisfaction. Using an integrative analysis, findings suggested that few factors that have high influence over customers are cleanliness, bathroom, and beds. Reviews laid higher influence on hotel's convenience to attractions, shopping, airports, and downtown. Also, food and beverage items play a part. Pantelidis [31] studied the impact of food characteristics over guest satisfaction. He considered 2,471 customer comments from 300 restaurants in London. Content analysis results showed the importance of food, and in particular, starters influence the experience. However, along with food, other determinants of guest experience are service, ambience, price, menu, and décor. He found that the results were consistent even under economic crisis, and that these experiences determine the longevity of business and customer's intention of loyalty.

An ecotourism satisfaction study was performed to determine the ecolodge stay experience [32]. Content analysis was used to analyze an ecotourism data obtained from Tripadvisor site. A two-step statistical procedure was implemented to classify the experience level into four categories, such as satisficrs, dissatisfies, critical, and neutral. The study considered the frequency of online reviews, expression of favorable attitude, and the overall satisfaction levels reported by the reviewers along with textual comments. Li et al. [33] took 42,668 online reviews from 778 hotels to study guest satisfaction. Content analysis was used to obtain the study results. The findings showed that transportation convenience, food and beverage management, convenience to tourist destinations, and value for money impact the guest experience of tourists from both luxury and budget hotels. Guests paid more consideration to bed, reception services, room size, and decoration.

Another study considering 1,345 customer feedbacks from 97 hotels in the Hangzhou area was performed to evaluate guest satisfaction [9]. Twenty-three attributes considered as the determinants of customer satisfaction were divided into four categories, such as satisfiers, dissatisfies, bidirectional, and neutrals. These attributes incorporated features such as room facilities, general hotel facilities, food quality, dining environment, price, location, and staff service. One-way ANOVA (Analysis of Variance) results showed positive influence of softness of beds, availability of western food, availability of 24-h reception, sound proofing of the room, and on-site parking. Xiang et al. [10] used big data analytics considering 60,648 reviews covering 10,537 hotels extracted from Expedia.com to study the guest experience towards hotel. Regression results show the importance of UGC for determining the guest behavior in hotels. Research findings emphasized on semantic differentiations in comparison to motivation and hygiene variables. Also, a strong association between experience and satisfaction was obtained.

An empirical study consisting of 176 respondents was conducted for measuring guest satisfaction [22]. The conceptual framework consisted of independent variables, such as empathy, paraphrasing, and speeds of response. ANOVA results showed support for the influence of the former two constructs but did not support the speeds of response. They also noticed the importance of negative feedbacks on the user decision. Xiang et al. [34] clustered an unstructured data considering 60,648 reviews covering 10,537 hotels extracted from Expedia.com, into two groups. One group consists of guest experience dimensions and the other group consists of satisfaction ratings. The correspondence analysis findings showed various types of salient traits of hotels that satisfied their customers, where guests had issues regarding cleanliness and maintenance-related factors.

Another guest satisfaction process was analyzed on 14,175 textual reviews extracted from Tripadvisor using text analytics approach [18]. The performance of five hotels in Singapore was checked to determine the satisfaction using various aspects, such as location, sleep quality, rooms, service quality, the value for money, and cleanliness. Also, sentiment analysis checked the review emotions. On the basis of UGC, results showed the importance of a good room and a hotel with a pool and good service, whereas ratings emphasized on rooms, the value for money, and location. Berezina et al. [13] used 2,510 reviews extracted from Tripadvisor to judge the key determinants of satisfaction and dissatisfaction. The text mining results showed that the variables such as place of business, room, furnishing, members, and sports were present in both positive and negative reviews. Moreover, research findings present that satisfied customers talk about intangible aspects of their stay, more than the dissatisfied ones, whereas unsatisfied customers mentioned tangible aspects of their stay.

A hybrid approach combining quantitative and qualitative web content analysis with Penalty-Reward Contrast Analysis (PRCA) was proposed to judge the satisfaction of guests with mobility challenges [11]. The PRCA approach classified 11 attributes that affect satisfaction into three factors, namely basic, performance, and excitement factors. The study findings reported that basic factors, including entrance accessibility, moving convenience, and information credibility should be met to avoid dissatisfaction. The performance factors involving shower accessibility, room settings, staff attitude and capability, access to the room, and public area accessibility had a neutral influence on satisfaction. The excitement factors including room quality, general lodging feature, and luggage and equipment support play a part in the longevity of business. Liu et al. [20] studied the impact of language differentiation of customers on the determinants of their satisfaction towards a hotel. They obtained 412,784 user-generated reviews on Tripadvisor for 10,149 hotels from five Chinese cities, which were analyzed using ANOVA technique along with perpetual map. The findings showed that tourists speaking assorted lingo, such as English, German, French, Italian, Portuguese, Spanish, Japanese, and Russian, vary significantly on their roles of various hotel attributes, namely rooms, location, cleanliness, service, and value in forming their overall satisfaction for hotels.

Another study was conducted to obtain the key determinants of guest satisfaction in case of a business trip [21]. A dataset comprising 1.6 million reviews covering 13,410 hotels was extracted for study purpose. Multilevel analysis showed that the

TABLE 12.1
ML and Hotel Online Reviews Literature

Author(s)	Purpose of the Study	Technique(s)
[27]	Sentiment of reviews	SVM
[35]	Sentiment of reviews	NB
[28]	Polarity of documents	TF-IDF
[36]	Sentiment classification of reviews	SVM
[37]	Hotel service quality	NB
[38]	Sentient classification	SVM
[39]	Opinion spam detection	Sparse additive generative model (SAGE)
[40]	Fake review classification	Logistic regression, RF, DTs, SVM, NB
[41]	Opinion mining	SVM and fuzzy domain ontology (FDO)
[42]	Comparative analysis of online review platforms	Topic modeling
[43]	Aspect-based sentiment analysis	Recurrent neural network (RNN) and SVM
Present study	Hotel guest satisfaction	SVM, RF, DTs, ANN, NB

overall satisfaction of guests is less in case of a business trip in comparison to a pleasure trip. The results were moderated by certain contextual factors, such as the traveler's leisure versus work orientation, economic and cultural characteristics of the destination, and the traveler's nationality. Zhao et al. [12] performed an econometric modeling to judge guest satisfaction using technical characteristics of textual reviews along with the customer's involvement in review committee on 127,629 reviews extracted from Tripadvisor. The modeling results showed that higher level of subjectivity, readability, and a longer textual review lead to low customer satisfaction, whereas a higher level of diversity and sentiment polarity of textual review leads to high customer satisfaction. Also, review involvement positively influenced overall satisfaction.

The work reported in this chapter aims to evaluate the guest satisfaction from a hotel review dataset obtained from Tripadvisor. In order to measure the satisfaction level, we make use of classifiers, such as NB, DT, RF, SVM, and ANN, to check their accuracy using performance measures. Some of the previous works using ML in the hotel industry are provided in Table 12.1.

12.3 METHODOLOGY

12.3.1 Data Description and Analysis

A text analytics research was conducted with the aim to analyze the hotel guest satisfaction represented through EWOM and its influence on overall hotel ratings available at TripAdvisor.com. We used the data of Tripadvisor.com, because it is world's biggest social media platform specific to e-services and also has more than 460 million EWOMs related to hotels, restaurants, and other e-services. We used TripAdvisor.com EWOMs dataset due to [44]. A total of 5,69,861 reviews were downloaded in 2015.

TABLE 12.2
Dataset Information

Categories	No. of Reviews
Satisfied guests	2,19,984
Unsatisfied guests	19,875
Total number of hotel reviews (N)	2,39,859
Total number of reviews	5,69,861

This dataset is also freely available on https://twin-persona.org. The consumer review enclosed both the textual experience as well as the overall rating provided by the guests to the hotel. The description of dataset is presented in Table 12.2.

This dataset contained reviews from various service categories, such as hotel, restaurants, and attractions, including 5,69,861 reviews for the year 2015. In this chapter, we have considered all the 2,39,859 reviews related to hotel services. These were further categorized as αsatisfied and unsatisfied hotel guests, based on a threshold value which is described in detail further. This gave us an aggregate of 2,19,984 satisfied and 19,875 unsatisfied hotel guests. Each EWOM contains the name of associated hotels, review text, review title, and overall ratings. The distribution of overall ratings versus the number of reviews related to our dataset is presented in Figure 12.1. The overall ratings vary from one to five and are denoted on x-axis, whereas the number of reviews is depicted on the y-axis. This shows the relative frequency of review volume at different overall ratings.

The main goal of this study was to analyze the EWOM text and how they impact on the guest satisfaction of hotels (i.e., overall ratings). The flow diagram of our analysis is given in Figure 12.2.

12.3.2 Data Cleaning

It included the processes consistent with previous studies [45,46], including punctuations, stop words, tokenization, stemming, and non-English words. We implemented

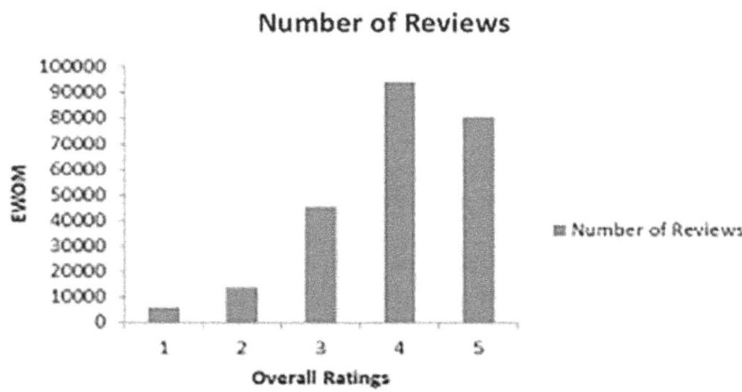

FIGURE 12.1 Distribution of overall ratings.

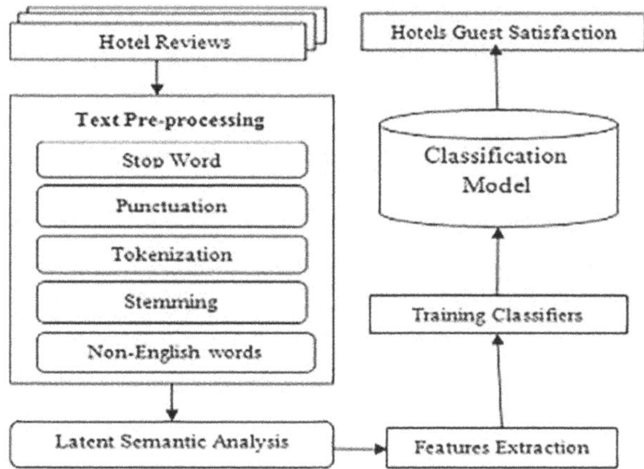

FIGURE 12.2 Research framework.

text preprocessing through the Natural Language Toolkit (NLTK) module using python programming.

12.3.3 Latent Semantic Analysis

On the basis of accessed review, the next task is to extract the keywords that influence guest satisfaction. Extracting key dimensions is an important practice, especially in the case of big data, to reduce the dimensionality of data by removing unimportant variables. Since the traditional dimensionality reduction methods were incapable of truly extracting implicit dimensions, the use of topic modeling technique has been popularized by extant researchers. This chapter adopts the probabilistic LSA to obtain the topics for guest satisfaction. The advantage of probabilistic LSA is that it avoids polysemy, i.e., it allows minimal overlapping occurrence of a word that has a situational meaning in the factored representation. Being an unsupervised ML method, LSA is efficient in handling big data as well as data spread across various time periods. It is a statistical model dependent on posterior probability, the working for which is defined as follows.

Let the number of words be N and the number of reviews be M. We assume that with each occurrence of a word $w \in W = \{w_1, w_2, \ldots, w_M\}$ in a document $d \in D = \{d_1, d_2, \ldots, d_N\}$, there exists an unobserved class variable $z \in Z = \{z_1, z_2, \ldots, z_M\}$.

The probability of obtaining an unobserved pair (d, w) can be represented as a joint probability distribution provided later:

$$P(d,w) = P(d)P(w \mid d) \tag{12.1}$$

where

$$P(w \mid d) = \sum_{z \in Z} P(w \mid z)P(z \mid d) \tag{12.2}$$

Also, the probability of selecting a document d is $P(d)$, probability of picking a latent class z is $P(z \mid d)$, and the probability of generating a word w is $P(w \mid d)$.

The earlier model assumes two conditions at grassroot level.

a. The observation pairs (d,w) are assumed to be independent; this corresponds to the "bag-of-words" approach.
b. Conditioned on the latent class z, words w are generated independent of the document d. Asserting the number of states to be smaller than the number of documents ($K \leq N$), z acts as a threshold variable in forecasting w conditioned on d.

Based on the maximum likelihood function defined in Equation (12.3), we get the probabilities required in Equations (12.1) and (12.2).

$$L = \sum_{d \in D} \sum_{w \in W} n(d,w) \log P(d,w) \tag{12.3}$$

where $n(d,w)$ denotes the term frequency, i.e., the number of times w occurred in d. Note that an equivalent symmetric version of the model can be obtained by inverting the conditional probability $P(z \mid d)$ by using Bayes' rule

$$P(d \mid w) = \sum_{z \in Z} P(z)P(w \mid z)P(d \mid z) \tag{12.4}$$

This chapter uses LSA to obtain and tag guest satisfaction attributes for all 2,39,859 hotel reviews accessed in our analysis. LSA does not make any assumption regarding the structure of text or the grammatical properties of the language. LSA extracted 20 topics and every topic contained seven words. These topics represent the important aspects related to travelers' satisfaction, depending on the frequency of occurrence associated with consumers' own hotel experiences.

12.3.4 Classifiers and Performance Measures

We used some popular ML techniques, namely DT, ANN, NB, SVM, and RF, to classify guests as satisfied or unsatisfied on the basis of key features of guest satisfaction expressed in review text. These classifiers help in training procedures to build the guest satisfaction predictive model. This is the first chapter that utilizes ML for evaluating a prognostic model for guest satisfaction of hotels, handled by Python programming. There are at most 20 extracted features, and the feature matrix will be $N \times 20$, where N denotes the volume of hotel reviews. The performance of five classifiers evaluated on the basis of tenfold cross-validation, f-measure, recall, and precision metrics. The mathematical definition of the performance evaluations metrics used in this chapter is given in Table 12.3.

The hotel guest satisfaction measured through overall ratings range from 0 to 1. A binary classification model was developed for predicting guest satisfaction and

TABLE 12.3
Classifiers' Performance Measures

Measure	Formula	Description
Precision	$\frac{TP}{TP+FP}$	Out of data points that were found to be relevant, actually were relevant
Recall	$\frac{TP}{TP+FN}$	Ability to find out all the relevant occurrences in the data
$F1$-measure	$\frac{\text{Precision} \cdot \text{recall}}{\text{Precision} + \text{recall}}$	A measure to seek a balance between precision and recall
Accuracy	$\frac{TP+TN}{TP+TN+FP+FN}$	To measure effectiveness of a classifier in prediction

obtaining satisfied guests from the collected review dataset. We divided the guest categories on the basis of overall rating into satisfied and unsatisfied guests on the basis of a bottleneck value. For our analysis, it is set at 0.60 (i.e., $3/5=0.60$). If customers gave three or greater overall ratings, then they come under the category of satisfied guest and, otherwise, vice versa. This shows that a customer who gave 60% or more ratings to the hotel is labeled as satisfied guest and, otherwise, unsatisfied guest.

12.4 EXPERIMENTAL RESULTS

Here, we quote the findings of dimension extraction for the hotel guest satisfaction. We then analyze the influence of these extracted features on the hotel guest satisfaction. The experiments comprised features extraction, their relative importance, and guest satisfaction prediction analysis.

12.4.1 Features Related to Guest Satisfaction

We apply an LSA to obtain and tag guest satisfaction attributes for all hotel's EWOM accessed in our analysis. The LSA extracted 20 topics, and every topic contained seven words based on their relative weights.

The features were named with the help of two industry experts dealing with hospitality sector. The feature name relied on the rational relationship between recurrent occurring words for a topic. Table 12.4 shows a topic "Natural Beauty," representing the words "beach (20.5%)," "ocean (3.4%)," and "pool (3.4%)"; all of which tops the list. We also tested the topic name through logical connections to other words in that particular topic. For example, other words such as "beautiful" and "view" also relate to the topic "Natural Beauty." If other words in the topic did not relate to the provided topic name, then the naming process was started afresh.

Figure 12.3 shows 20 most important features (topics) extracted from 2,39,859 hotel reviews. Out of 20, two topics were shown the overall guest perception towards the hotel: Style and dećor and satisficing. Style and dećors shows that a customer feels

TABLE 12.4
Examples of Identified Topic Labels

Topic	Relative Weight (%)	Topic	Relative Weight (%)
Topic 1: Natural Beauty		**Topic 2: Recommendation**	
Beach	20.5	Recommend	14.8
Ocean	3.4	Best	12.6
Pool	3.4	Hotel	3.6
Resort	3.3	Staff	1.8
Beautiful	1.8	Service	1.6
Kids	1.6	Modern	1.3
View	1.3	Housekeeping	1.3

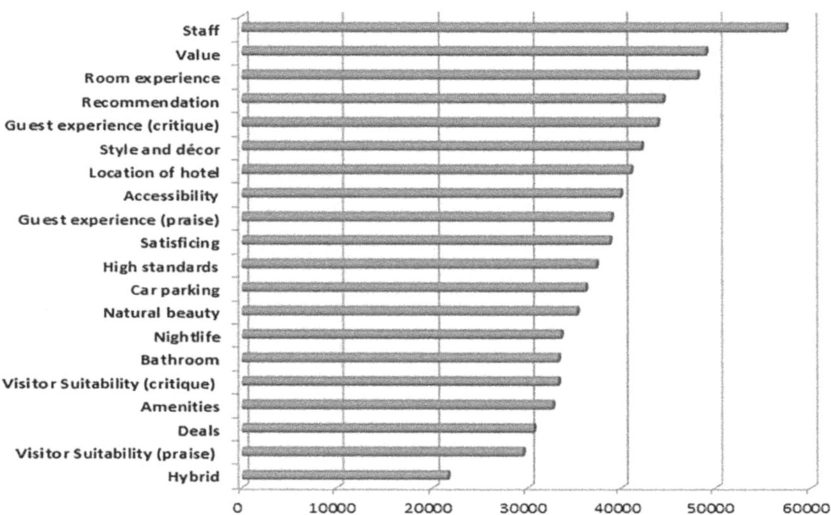

FIGURE 12.3 Extracted features.

good about the hotel's actual appearance. It shows that the selected hotel delivers a clean and beautiful place as they promised. Satisficing shows that customers found satisfactory services by the hotel, which met their expectations towards that hotel. The features that showed the level of satisfaction and dissatisfaction of hotel guests are guest experience (critique), guest experience (praise), and recommendation. One of the topics was labeled as "Hybrid," as it incorporated more than one group of words that represents very different levels of hotel guest experiences. The other topics show 14 specific aspects related to service quality (e.g., staff, room experience, and bathroom). Therefore, all these 20 features can influence the hotel guest satisfaction. Although the relative percentage is not fixed, it can be different with respect to other hotels, investors, and owners. However, hotel managers and practitioners should focus on these features related to hotel guest satisfaction. It can enhance the level of guest satisfaction towards that hotel.

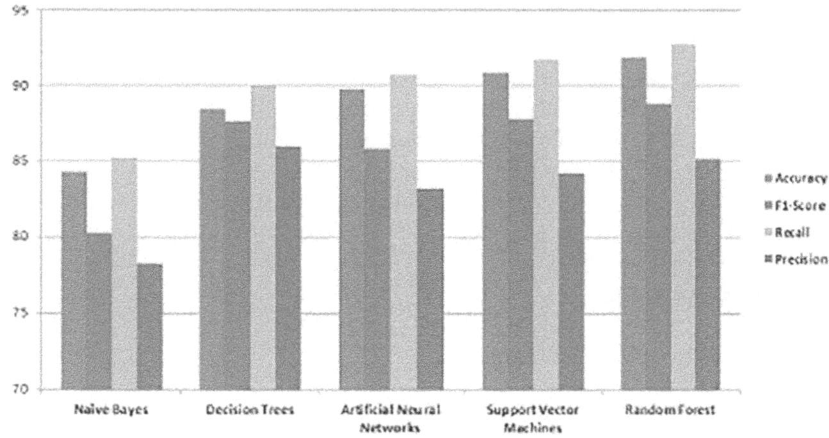

FIGURE 12.4 Prediction performance using guest satisfaction dimensions.

12.4.2 Hotel Guest Satisfaction Prediction

Initially, we constructed the predictive models, and then their accuracies are compared using tenfold cross-validation technique. Five ML classifiers learned 20 extracted topics (nightlife, value, amenities, natural beauty, guest experience (critique), recommendation, staff, guest experience (praise), location of the hotel, accessibility, car parking, visitor suitability (critique), satisficing, style and décor, bathroom, deals, visitor suitability (praise), hybrid, room experience, and high standards). The results of the experiments using 20 extracted features are provided in Figure 12.4. The RF classifier performed best with 91.94% accuracy, 88.8% f-measure, 92.75% recall, and 85.18% precision. These results clearly provide good overall performance and validate the measured guest satisfaction features in terms of accuracy, recall, precision, and f-measure for the guest satisfaction level in hotels.

12.5 DISCUSSIONS AND CONCLUSION

The series of experiments and findings gives strong empirical support to our quest. Our findings are robust and consistent across one real-life dataset, 20 extracted features, and three evaluation metrics (f-measure, recall, and precision). First, this chapter proposes a novel technique to extract features related to guest satisfaction or dissatisfaction from a big hotel review dataset of Tripadvisor.com. For feature extraction, the LSA of hotel reviews revealed sensible topics along with their relative weight determined through conversational power.

Second, this study suggested a new guest satisfaction prediction model that utilized 20 types of features, namely nightlife, value, amenities, natural beauty, guest experience (critique), recommendation, staff, guest experience (praise), the location of hotel, accessibility, car parking, visitor suitability (critique), satisficing, style and décor, bathroom, deals, visitor suitability (praise), hybrid, room experience, and high standards related to hotel guest satisfaction. The most exciting add on is the influence of these features on the hotel guest satisfaction prediction. Our proposed hotel guest prediction model is very important to identify the features that aid marketers to judge the guest satisfaction level. The suggested predictive model found predictive f-measure above 80%with respect to each classifier using top-20 hotel guest satisfaction features (Figure 12.4). It also obtained a predictive f-measure above 85% for all other classifiers except NB. The RF classifier is noted to be a good predictive classifier that outperformed the others as shown in Figure 12.4. The relative importance of each extracted topic is also examined.

12.6 IMPLICATIONS

12.6.1 Theoretical Implications

Much of the previous studies in the hospitality sector focusing on customer (guest) satisfaction were conducted empirically or through econometric modeling. Compared with the earlier studies that made use of primary datasets obtained from survey questionnaires and focus groups (in case of empirical studies) or using technical aspects of UGC, such as length, sentiments, valence, and many more, this study uses topic model to gain insights into what guests talk about in their reviews, and based on the obtained topics, we judge the satisfaction level with the help of ML classifiers.

One of the research findings are extracting the topics that represent the collection of buzzwords that reviewers talk about. The topic modeling technique adopted in this chapter is LSA. The 20 most important features (topics) extracted from 2,39,859 hotel reviews were nightlife, value, amenities, natural beauty, guest experience (critique), recommendation, staff, guest experience (praise), the location of hotel, accessibility, car parking, visitor suitability (critique), satisficing, style and décor, bathroom, deals, visitor suitability (praise), hybrid, room experience, and high standards. These are consistent with previous works in tourism literature.

Since the online reviews are available in abundance and represent information in open structural form, it becomes complex to analyze them. This study tackles these shortcomings by using five supervised ML techniques, namely RF, DTs, NB, SVM, and ANNs to classify the reviews into satisfiers and dissatisfiers. The RF classifier delivers the best performance when compared with other four classifiers, with performance values having 91.94% accuracy, 88.8% f-measure, 92.75% recall, and 85.18% precision. This is consistent with recent studies conducted using ML.

12.6.2 Managerial Implications

The study comes out with few significant implications for managers. It suggests hoteliers and investors the importance of latent dimensions of guest satisfaction from UGCs available on their sites. The study makes use of text mining methodologies that help the practitioners in understanding the linguistic aspects of online reviews and how these features impact the overall satisfaction of the guests. Therefore, apart from concentrating on the perceptions made by guests through textual reviews and improving their services accordingly, the marketers must also take note of the obtained latent features helpful in achieving the maximum satisfaction level of customers. Recent statistical reports project the importance of EWOM and how these comments influence the pre- and postpurchasing behavior of customers. Therefore, managers must focus on maintaining a platform where previous and experienced customers can post their feedbacks regarding the service provided by the hotels and take actions in case a complaint is reported.

The first thing that online customers encounter while shopping is the firm's web page. The web site is the key for a firm to reach a wider audience not only within its country but also across borders. The researchers have stressed on system quality, content quality, and service quality for the popularity of a platform among browsers, which the marketers must focus in order to enhance their site traffic. The online service providers must have a web site that is compatible to the requirements of online shoppers, such as a safe and secure transaction medium, a platform that is easily accessible, possibility to book their services easily and at affordable prices, and an approachable customer support system. In addition to these facilities, the customers also expect a platform that is available 24×7, displays correct and precise information regarding their services and the underlining conditions, and enables the purchasers to compose their service package as per their personalized needs.

The digital form of word-of-mouth, i.e., EWOM, is the process of spreading the message related to an object by near and dear ones as well as acquaintances. Over the years, researchers have noticed the influence of social stimuli in the decision-making process by a shopper. Strong ties are mainly related with the people who are close and in continuous interaction with the customer like his friends or family. These are the reference group for the purchaser, and he or she looks forward for the comments provided by them related to a product/service before making a purchasing decision. Therefore, the service providers must keep track of the feedbacks provided by their customers and try to rectify their complaints, if any. This will help to retain its customer base and also add potential purchasers to their base. Moreover, the firm must be active on social platforms to maintain a continuous dialog between the service provider and customers. This will enhance the value perceived by the potential customer for the firm as well as its offerings.

12.7 LIMITATIONS AND FUTURE SCOPE

Though the research is well documented and systematic, it encounters some limitations. First, the study considers studying only hotel guest satisfaction; however, future studies may be conducted on the same lines for other sectors in hospitality and tourism industry. Second, the results are applicable for only a single dataset. To get more detailed insights, future studies may consider more datasets. Third, the results are a representation of experienced goods and findings may not be generalized for all product categories. The analysis ignored infrequent and rare words, which could be helpful for determining expected consumer preferences. This study may further be extended for electronic commerce applications.

REFERENCES

1. Aggarwal, A.G. and Aakash, N.A., Multi-criteria-based prioritisation of B2C e-commerce website. *International Journal of Society Systems Science*, 2018. **10**(3): pp. 201–222.

2. Memon, Q. and Khoja, S., Semantic web for program administration. *International Journal of Emerging Technologies in Learning*, 2010. **5**(4): pp. 31–40.
3. Tandon, A., Sharma, H., and Aggarwal, A.G., Assessing travel websites based on service quality attributes under intuitionistic environment. *International Journal of Knowledge-Based Organizations (IJKBO)*, 2019. **9**(1): pp. 66–75.
4. Casaló, L.V., Flavián, C., and Guinalíu, M., The role of satisfaction and website usability in developing customer loyalty and positive word-of-mouth in the e-banking services. *International Journal of Bank Marketing*, 2008. **26**(6): pp. 399–417.
5. Memon, Q. and Khoja, S., Semantic web approach for program assessment. *International Journal of Engineering Education*, 2009, **25**(5): pp. 1020–1028.
6. Statista. 2018 [cited 2018 August]; Available from: www.statista.com/statistics/289770/india-retail-e-commerce-sales/.
7. Hao, J.-X., et al., A genetic algorithm-based learning approach to understand customer satisfaction with OTA websites. *Tourism Management*, 2015. **48**(June): pp. 231–241.
8. Ali, F., Hotel website quality, perceived flow, customer satisfaction and purchase intention. *Journal of Hospitality and Tourism Technology*, 2016. **7**(2): pp. 213–228.
9. Zhou, L., et al., Refreshing hotel satisfaction studies by reconfiguring customer review data. *International Journal of Hospitality Management*, 2014. **38**: pp. 1–10.
10. Xiang, Z., et al., What can big data and text analytics tell us about hotel guest experience and satisfaction? *International Journal of Hospitality Management*, 2015. **44**: pp. 120–130.
11. Zhang, Y. and Cole, S.T., Dimensions of lodging guest satisfaction among guests with mobility challenges: A mixed-method analysis of web-based texts. *Tourism Management*, 2016. **53**: pp. 13–27.
12. Zhao, Y., Xu, X., and Wang, M., Predicting overall customer satisfaction: Big data evidence from hotel online textual reviews. *International Journal of Hospitality Management*, 2019. **76**: pp. 111–121.
13. Berezina, K., et al., Understanding satisfied and dissatisfied hotel customers: Text mining of online hotel reviews. *Journal of Hospitality Marketing & Management*, 2016. **25**(1): pp. 1–24.
14. Shopify. 2018 [cited 2018 August]; Available from: www.shopify.com/enterprise/global-ecommerce-statistics.
15. Mishra, A., et al., Adolescent's eWOM intentions: An investigation into the roles of peers, the Internet and gender. *Journal of Business Research*, 2018. **86**: pp. 394–405.
16. Fan, Z.-P., Che, Y.-J., and Chen, Z.-Y., Product sales forecasting using online reviews and historical sales data: A method combining the Bass model and sentiment analysis. *Journal of Business Research*, 2017. **74**: pp. 90–100.
17. Aakash, A. and Aggarwal, A.G., Role of EWOM, product satisfaction, and website quality on customer repurchase intention, in *Strategy and Superior Performance of Micro and Small Businesses in Volatile Economies*. Editors: João Conrado de Amorim Carvalho and Emmanuel M.C.B. Sabino, 2019, IGI Global. pp. 144–168.
18. Hargreaves, C.A., Analysis of hotel guest satisfaction ratings and reviews: an application in Singapore. *American Journal of Marketing Research*, 2015. **1**(4): pp. 208–214.
19. Stringam, B.B. and Gerdes Jr, J., An analysis of word-of-mouse ratings and guest comments of online hotel distribution sites. *Journal of Hospitality Marketing & Management*, 2010. **19**(7): pp. 773–796.
20. Liu, Y., et al., Big data for big insights: Investigating language-specific drivers of hotel satisfaction with 412,784 user-generated reviews. *Tourism Management*, 2017. **59**: pp. 554–563.
21. Radojevic, T., et al., The effects of traveling for business on customer satisfaction with hotel services. *Tourism Management*, 2018. **67**: pp. 326–341.

22. Min, H., Lim, Y., and Magnini, V.P., Factors affecting customer satisfaction in responses to negative online hotel reviews: The impact of empathy, paraphrasing, and speed. *Cornell Hospitality Quarterly*, 2015. **56**(2): pp. 223–231.
23. Wallach, H.M., Topic modeling: Beyond bag-of-words. In *Proceedings of the 23rd international conference on Machine learning*. 2006. ACM. Pittsburgh, Pennsylvania, USA
24. Crossley, S., Dascalu, M., and McNamara, D., How important is size? An investigation of corpus size and meaning in both latent semantic analysis and latent dirichlet allocation. In *The Thirtieth International Flairs Conference*. 2017. Marco Island, FL, USA
25. Hofmann, T., Probabilistic latent semantic indexing. In *ACM SIGIR Forum*. 2017. ACM.
26. Dybowski, T. and Adämmer, P., The economic effects of US presidential tax communication: Evidence from a correlated topic model. *European Journal of Political Economy*, *55*(C), 511–525, 2018.
27. Zheng, W. and Ye, Q., Sentiment classification of Chinese traveler reviews by support vector machine algorithm. In *Intelligent Information Technology Application, 2009. IITA 2009. Third International Symposium on*. 2009. IEEE., Nanchang, China.
28. Shi, H.-X. and Li, X.-J., A sentiment analysis model for hotel reviews based on supervised learning. In *Machine Learning and Cybernetics (ICMLC), 2011 International Conference on*. 2011. IEEE., Guilin
29. Chu, S.-C. and Kim, Y., Determinants of consumer engagement in electronic word-of-mouth (eWOM) in social networking sites. *International Journal of Advertising*, 2011. **30**(1): pp. 47–75.
30. Saha, G.C. and Theingi, Service quality, satisfaction, and behavioural intentions: A study of low-cost airline carriers in Thailand. *Managing Service Quality: An International Journal*, 2009. **19**(3): pp. 350–372.
31. Pantelidis, I.S., Electronic meal experience: A content analysis of online restaurant comments. *Cornell Hospitality Quarterly*, 2010. **51**(4): pp. 483–491.
32. Lu, W. and Stepchenkova, S., Ecotourism experiences reported online: Classification of satisfaction attributes. *Tourism management*, 2012. **33**(3): pp. 702–712.
33. Li, H., Ye, Q., and Law, R., Determinants of customer satisfaction in the hotel industry: An application of online review analysis. *Asia Pacific Journal of Tourism Research*, 2013. **18**(7): pp. 784–802.
34. Xiang, Z., Schwartz, Z., and Uysal, M., What types of hotels make their guests (un) happy? Text analytics of customer experiences in online reviews, in *Information and Communication Technologies in Tourism 2015*, L. Tussyadiah and A. Inversini, Editors. 2015, Springer. pp. 33–45.
35. Baharudin, B. Sentence based sentiment classification from online customer reviews. In *Proceedings of the 8th International Conference on Frontiers of Information Technology*. 2010. ACM., Islamabad
36. Yin, P., Wang, H., and Zheng, L., Sentiment classification of Chinese online reviews: Analysing and improving supervised machine learning. *International Journal of Web Engineering and Technology*, 2012. **7**(4): pp. 381–398.
37. Duan, W., et al., Mining online user-generated content: Using sentiment analysis technique to study hotel service quality. In *System Sciences (HICSS), 2013 46th Hawaii International Conference on*. 2013. IEEE. Wailea, Maui, HI USA
38. Wang, H., et al., Text feature selection for sentiment classification of Chinese online reviews. *Journal of Experimental & Theoretical Artificial Intelligence*, 2013. **25**(4): pp. 425–439.
39. Li, J., et al., Towards a general rule for identifying deceptive opinion spam. In *Proceedings of the 52nd Annual Meeting of the Association for Computational Linguistics* (Volume 1: Long Papers). 2014. Baltimore, Maryland

40. Banerjee, S., Chua, A.Y., and Kim, J.-J., Using supervised learning to classify authentic and fake online reviews. In *Proceedings of the 9th International Conference on Ubiquitous Information Management and Communication*. 2015. ACM.
41. Ali, F., Kwak, K.-S., and Kim, Y.-G., Opinion mining based on fuzzy domain ontology and support vector machine: A proposal to automate online review classification. *Applied Soft Computing*, 2016. **47**: pp. 235–250.
42. Xiang, Z., et al., A comparative analysis of major online review platforms: Implications for social media analytics in hospitality and tourism. *Tourism Management*, 2017. **58**: pp. 51–65.
43. Al-Smadi, M., et al., Deep recurrent neural network vs. support vector machine for aspect-based sentiment analysis of Arabic hotels' reviews. *Journal of computational science*, 2018. **27**: pp. 386–393.
44. Roshchina, A., Cardiff, J., and Rosso, P., TWIN: personality-based intelligent recommender system. *Journal of Intelligent & Fuzzy Systems*, 2015. **28**(5): pp. 2059–2071.
45. Malik, M. and Hussain, A., An analysis of review content and reviewer variables that contribute to review helpfulness. *Information Processing & Management*, 2018. **54**(1): pp. 88–104.
46. Singh, J.P., et al., Predicting the "helpfulness" of online consumer reviews. *Journal of Business Research*, 2017. **70**: pp. 346–355.

13 Sentiment Analysis for Decision-Making Using Machine Learning Algorithms

Mohamed Alloghani
Liverpool John Moores University
Abu Dhabi Health Services Company (SEHA)

Thar Baker, Abir Hussain, and Mohammed Al-Khafajiy
Liverpool John Moores University

Mohammed Khalaf
Almaaref University College

Jamila Mustafina
Kazan Federal University

CONTENTS

13.1 INTRODUCTION

Since time immemorial, people have always been curious to understand their surroundings. The advent of data mining brought with "Opinion Mining," sentimental analysis (SA), and other techniques have improved the quest to understand complex textual circumstances and the environment. In the context of natural language processing (NLP), opinion mining and SA refer to algorithms that compute and identify critical patterns in opinions, sentiments, and subjectivity from written texts [1]. The technological advancement and subsequent development in SA techniques have promoted opinion as for the first-class attribute with relatable constructs. Machine learning (ML) algorithms have proven efficient and effective in different classification and prediction tasks, including the context of document analysis. Whether using supervised or unsupervised learning techniques, with proper improvisations and modifications, these algorithms can be used to analyze and collate negative and positive sentiments in documents and unit texts [1,2]. Both supervised and unsupervised learning techniques can detect polarity in sentimental reviews, although their deployment uses different approaches, for example, supervised learning is only applicable when training subsets are available, while unsupervised learning suits datasets with linguistic resources but missing a training subset. The basis of SA is the assumption that opinions, attitudes, and emotions are subject to impressions that have innate binary opposition [2]; in this case, opinions are expressed as either like or dislike, good or bad, positive or negative, among others. The analysis of such sentiments uses either NLP, statistics, or ML techniques in the characterization of sentiments embedded within a text unit. The concern of SA is to extract a specific content from the provided text. It entails information retrieval, which involves effective techniques for discarding content that is subjective. Some SA tasks also recognize and isolate opinion-inclined queries. Above all, SA summarizes multiple perspectives and returns an actionable stance regarding the issue of interest.

This technique is applicable in business intelligence and interindustry business operations. In business intelligence, SA techniques permit search on opinions relating to a product of interest. Information on prices and customer satisfaction from a consensus point of view can help in obtaining accurate information, without necessarily subscribing to a service or creating a profile. It is this feature that marks a significant difference between SA and other data mining techniques. A majority of scholarly articles on SA have deployed different algorithms at a document level. Also, most articles distinguish negative and positive sentiments in such texts in the form of reviews. Nonetheless, the current development in NLP has brought with it more functionalities, including those that support sentence-level analysis of multiperspective documents.

SA is a classification process with three distinct levels. The first is a document-level SA that classifies the unit text as expressing either negative or positive sentiments. The second is a sentence-level SA that identifies and extracts sentiments found in each sentence. In the sentence-level analysis, it is mandatory to establish whether a sentence is subjective or not. For subjective sentences, this SA technique establishes whether it expresses negative or positive sentiments [1,2]. However, it is debatable whether sentences that bear sentiments are necessarily subjective. The first and second levels of SA are indistinguishable, although informed and detailed analysis is obtained using aspect-level SA, which classifies sentiments based on the object or topic of interest but using an attribute. Hence, aspect-level SA first identifies an object, for example, a movie and its identifiers, including genre and characters, among others. The approach is necessary because different people have different preferences, especially regarding some specific attributes [2]. In the context of movies, action diehard fans may rate a movie differently based on their opinion on stunts and how good or bad they were pulled.

SA techniques alongside other NLP algorithms are becoming more relevant with each emerging Big Data concept. For a long time, Big Data was associated with numerical or quantitative, but this will not be the case with each new application field. The dataset used in the research qualifies as big data because it contains over 465,000 tag applications, although the movie reviews in the data amount to 12 million in tag genome data with comments and scores for over 27,000 movies. Big Data analytics relies on advanced tools to identify patterns, reveal trends and correlations, and such tools are used in different decision-making instances. Specifically, Big Data has applications in customer, fraud, compliance, and operational analytics. In customer analytics, it intends to facilitate the processes of product improvement, value addition, and customer satisfaction strategies, and as such is applicable in the movie industry. SA of the big data in the movie industry can help in identifying trends and patterns that can aid in movie creation and budgeting.

13.2 LITERATURE REVIEW

The SA techniques rely on either ML techniques or lexicon-based models. ML approaches consist of unsupervised and supervised learning algorithms. Supervised learning requires a labeled dataset with training and validating subsets [3]. The algorithms learn the data using the training set, make predictions, and test their deviations from the values in the testing subset. The algorithms include decision trees, linear discriminants, and rule-based models among others [4–11]. The linear classifiers consist of support vector machines (SVM) and neural networks (NN), while probabilistic classifiers consist of maximum entropy, Bayesian network, and naïve Bayes. The classification in the article relied on SVM and naïve Bayes. Arguably, these are the two most commonly used techniques in SA [12].

Both ML and lexicon-based techniques have their merits and demerits. For instance, ML models require labeled training subsets that may not be necessarily available. Conversely, the lexicon-based models utilize language-dependent resources; otherwise, these techniques return lower recall measures. SA analysis uses NLP, textual investigation, and computed linguistics for the identification and extraction of subjective information within the sourced material [13]. That is, SA techniques have proven effective in emotion detection (ED), transfer learning (TL), and building resources (BR). All the application domains involve extraction and analysis of implicit or explicit emotions and the creation of lexical corpora [14]. The cross-domain learning or the TR specializes in lexical annotation based on polarity. In the polarity classification, the algorithm searches for comments or reviews as regards to whether they express positive or negative sentiments about the subject [15].

In other applications, SA techniques inspire sentiment classification, opinion summarization, and feature-based sentiment classification. Most of the SA studies have been done using survey data, or rather, they have been applied in the analysis of surveys besides the conventional statistical methods. The surveys deploy different SA techniques to solve different classification problems, although they demonstrate the versatility of SA as an analysis technique. Most of the current research prospects seek to improve SA, and with such needs, researchers are looking into the possibilities of integrating or improving SA using collaborative filtering (CF) techniques and NLP.

CF is a group of model-based and memory-based algorithms used in recommender frameworks. The framework has provisions for using the rating matrix of a training data subset that consists of binary values. It is the binary options that are likened to sentiments gathered on products or services [16]. Besides the rating matrix commonly associated with collaborative filters, the comments that accompany the ratings often go unnoticed, because the emphasis is on the rating. At the same time, most of the CF techniques are beyond laboratory work, as most of them are already integrated into live systems to complete different recommendations tasks. As such, using CF to improve SA is a mutually benefiting prospect, because CF recommenders can use SA techniques for polarity classification, while SA can leverage the robust CF framework to improve its framework and applicability. In particular, CF systems grapple with cold start problem, because new users require some reference similarity index to generate recommendation, while SA can search, index, and annotate the information in the database using polarity classifiers, without requiring the historical information about the user [17].

In application, data-collated recommendation application also deals with high sparsity, which undermines the interpretation of metadata. However, SA techniques can analyze and interpret the sparse and nonzero data as negative indicators of users' sentiments. In addition, the current data frameworks gather user ratings without a distinction for positive or negative sentiments.

For the purpose of implementing ML algorithms for analyst comments and reviews, it is imperative to select and extract features. In the case of MovieLens data, the features are the tags, and they are stored in the data frame as vectors defined as $\vec{F} = (f_1, f_2, \ldots, f_n)$, with each feature in the tag space representing specific tags relatable to the user [18,19]. In this case, the binary values are used because the ratings reflect the intensity of the user sentiments about the rated movie, although the same intensity indicator lacks in the movie reviews.

Some articles have used the *Bag-of-word* model to select feature for inclusion in the SA model [19]. The model extracts models from unordered databases, and the features (all words or tags) form the dictionary used for SA learning and predictions. For instance, a feature space defining tags for positive feedback is represented by the following models [20].

$$\vec{F}_1 = \{\text{'tag1'}: 1, \text{'tag2'}: 1, \text{'tag3'}: 1\} \tag{13.1}$$

The selection of the feature occurs when different users apply similar tags to review the same movie so that it features the space defined as follows [21]:

$$\vec{F}_2 = \{\text{'tag2'}:1, \text{'tag5'}:1, \text{'tag6'}:1\} \quad (13.2)$$

For example, in the two instances, the "tag2" is used to review the same movie so that a user similarity exists with respect to that tag for that item.

Those tags identified as belonging to the positive class have a higher similarity [22]. However, missing or omitted information is identified and grouped in the negative class, and this is the tenet for All Missing as Negative assumption used in most studies despite it is the biases of recommendations [23]. However, the absence of values does not improve model performance as the existence of the positive. As such, it is important to allocate lower weights to the negative tags. Most weighing models assume that the missingness of the data includes negative ratings with equal probabilistic distribution across users and movies [24]. In this case, the weights are uniformly assigned using a distribution function defined as $\delta \in [0, 1]$, and it represents negative tags [25]. Similarly, a weighting system that considers additional positive tags can also be adopted because the probability of a user not liking many movies tends to be high. In this case, the nonexistent information does not have any impact on the model. Additionally, such a weighting system assumes that the missing information, even though include positive tags, is inconsequential compared with the negative sentiments [26]. The weighting approaches are summarized in the following table.

Table 13.1 summarizes the low-rank approximation computation for the ranking matrices. The ranks are based on uniform (consistent), user-oriented (learning), and item-oriented (movie learning) schemes. The consistent scheme fixes the missing entries while awarding high confidence to the user based on their ratings. Further, movie leaning uses probability with the assumption that a movie with less positive reviews has higher instances of negative ratings and hence higher probabilities.

13.2.1 Related Studies and Techniques

Regarding the implementation of SA in polarity classification, this chapter used MovieLens data and, as such, its appraised user sentiments on the movies. Even though several studies have used similar SA techniques, very few have analyzed user reviews on movies. As shown in the summary, these studies use different data

TABLE 13.1
Weighted Alternate Least Squares

	Positive Tags	Negative Tags
Consistent	$W_{ij} = 1$	$W_{ij} = \delta$
User leaning	$W_{ij} = 1$	$W_{ij} = \Sigma j \cdot R_{ij}$
Movie leaning	$W_{ij} = 1$	$W_{ij} = m - \Sigma j \cdot R_{ij}$

TABLE 13.2
Summary of SA Studies and Techniques

Article	Year	Algorithms Used	Polarity	Data Source	Dataset Source (s)
[27]	2016	Classification-based	G	Web forms	IDM movie review
[28]	2016	Log-likelihood ratio (LLR) and delta TF-IDF	G	Movie review pages	4,000 Chinese movie reviews
[29]	2016	MapReduce	G	Hadoop platform	Hadoop database
[30]	2016	Naive Bayes, SVM	G	Tweet movie reviews	Twitter
[31]	2017	GINI index and SVM	Neg/pos	Movie reviews	N/A

sources, and comparison may be difficult because of the subjective nature of opinions depending on the subject (Table 13.2).

Nonetheless, Kim et al. [32] conducted a comparative study in which they established that NN and SVM are more accurate than naïve Bayes classier. Based on the study, NN had an accuracy of 70.72%, while SVM had 70.63% accuracy. In another paper, Kalaivani and Shunmuganathan [33] used SA to classify the user review of movies using SVM, naïve Bayes, and *k*NN (k- nearest neighbor) algorithms. According to the study, SVM outperformed both NN and *k*NN and recorded over 80% accuracy in making the predictions. The SA task is time consuming in cases where the target variable has longer *n*-gram lengths. However, texts or reviews with shorter *n*-gram lengths also result in unbalanced positive, negative, and neutral corpora. Consequently, moving forward and considering the influence of Big Data analytics in different NLP applications, it is important to develop techniques for handling the computational memory problem. The R package Big Data and its dependencies are a step towards handling ML problems, especially for text-based data.

Sahu and Ahuja [27] conducted SA analysis of an IDM (movie review based on a structured N-gram) and concluded that their approach had a classification accuracy of 88.95%. Chu et al. [28] also conducted an SA of movie reviews from Chinese databases and established that despite favoritism toward SVM and naïve Bayes as the best performing SA analysis techniques, LLR performs better and has lower misclassification error margins. Furthermore, Amolik et al. [30] conducted an SA of movie reviews based on comments retrieved from Twitter using SVM and NB (Naive Bayes). The study classified each of the reviews as positive, negative, or neutral, and established that NB and SVM have accuracies of 75% and 61%, respectively. In their conclusion, the authors asserted that the accuracy of an SVM algorithm increased with an increase in the size of the training subset. Finally, Tripathy Agrawal and Rath [34] implemented logistic regression, SVM, and NB to conduct an SA of movie reviews obtained from different platforms and claimed that logistic regression classifier is more accurate than SVM and NB. However, SVM performed better than NB in that analysis.

It is clear from these studies that SVM is more accurate, but not necessarily precise, compared with NB algorithm. Additionally, the results of the studies established an accuracy that ranges between 75% and 88.95%, and based on this general trend, this study obtained a lower accuracy. The lower accuracy can be ascribed to the size of the training data, data sparsity, and cold start problem. Nonetheless, the algorithm had the highest sensitivity (88%) compared with the rest of the studies and, hence, had the highest true positive (TP) or probability of detection.

13.3 METHODS

SA techniques, including binary and multiclass classification, classify documents or corpus reviews as either positive or negative. However, multiclass sentiment classification algorithm divides user reviews into strong positive, positive, neutral, negative, or strong neutral [35]. In general, binary SA classification suits comparison of documents or corpus.

The MovieLens dataset contains tags stored in unformatted text format. As such, the data requires transformation before loading for ML algorithms [35,36]. The data transformation involves the ratings, but the target attributes are the movie tags and user comments. The data was processed and tasks such as feature selection, comment segmentation for corpus creation, and transformation of rating vectors use term frequency-inverse document frequency. The transformation of ratings alongside the creation of the corpus from the comments and reviews yielded the data used to complete the analysis. The processed data, corpus, was loaded, and naïve Bayes and SVM were used to classify the data and obtain sentiments. The visual program that was implemented to obtain the results of the study is as shown in Figure 13.1.

Figure 13.1 illustrates the process of implementing SA. It consists of data importation and preparation, and implementation of both naïve Bayes and support vector SA algorithms. The imported MovieLens dataset contains text-based comments and

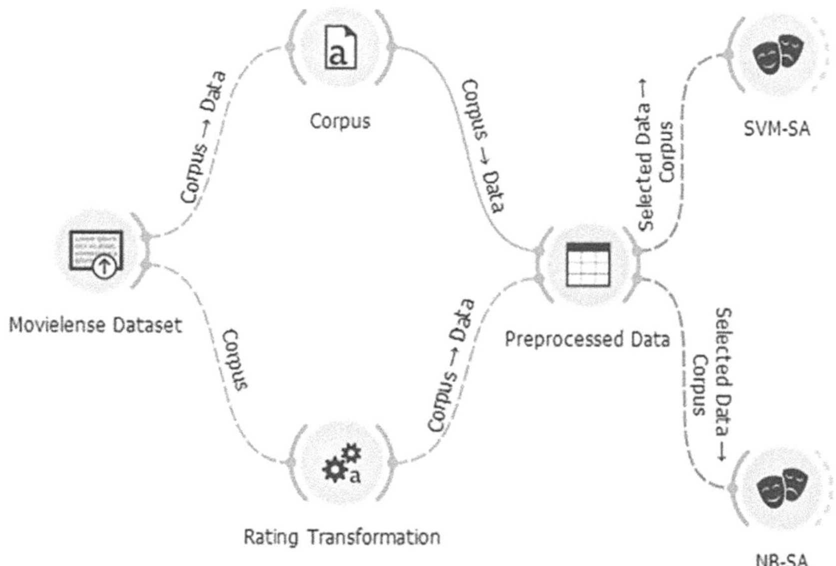

FIGURE 13.1 The visual program of SA implemented in the study.

numerical ratings; both are converted as corpus and merged to a data table and used as input for NB and SVM SA analysis.

13.3.1 Naïve Bayes Classifier

The classifier is of probabilistic nature, and it relies on Bayesian probability theory. It is the simplest and easiest algorithm to implement. The classification model calculates the posterior probability of the features of the reviews. The posterior probability is based on the distribution of features that Equations (13.1) and (13.2) illustrate. In application, naïve Bayes implements an algorithm that omits the position of the tag in the rating matrix. The Bayes theorem for predicting the classes in which each tag belongs is as follows [37]:

$$P(\text{tab} \mid \text{feature}) = \frac{P(\text{tag}) \times P(\text{feature} \mid \text{tag})}{P(\text{feature})} \tag{13.3}$$

$P(\text{tag})$ refers to the probability of the tag in the rating matrix, and it shows the likelihood of a random user giving the review. $P(\text{feature} \mid \text{tag})$ refers to the prior probability for the algorithm, which classifies a feature as a tag [37]. However, the naïve algorithm must meet the requirement of independence among the features so that Equation (13.3) becomes [37]

$$P(\text{tab} \mid \text{feature}) = \frac{P(\text{tag}) \times P\left(f_1 \middle| \text{tag} \times \ldots \times P(f_n \mid \text{tag})\right)}{P(\text{feature})} \tag{13.4}$$

Equation (13.4) and the improved version of Equation (13.3) solve the problem of the prediction of positive classes, which is being higher than that of negative classes [37]. Despite the improvement, the average accuracy of the model reduces, but both precision and recall improve using the new model.

13.3.2 Support Vector Machine

SVM is a nonprobabilistic positional notation additive classifier represented in this research paper by every evaluation in the form of a variable information component in the space. The process utilized in the analysis of full vectorized data and the major design behind training the representation is to search for a hyperplane denoted by $\vec{w}$. The SVM algorithm selects smaller numbers of important boundary instances (support vectors) from each tag and creates linear discriminant functions that separate the features [38,39]. Such an approach overcomes the limitations of linear boundaries and allows for the inclusion of additional nonlinear terms in the LDF, and hence the possibility of developing quadratic, cubic, or any higher-order decision-making boundaries in the model. Collecting textual information vectors defined as being optimally split up by the hyperplane is simply referenced when there is no fault and when the detachment between the nearest points of every class and hyperplane is complete. Following the preparation of the model, experimental evaluation can be mapped into a similar space and can be expected to be assigned to a class based on which region the hyperplane rests. According to Equation (13.5), $c_{j\in\{-1,1\}}$ is divided according to (positive, negative) for a file (denoted as d_j). The mathematical statement of $\vec{w}$ is provided by [40]

$$\vec{w} = \sum a_j c_j \vec{d_j} \ , \ a_j \geq 0 \tag{13.5}$$

The dual optimization challenge provides the values for a_j's. The entire $\vec{d_j}$ and a_j are higher than zero and is defined as support vectors not only because they are document based but also because they contribute to $\vec{w}$. Other research studies have shown that naïve Bayes with local weightings outperforms both naïve Bayes itself and *k*NN. It also compares favorably with more sophisticated ways of enhancing naïve Bayes by relaxing its intrinsic independence assumption [41]. Locally weighted learning only assumes independence within a neighborhood, not globally in the whole instance space as a standard naïve Bayes does.

13.4 DATA ANALYSIS AND ALGORITHM INITIALIZATION

Accounts on the setup of the analysis, including data preparation and model initialization, evaluation measures, and planned approach are presented.

13.4.1 Experimental Setup

The MovieLens dataset was retrieved from www.movielens.org, and it contains 100,004 ratings and 1,296 tags for 9,125 movies. The data reviews were written by 671 users between January 1995 and October 2016, and all the users were included in the analysis. However, the qualification criterion included users' tags with at least 20 movies reviewed, all users were identified using a unique ID without more demographic information.

The dataset contains links, movies, tags, and tag files, and each contains specific information. Using the two approaches, SA analysis was applied to the tags. As for feature selection, K-means classifier was used instead of the other conventional methods. Ideally, the tags were predicted using an algorithm that is verifiable and can be evaluated. In other terms, our expectation is that users with the same taste will normally rate movies with a high relationship [42].

13.4.2 Data Preparation and Model Initialization

The Bayesian network classifier takes only shorter time and requires a minimal measure to train data for estimation of the necessary factors to be classified. Its conditional probability simplicity and solid assumptions make it satisfactory in multiple domains. For instance, it avails practical learning combined with past knowledge and observed data [42]. The naive probability was computed for the tags. csv document with feature identified by the K-means classifier as the most common tag for high-rated movies. The tags conform to the independence assumption. As for the analysis procedure, the data was loaded into R, and the necessary packages are loaded as well. As part of data management, the necessary variable and data type specifications were performed. The loaded data contained 100,005 ratings, with timestamp, genre, and tags as the only variables. The tag variable was the only variable used in learning and classification. The tags were categorized into positive and negative followed by being appended to each other in accordance and stored into already-created empty lists. The modeling scheme designated 75% of data to a training set and the rest to validation.

13.4.3 Evaluation Measures

The performance metrics used to evaluate the two algorithms were based on the classical confusion matrix experiment and computations. The matrix presented in Table 13.3 contains counts of sentiment instances correctly classified and misclassified. The performance of the models was evaluated based on TP, True Negative (TN), False Positive (FP), and False Negative (FN). The TP is the count of the number of correctly predicted sentiments [43]. That is, the model predicts the actual positive sentiments correctly. The number of actual positive sentiments predicted as negative constitute the FN measure, while the number of negative sentiments predicted or classified as positive sentiments constitute the FP measure [43]. Finally, the number of negative sentiments correctly predicted is the TN measure.

From the confusion matrix table, metrics such as precision, recall, *F*-measure, and accuracy can be computed. The definition and computational formula for each of the metrics are discussed as follows.

Precision: It refers to the ratio between the number of correctly predicted positive sentiments to the sum of correctly predicted positive sentiments and the negative sentiments predicted as positive sentiment [43]. The computation equation is shown as follows.

$$\text{Precision} = \frac{\text{TP}}{\text{TP} + \text{FP}} \tag{13.6}$$

Recall: It is a measure of sensitivity, because it refers to the ratio between correctly classified positive sentiments and the sum of positively classified sentiments and positive sentiments incorrectly classified as negative sentiments [43]. The computation equation is shown as follows.

$$\text{Recall} = \frac{\text{TP}}{\text{TP} + \text{FN}} \tag{13.7}$$

TABLE 13.3
Confusion Matrix and Model Sentiment Prediction Elements

		Predicted Sentiment Class	
		Positive	**Negative**
Actual Sentiment Class	**Positive**	TP	FN
	Negative	FP	TN

Recall and precision tend to contradict each other, because models with higher precision tend to be less sensitive.

F-measure: It is a single value measure that combines recall and precision, and it is also referred as the F-1 measure [42]. The computation equation is as shown below.

$$\text{F-Measure} = \frac{2 * \text{Precision} * \text{Recall}}{\text{Precision} + \text{Recall}} \tag{13.8}$$

Accuracy: It is the classification accuracy of the model and is a ratio between the sum of TP and TN and the total number of sentiments within the dataset [43]. The computation equation is shown as follows.

$$\text{Accuracy} = \frac{\text{TP} + \text{TN}}{\text{TP} + \text{TN} + \text{FP} + \text{FN}} \tag{13.9}$$

13.4.4 Planned Approach

The implementation of SA analysis in analyzing the reviews in the Movielens data consisted of several preprocessing activities, such as handling of missing data or sparsity, removal of special characters applied, such as "!," "@," and blank spaces needed, and creation of a corpus based on unsupervised learning to label the data. Other phrases such as "wow" and "ooh" as well as repetitive characters were also removed. The dataset consists of numerical attributes and string attributes. The numerical attributes consist of five-star ratings, while string attribute contains the comments and reviews on the movies. Studies that have used SA to conduct phrase-level categorization or polarity classification most uses Bag-of-word model to select features when dealing with large datasets [44]. However, the approach used in this article relied on a k-means algorithm to predict tags or comments that were common among users. The use of the two supervised learning algorithms permitted control of the qualifying polar phrases, and the predicted words from K-means were used in constructing the count vectorizer matrix.

13.5 RESULTS AND MODEL EVALUATION

The preliminaries result of the preprocessing stages, including scatterplot of user ID against movie ID and term frequencies for words from the corpus, are presented as follows. The scatterplot of the users and movies rated and reviewed is as shown in the following figure.

Figure 13.2 shows sparsity that the rest of the analysis considers as negative reviews. It is a scatterplot of the number of users against the number of movies rated and presumably reviewed. The colored (in pdf version of this chapter) dots represent the reviews and human being's visualization of a sparse matrix data, and the spaces represent nonreviewed or commented movies. During SVM analysis, the process of vectorization considers the white spaces as negative reviews.

The statistics of the vectorized form of the reviews in Figure 13.3 shows the minimum, maximum, and confidence intervals for both the positive and negative sentiments. The predicted sentiments are considered as binomial attributes, with the least

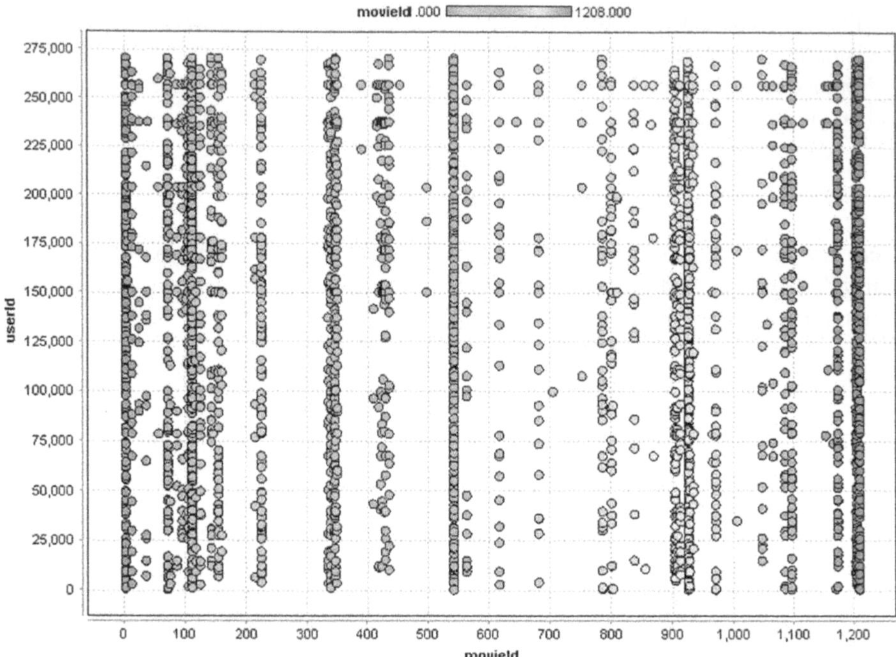

FIGURE 13.2 Scatterplot of movies reviewed against the number of reviews.

Name	Type	Missing	Statistics		
Sentiment	Bionomial	0	Least Positive (0)	Most Negative (1)	Negative Values (1)
CI (Negative)	Real	0	0.587 (Min)	0.587 (Max)	0.587 (Average)
CI (Positive)	Real	0	0.413 (Min)	0.413(Max)	0.413 (Average)

FIGURE 13.3 Summary statistics of the SA SVM model output.

positive sentiment being 0 and the most negative being 1. The maximum, minimum, and average real-valued ratings of the negative sentiments have the same value of 0.587. Similarly, the maximum, minimum, and average real-valued ratings of the negative sentiments have the same value of 0.413. The negative comments dominate the dataset partly because of the inclusion of the white space (lack of reviews) as negative reviews. However, both positive and negative sentiments are representable using a single value.

The confusion matrix associated with the model is as shown in the following table. The consequence of the inclusion of the whitespace is a higher prediction precision of negative reviews.

From the earlier table, the SVM model had an accuracy of 61% ± 10.44%, which is lower than the values that other researchers have established. Furthermore, the model had a precision of 59.05% ± 7.34%, a recall of 88.73% ± 9.04%, an f-measure of 70.61% ± 6.91%. Hence, the model has a sensitivity of 88.73% ± 9.04% and a

TABLE 13.4
Confusion Matrix Retrieved from the SA SVM Model

	True Negative	True Positive	Class Precision
Pred. Negative	29	12	70.73%
Pred. Positive	66	93	58.49%
	30.53%	88.57%	

TABLE 13.5
Token Count Matrix

Attribute	No. 1	No. 2	No. 3	No. 4	No. 5
Phrase 1	1	1	1	0	0
Phrase 2	1	1	0	1	0
Phrase 3	1	1	0	0	1

specificity of 30.67% ± 18.11%. The token matric model from the tokenization and vectorization of sentiments in the reviews is presented in Table 13.5. The tokenization is based on the frequency of occurrence of positive, neutral, and negative sentiments based on the Count Vectorizer of "great", "fine," and "awful" phrases.

The generated token matrix is 3 × 5 for the three phrases and five sentiment classes (very negative and positive, somewhat negative and positive, and neutral). The counts are an indication of the term frequency for each word used in the corpus or grouped reviews. For instance, for a corpus consisting of 100,000 text phrases with word "awesome" appearing 100 times, it suffices to conclude that the term frequency for "awesome" is 0.001. The term frequency refers to the proportion representing the number of times a word appears in a corpus. The term frequency for the two algorithms is shown in Table 13.6 and Figure 13.4. From Table 13.6, it is apparent that the term frequency increased from 2,000 tags onwards, and the two models have an almost equal term frequency at the corresponding number of tags.

From Table 13.6 and Figure 13.4, it is clear that SVM performed better than NB in terms of term frequency. In both cases, the maximum number of term frequency is obtained when the number of tags is less than 500.

Despite the similarity in the profiles of the algorithms in Figure 13.4, SVM had a maximum term frequency of 4.2%, while the NB algorithm had a term frequency of 4%, and, as such, SVM is more suitable in predicting the sentiment classes in the dataset. Further assessment of the algorithms required information on historical movie production trends and genres. Figure 13.5 shows that movie production grew exponentially between the early 1980s and 2016. The trend is of importance because movie production has undergone a tremendous technological development. Such developments hold sway on the sentiments toward movies.

TABLE 13.6
Accuracy Comparison of Datasets

	Term Frequency		
No. of Experiments	**Number of Tags in the Training Set**	**Naïve Bayes Rating**	**SVM Rating**
1	100	3.295	3.37
2	200	4.035	4.2
3	500	4.008333	3.965
4	1,000	3.541	3.519
5	1,500	2.94	2.935
6	2,000	2.599	2.608
7	2,500	2.384	2.368
8	3,000	3.338677	3.37976
9	4,000	3.4335	3.4285
10	4,500	3.655	3.663

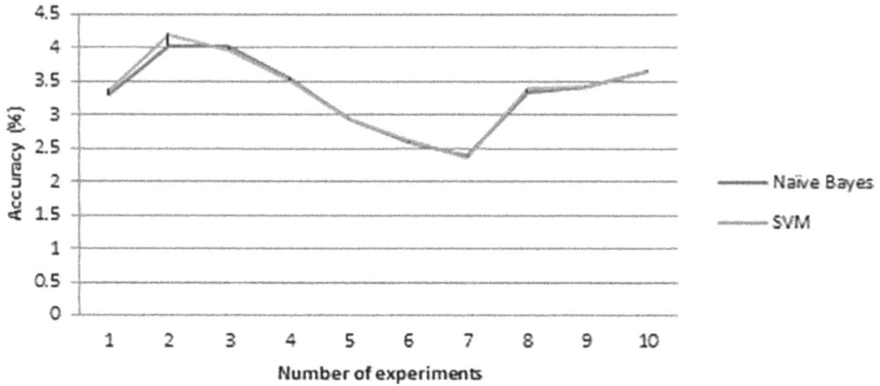

FIGURE 13.4 Graphic representation of accuracy in the experiments.

Besides the history, the demand for movies is driven by genre popularity, as shown in Table 13.7. Based on the table, Drama, Comedy, and Thriller are the most popular genres.

However, the distribution curves for the genres suggest that action genre tends to be more popular over time compared with other genres.

The graphical data in Figure 13.6 reflects a sentiment distribution of "thoughtful" tagged movies as Sci-Fi, being less preferred compared with "lovely" tagged movies that are associated with drama genre movies.

The visualization in Figure 13.7 suggests that Lovely and Sad tags were the most frequently used words in the review. The Lovely comments were associated with sentiments such as funny and revealing, while the sad tag was associated with

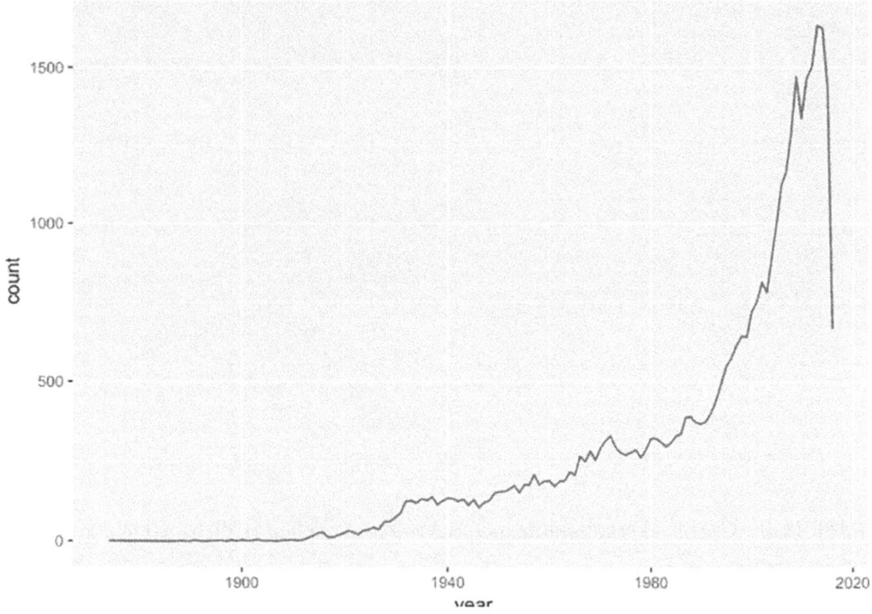

FIGURE 13.5 The number of movies produced over the years.

TABLE 13.7
Tabulation of Movie Genre Popularity Distribution

Genre	Count
Drama	23,152
Comedy	16,712
Thriller	11,320
Romance	10,759
Action	10,370
Horror	9,180
Crime	9,094
Documentary	1,190
Adventure	6,801
Sci-Fi	1,420

dangerous and scary sentiments. The *n*-gram in the figure summarizes the distribution of sentiments associated with the frequencies presented in Figure 13.7.

Figure 13.8 shows the distribution of review sentiments based on the polarity of tags. The figure asserts that reviews or comments with shorter n-gram lengths tend to be mostly neutral, while those with longer *n*-gram lengths tend to be either negatively or positively polarized.

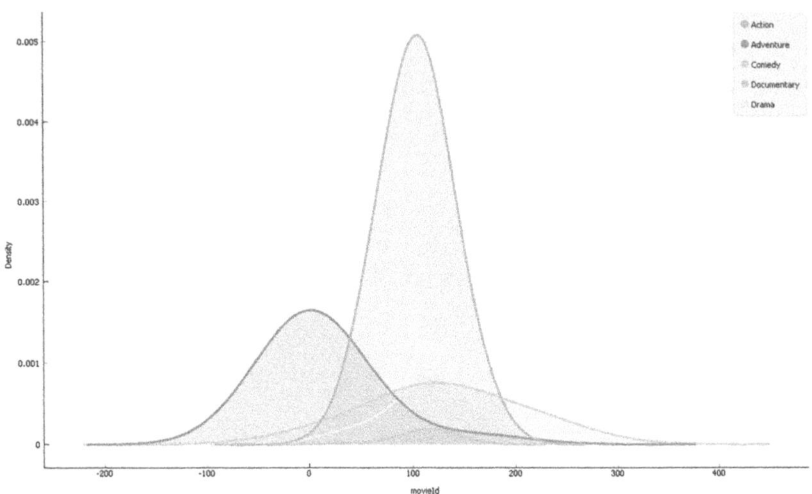

FIGURE 13.6 Graphical representation of movie genre popularity distribution.

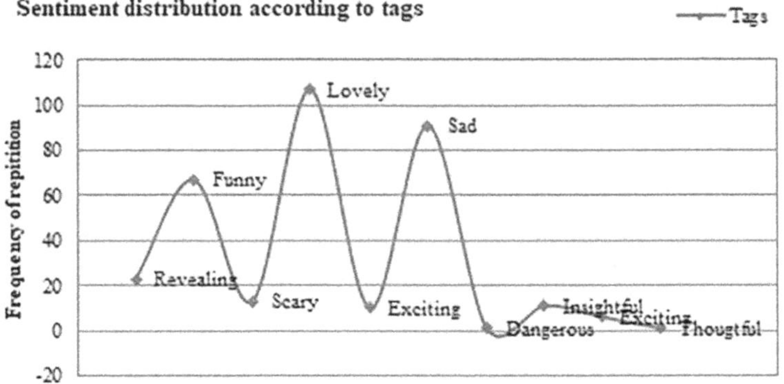

FIGURE 13.7 Sentiment distribution-based user tags.

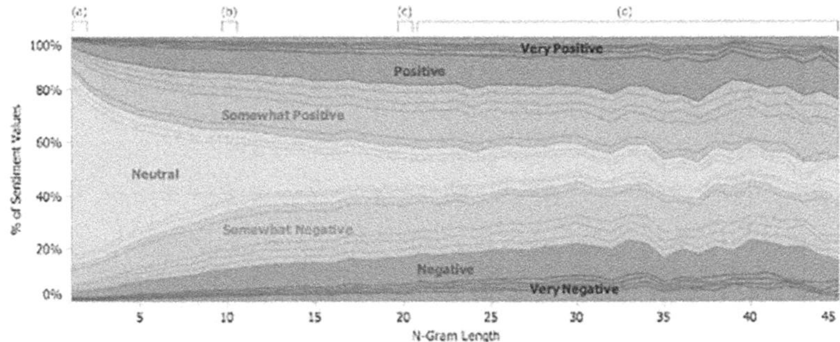

FIGURE 13.8 Illustration the profile of both positive and negative tags.

13.6 CONCLUSION AND FUTURE WORKS

Of the two algorithms, SVM algorithm outperforms other classifiers, since it is competent in predicting the reaction of a review. In addition to the automation of feature extraction, SA can also aid in a variety of predictive activities. For instance, the deployment of either a deep learning or ML algorithm in the classification movie reviews unveils patterns and sentiments that can benefit the movie industry. However, the implementation of such an algorithm is dependent on the computing power and size of the data. For Big Data, the process of learning can be quite long. Most ML algorithms isolate and learn only influential attributes for predictive decision-making. For instance, decision trees identify the most promising feature to form the node and build branches around. Theoretically, irrelevant attributes reduce the performance of the model. As established in the literature, in most cases, both SVM and naïve Bayes algorithms are the most commonly implemented algorithms. SVM method is probably the most efficient SA classification algorithm today. However, future research prospects should consider the influence of impacts of NLP on SA techniques. Additionally, it would be prudent to appraise the performance of the models under different feature selection models. Given the existence of Big Data, feature selections will increasingly become as important as SA classification. Finally, studies should investigate the feasibility of developing a hybrid recommender model that uses model-based, memory-based, SA algorithm to help deal with data sparsity and cold start problems.

13.7 ACKNOWLEDGMENT

The data sources used in this research paper are retrieved from www.movielens.org. We are grateful for the dataset and would like to thank them for making such an open dataset available online, which helped us to achieve the objectives and results of the study.

REFERENCES

1. Perea-Ortega JM, Martinez-Cámara E, Martn-Valdivia M-T, Ureña-López LA (2013) Combining supervised and unsupervised polarity classification for non-English reviews. In: *International Conference on Intelligent Text Processing and Computational Linguistics*, pp. 63–74. Samos, **Greece**
2. Pang B, Lee L, et al. (2008) Opinion mining and sentiment analysis. *Found Trends® Inf Retr* 2:1–135.
3. Devika MD, Sunitha C, Ganesh A (2016) Sentiment analysis: A comparative study on different approaches. *Procedia Comput Sci* 87:44–49.
4. Aljaaf AJ, Al-Jumeily D, Hussain AJ, Fergus P, Al-Jumaily M, Abdel-Aziz K (2015) Toward an optimal use of artificial intelligence techniques within a clinical decision support system. In: *2015 Science and Information Conference (SAI)*, London, pp. 548–554.
5. Keight R, Aljaaf A, Al-Jumeily D, Hussain A, Özge A, Mallucci C (2017) An intelligent systems approach to primary headache diagnosis. In: *Intelligent Computing Theories and Application - 13th International Conference*, ICIC 2017, Liverpool, UK, August 7–10.

6. Aljaaf AJ, Hussain AJ, Fergus P, Przybyla A, Barton GJ (2016) Evaluation of machine learning methods to predict knee loading from the movement of body segments. In: *2016 International Joint Conference on Neural Networks (IJCNN)*, Vancouver, BC, pp. 5168–5173.
7. Aljaaf AJ, et al. (2018) Early prediction of chronic kidney disease using machine learning supported by predictive analytics. In: *2018 IEEE Congress on Evolutionary Computation (CEC)*, Rio de Janeiro, pp. 1–9.
8. Aljaaf AJ, Al-Jumeily D, Hussain AJ, Fergus P, Al-Jumaily M, Radi N (2015) Applied machine learning classifiers for medical applications: Clarifying the behavioural patterns using a variety of datasets. In: *2015 International Conference on Systems, Signals and Image Processing (IWSSIP)*, London, pp. 228–232.
9. Aljaaf AJ, Al-Jumeily D, Hussain AJ, Fergus P, Al-Jumaily M, Radi N (2015) A systematic comparison and evaluation of supervised machine learning classifiers using headache dataset. In: Huang DS, Han K (eds.) *Advanced Intelligent Computing Theories and Applications*. ICIC 2015. Lecture Notes in Computer Science, vol. 9227. Springer, Cham.
10. Aljaaf AJ, Al-Jumeily D, Hussain AJ, Lamb D, Al-Jumaily M, Abdel-Aziz K (2014) A study of data classification and selection techniques for medical decision support systems. In: Huang DS, Jo KH, Wang L (eds.) *Intelligent Computing Methodologies*. ICIC 2014. Lecture Notes in Computer Science, vol. 8589. Springer, Cham.
11. Al-Kassim Z, Memon Q (2017) Designing a low-cost eyeball tracking keyboard for paralyzed people. *Computers & Electrical Engineering* 58:20–29.
12. Medhat W, Hassan A, Korashy H (2014) Sentiment analysis algorithms and applications: A survey. *Ain Shams Eng J* 5:1093–1113.
13. Hemmatian F, Sohrabi MK (2017) A survey on classification techniques for opinion mining and sentiment analysis. *Artif Intell Rev* 1–51.
14. Bhadane C, Dalal H, Doshi H (2015) Sentiment analysis: Measuring opinions. *Procedia Comput Sci* 45:808–814.
15. Guzman E, Maalej W (2014) How do users like this feature? A fine grained sentiment analysis of app reviews. In: *2014 IEEE 22nd International Requirements Engineering Conference (RE)*, pp. 153–162. Karlskrona.
16. Araque O, Corcuera-Platas I, Sanchez-Rada JF, Iglesias CA (2017) Enhancing deep learning sentiment analysis with ensemble techniques in social applications. *Expert Syst Appl* 77:236–246.
17. Galvis Carreño LV, Winbladh K (2013) Analysis of user comments: An approach for software requirements evolution. In: *Proceedings of the 2013 International Conference on Software Engineering*, pp. 582–591. San Francisco, CA, USA.
18. Taboada M, Brooke J, Tofiloski M, et al. (2011) Lexicon-based methods for sentiment analysis. *Comput Linguist* 37:267–307.
19. Prabowo R, Thelwa M (2009) Sentiment analysis: A combined approach. *Journal of Informatics*. Vol. (2), 143–157.
20. Pappas N, Popescu-Belis A (2013) Sentiment analysis of user comments for one-class collaborative filtering over ted talks. In: *Proceedings of the 36th International ACM SIGIR Conference on Research and Development in Information Retrieval*, pp. 773–776. Dublin, Ireland.
21. Siersdorfer S, Chelaru S, Nejdl W, San Pedro J (2010) How useful are your comments?: Analyzing and predicting youtube comments and comment ratings. In: *Proceedings of the 19th International Conference on World Wide Web*, pp. 891–900.
22. Ahmad M, Aftab S (2017) Analyzing the performance of SVM for polarity detection with different datasets. *Int J Mod Educ Comput Sci* 9:29.
23. Serrano-Guerrero J, Olivas JA, Romero FP, Herrera-Viedma E (2015) Sentiment analysis: A review and comparative analysis of web services. *Inf Sci (Ny)* 311:18–38.

24. Hu N, Bose I, Koh NS, Liu L (2012) Manipulation of online reviews: An analysis of ratings, readability, and sentiments. *Decis Support Syst* 52:674–684.
25. Kaur D (2017) Sentimental analysis on Apple Tweets with machine learning technique. *Int J Sci Eng Comput Technol* 7:76.
26. Wilson T, Wiebe J, Hoffmann P (2005) Recognizing contextual polarity in phrase-level sentiment analysis. In: *Proceedings of Human Language Technology Conference and Conference on Empirical Methods in Natural Language Processing.* Vancouver, British Columbia, Canada.
27. Sahu TP, Ahuja S (2016) Sentiment analysis of movie reviews: A study on feature selection and classification algorithms. In: *International Conference on Microelectronics, Computing and Communication*, MicroCom.
28. Chu CH, Wang CA, Chang YC, et al. (2017) Sentiment analysis on Chinese movie review with distributed keyword vector representation. In: *TAAI 2016-2016 Conference on Technologies and Applications of Artificial Intelligence, Proceedings*, pp. 84–89.
29. Gupta P, Sharma A, Grover J (2016) Rating based mechanism to contrast abnormal posts on movies reviews using MapReduce paradigm. In: *2016 5th International Conference on Reliability, Infocom Technologies and Optimization, ICRITO 2016: Trends and Future Directions*, pp. 262–266.
30. Amolik A, Jivane N, Bhandari M, Venkatesan M (2016) Twitter sentiment analysis of movie reviews using machine learning technique. *Int J Eng Technol* 7:2038–2044. doi:10.5120/ijca2017916005.
31. Manek AS, Shenoy PD, Mohan MC, Venugopal KR (2017) Aspect term extraction for sentiment analysis in large movie reviews using Gini Index feature selection method and SVM classifier. *World Wide Web* 20:135–154. doi:10.1007/s11280-015-0381-x.
32. Kim Y, Kwon DY, Jeong SR (2015) Comparing machine learning classifiers for movie WOM opinion mining. *KSII Trans Internet Inf Syst* 9:3178–3190. doi:10.3837/tiis.2015.08.025.
33. Kalaivani P, Shunmuganathan K, Index MI, et al. (2013) Sentiment classification of movie reviews by supervised machine learning approaches. *Indian J Comput Sci Eng.* doi:10.1016/j.proeng.2014.03.129.
34. Tripathy A, Agrawal A, Rath SK (2015) Classification of sentimental reviews using machine learning techniques. In: *Procedia Computer Science*, pp. 821–829. Delhi, India.
35. Liu B (2015) *Sentiment Analysis: Mining Opinions, Sentiments, and Emotions.* Doi: 10.1017/CBO9781139084789.
36. Qiu G, He X, Zhang F, et al. (2010) DASA: Dissatisfaction-oriented advertising based on sentiment analysis. *Expert Syst Appl* 37:6182–6191.
37. Dey L, Chakraborty S, Biswas A, et al. (2016) Sentiment analysis of review datasets using naive bayes and k-nn classifier. arXiv Prepr arXiv161009982.
38. Di Caro L, Grella M (2013) Sentiment analysis via dependency parsing. *Comput Stand Interfaces.* doi:10.1016/j.csi.2012.10.005.
39. Memon Q (2019) On assisted living of paralyzed persons through real-time eye features tracking and classification using Support Vector Machines. *Medical Technologies Journal*, 3(1):316–333.
40. Fan T-K, Chang C-H (2011) Blogger-centric contextual advertising. *Expert Syst Appl* 38:1777–1788.
41. Moreo A, Romero M, Castro JL, Zurita JM (2012) Lexicon-based comments-oriented news sentiment analyzer system. *Expert Syst Appl* 39:9166–9180.
42. Lane PCR, Clarke D, Hender P (2012) On developing robust models for favourability analysis: Model choice, feature sets and imbalanced data. *Decis Support Syst* 53:712–718.

43. Hahsler M (2011) Recommenderlab: A Framework for Developing and Testing Recommendation Algorithms. https://cran.r-project.org/web/packages/recommenderlab/vignettes/recommenderlab.pdf (20.09.2017).
44. Yu B (2008) An evaluation of text classification methods for literary study. *Lit Linguist Comput* 23:327–343.

14 Deep Learning Model
Emotion Recognition from Continuous Action Video

R. Santhosh Kumar and M. Kalaiselvi Geetha
Annamalai University

CONTENTS

14.1 INTRODUCTION

Understanding human emotions is a key area of research, since recognizing emotions may provide a plethora of opportunities and applications; for instance, friendlier human–computer interactions with an enhanced communication among humans, by refining emotional intelligence [1]. Recent research on experimental psychology demonstrated that emotions are important in decision-making and rational thinking. In day-to-day communication, human beings express different types of emotions. The human communication includes verbal and nonverbal communication. Sharing of wordless clues or information is called as nonverbal communication. This includes visual cues such as body language (kinesics) and physical appearance [2]. Human emotion can be identified using body language and posture. Posture gives information that is not present in speech and facial expression. For example, the emotional state of a person from a long distance can be identified using human posture. Hence, human emotion recognition (HER) through nonverbal communication can be achieved by capturing body movements [3]. The experimental psychology demonstrated how qualities of movement are related to specific emotions: for example, body turning towards is typical of happiness, anger, surprise; the fear brings to contract the body; joy may bring movements of openness and acceleration of forearms; fear and sadness make the body to turn away [4]. Emerging studies show that people can accurately decode emotions and cues from other nonverbal communications and can make inference about the emotional states of others. A certain group of body actions is called as gestures. The action can be performed mostly by the head, hands, and arm. These cues together convey information of emotional states and the content of interactions. With the support from psychological studies, identifying emotions from human body movement has plenty of applications. Suspicious action recognition to alarm security personnel, human–computer interaction, healthcare, and to help autism patients are few of the application areas of automatic emotion recognition through body cues [5–6]. However, in the surveillance environment, facial view is not clear when the camera is too far from humans. This type of issue can be rectified by capturing body movements (head, hands, legs, center body) for recognizing human emotion. The emotions are hard to identify from complex body-motion patterns. The challenge is based on the cultural habit that humans can express different expressions for the same emotion. With the help of deep features, the human emotion can be predicted easily using feedforward deep convolution neural network (FDCNN).

The following chapter is designed as follows. Section 14.2 briefly summarizes the related works. Section 14.3 explains the proposed work. Section 14.4 provides the experimental results. Finally, conclusions and future work are given.

14.2 RELATED WORKS

A system for automatic emotion recognition is developed using gesture dynamic's features from surveillance video and evaluated by supervised classifiers (dynamic time wrapping, support vector machine (SVM), and naïve Bayes) [7]. A framework is proposed to synthesize body movements based on high-level parameters and is represented by the hidden units of a convolutional autoencoder [8]. A system for recognizing the affective state of a person is proposed from face-and-body video using space–time interest points in video and canonical correlation analysis (CCA) for fusion [9]. A comprehensive survey of deep learning and its current applications in sentiment analysis is performed [10]. Recent works on high-performance motion data is described, and relevant technologies in real-time systems are proposed [11–12].

The deep learning algorithm to develop a novel structure in large data sets by using the backpropagation algorithm and processing images, video, speech, and audio for emotion recognition are developed [13]. A self-organizing neural architecture was developed for recognizing emotional states from full-body motion patterns [14]. A system for emotion recognition on video data is developed using both convolutional neural network (CNN) and recurrent neural network (RNN) [15]. The emoFBVP (face, body gesture, voice and physiological signals) database of multimodal (face, body gesture, voice, and physiological signals) recordings of actors enacting various emotional expressions are predicted [16]. A model with hierarchical feature representation for nonverbal emotion recognition and the experiments show a significant accuracy improvement [17–18]. The novel design of an artificially intelligent system is proposed for emotion recognition using promising neural network architectures [19]. A novel system for Emotion Recognition in the Wild (EmotiW) is developed using hybrid CNN–RNN architecture to achieve better results over other techniques [20]. A new emotional body gesture is developed to differentiate culture and gender difference framework for automatic emotional body gesture recognition. A new emotional body gesture is developed to differentiate culture and gender difference framework for automatic emotional body gesture recognition [21]. A novel approach for recognition of facial emotion expressions in video sequences proposes an integrated framework of two networks: a local network and a global network, which are based on local enhanced motion history image (LEMHI) and CNN-LSTM cascaded networks, respectively [22]. The deep learning algorithm for face detection achieves the state-of-the-art detection performance on the well-known FDDB face detection benchmark evaluation. In particular, we improve the state-of-the-art faster RCNN (Region-convolutional neural network) framework by combining a number of strategies, including feature concatenation, hard negative mining, multiscale training, model pretraining, and proper calibration of key parameters [23]. The author proposed architecture-based deep neural nets for expression recognition in videos, using an adaptive weighting scheme coping with a reduced size-labeled dataset [24]. The proposed technique uses movements of the human body for identification, particularly movement of the head, shoulders, and legs using CNN [25]. The challenging task of detecting salient body motion in scenes with more than one person is addressed, and a neural architecture that only reacts to a specific kind of motion in the scene is proposed: a limited set of body gestures [26]. The recent advances in CNNs and achievement of better performance in different applications like natural language processing, speech recognition, emotion recognition, activity recognition, and handwritten recognition are discussed [27]. Computer vision applications like emotion recognition, action recognition, image and video classification are experimented using dictionary learning-based approaches. From the large number of samples, representative vectors are learned and used in this concept [28]. The researchers developed a framework for human action recognition using dictionary learning methods [29]. Based on the hierarchical descriptor, the proposed method for human activity recognition outperforms the state-of-the-art methods. For a visual recognition, a cross-domain dictionary learning-based method was developed [30–31]. An unsupervised model was developed for cross-view human action recognition [32] without any label information. The coding descriptors of locality-constrained linear coding (LLC) [33] are generated by a set of low-level trajectory features for each action. The CNN is the most frequently used model from among the supervised category. The CNN [34] is a type of deep learning model that has shown better performance at tasks such as image classification, pattern recognition, human action recognition, handwritten digit classification, and HER. The multiple hidden layers present in the hierarchical learning model are used to transform the input data into output categories. The mapping back of different layers of CNN is called as deconvolutional networks (Deconvnets). The objects in the images are represented and recognized using a deep CNN model. The RNN is the other popular model of supervised category. The skeleton-based action and emotion recognition using RNNs are developed by this author [35].

14.3 LEARNING

Artificial intelligence is executed by machine learning technique. Learning is considered to be an important aspect for intelligent machines. In artificial intelligence research, a machine teaches to detect various patterns using machine learning patterns. Conventional machine learning techniques have the expertise to design a feature extractor that transformed the raw data into a feature vector to detect or classify patterns in the input data. Deep learning is a dedicated form of machine learning. The technique that instructs computers to do some operation and behave like humans is done by machine learning. From the input data, a machine learning task starts the feature extraction process. The features are fed to a model that classifies the objects in the image. Learning feature hierarchies are produced by combining low-level and high-level features [36]. In the deep learning model, features are automatically extracted from input data. Learning features are automatically set by several levels of abstraction. The numerous applications to machine learning techniques are increasing at an enormous rate. Figure 14.1 explains the learning-based approaches. It has the following two types of approaches.

Deep learning has emerged as a popular approach within machine learning. The traditional handcrafted feature-based approach and deep learning-based

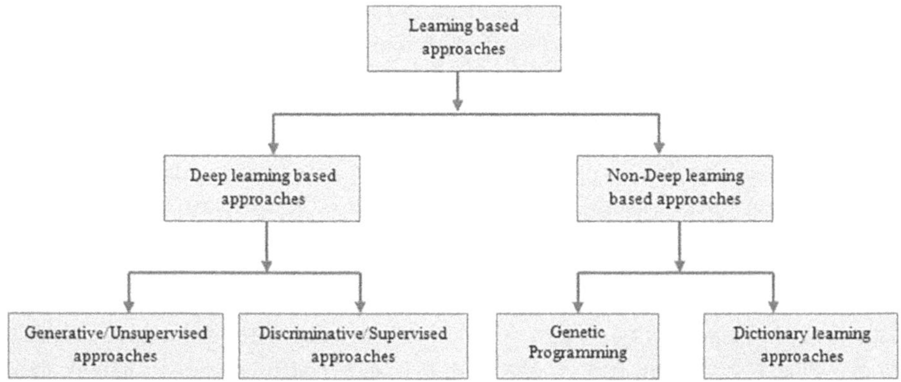

FIGURE 14.1 Learning-based approaches.

approach are the two major approaches in HER problem [37]. Several limitations of handcrafted approach fail to learn features automatically from the input data, as well as deep representation of data in classifiers. Alternatively, deep learning-based approach plays the concept of end-to-end learning by using the trainable feature extractor followed by a trainable classifier. The multiple layers of features are automatically extracted from raw data. This algorithm develops a multilayer representation of different patterns in the input data, where each successive layer is responsible for learning increasingly complex features [38]. The lower layers extract high-level features from the input data; thus, the representation increased the abstraction level at each consecutive layer. The need of handcrafted feature detectors and descriptors are eliminated by this automatic learning ability of deep learning models. In many visual categorization tasks, deep learning models have shown higher performance than traditional handcrafted feature-based techniques [39].

Deep learning models such as CNNs, deep belief networks (DBNs), deep RNNs, and deep Boltzmann machines have been successfully employed for many visual categorization tasks. Among these models, DBN is an unsupervised probabilistic graphical model capable of learning from the input data without prior knowledge. This model can also be trained in a semisupervised or unsupervised fashion, which is quite helpful for labeling data or dealing with unlabeled data. Learning-based approaches are divided into two approaches: nondeep learning approach and deep learning approach [40].

14.3.1 Nondeep Learning-Based Approaches

Dictionary learning method and genetic programming approach belong to nondeep learning-based methods.

14.3.1.1 Dictionary Learning Approaches

Computer vision applications like action recognition and image and video classification can be experimented using dictionary learning-based approaches [41]. From a large number of samples, the representative vectors are learned and used in this concept. Guha and Ward [42] developed a framework for human action recognition using dictionary learning methods. Based on the hierarchical descriptor, the proposed method [43] for human activity recognition outperforms the state-of-the-art methods. For visual recognition, a cross-domain dictionary learning-based method was developed [44]. An unsupervised model was developed by Zhu and Shao for cross-view human action recognition [45] without any label information. The coding descriptors of LLC [46] are generated by a set of low-level trajectory features for each action.

14.3.1.2 Genetic Programming Approaches

The unknown primitive operations can improve the accuracy performance of the HER task by using genetic programming techniques. Nowadays, this type of approach is being introduced for emotion recognition [47]. In this approach, the spatiotemporal motion features are automatically learned for action recognition. Filters such as 3D

Gabor filter and wavelet filter [48] have evolved for this motion feature. Similarly, a valuable set of features were learned for emotion recognition.

A probabilistic graphical model indicates the dependencies and random variables in a directed acyclic graph form. Different variations of Bayesian network have been introduced, such as conditional Bayesian networks, temporal Bayesian networks, and multientity Bayesian network (MEBN). In the work of Zhang et al. [49], an interval temporal Bayesian network (ITBN) was introduced for the recognition of complex human activities. To evaluate the performance of the proposed method, a cargo loading dataset was considered for experimentations and evaluations. Khan et al. [50] proposed another method for action detection using dynamic conditional Bayesian network, which also achieved the state-of-the-art results. In Park et al. [51], MEBN was used for predictive situation awareness (PSAW) using multiple sensors. These networks are robust for reasoning the uncertainty in the complex domains for predicting and estimating the temporally evolving situations.

14.3.2 Deep Learning-Based Approaches

For all types of datasets, there are no excellent handcrafted feature descriptors. To overcome this problem, the features are directly learning from raw data. Learning multiple levels of representation in data, such as speech, images/videos, and text is more advantageous in deep learning. These models do automated feature extraction, classification and process the images as raw data. These models have multiple processing layers in this work. There are three types of approaches in deep learning models [52,53]:

1. Generative/unsupervised approach (restricted Boltzmann machines (RBMs), DBNs, deep Boltzmann machines (DBMs), and regularized autoencoders);
2. Discriminative/supervised approach (CNNs, deep neural networks (DNNs) and RNNs);
3. Hybrid models (a characteristic combination of both approaches)

14.3.2.1 Generative/Unsupervised Approaches

In an unsupervised deep learning approach, the class labels are not required for the learning process. These types of approaches are specifically useful when labeled data are relatively unavailable. A remarkable surge in the history of deep models was triggered by the work of Hinton et al. [54], who developed a feature reduction technique and highly efficient DBN. In an unsupervised pretraining learning stage, a backpropagation method is used for fine-tuning. These types of deep learning approaches are used for many applications like object identification, image classification, speech classification, activity, and emotion recognition.

An unsupervised feature learning model from video data was proposed in the work of Le et al. [55] for human action recognition. The authors used an independent subspace analysis algorithm to learn spatiotemporal features, combining them with deep learning techniques such as convolutional and staking for action representation and recognition. DBNs trained with RBMs were used for HER [56]. This method performs better than the handcrafted learning-based approach on two public datasets. Learning the features continuously without any labels from the streaming video is a challenging task. Hasan and Roy-Chowdhury [57] addressed this type of problem using an unsupervised deep learning model. Most of the action datasets have been recorded under a controlled environment; besides, action recognition from unconstrained videos is a challenging task. A method for human action recognition from unconstrained video sequences was proposed by Ballan et al. [58] using DBNs. Unsupervised learning played a pivotal role in reviving the interests of researchers in deep learning.

14.3.2.2 Discriminative/Supervised Approaches

The CNN is the most frequently used model from the supervised category. CNN [56] is a type of deep learning model that has shown better performance at tasks such as image classification, pattern recognition, human action recognition, handwritten digit classification, and HER. The multiple hidden layers present in the hierarchical learning model are used to transform the input data into output categories. Its architecture consists of three main types of layers:

a. Convolutional layer
b. Pooling layer
c. Fully connected layer

The mapping back of different layers of CNN is called as Deconvnets. The objects in the images are represented and recognized using deep CNN models [59]. This author proposed spatial and temporal streams of CNN for action and emotion recognition. These two streams combined and outperformed better results than the other methods. RNN is the other popular model among the supervised category. The skeleton-based action and emotion recognition using RNNs are developed by this author [60]. The five parts of the human skeleton were separately fed into five subnets. The output from the subnets were combined and fed into a single layer for final demonstration. For the training process, deep learning-based model need a large size of video data. Collecting and annotating large size of video data require enormous computational resources. An outstanding accuracy has been achieved in many application fields.

14.3.3 Convolutional Neural Network

The CNN has six components [27]:

14.3.3.1 Convolutional Layer

The process of convolution has four Steps. Figure 14.2 shows the architecture of an FDCNN model.

- Line up the feature and image.
- Multiply each image pixel by its corresponding feature pixel.
- Add the values and find the sum.
- Divide the sum by the total number of pixels in the feature.

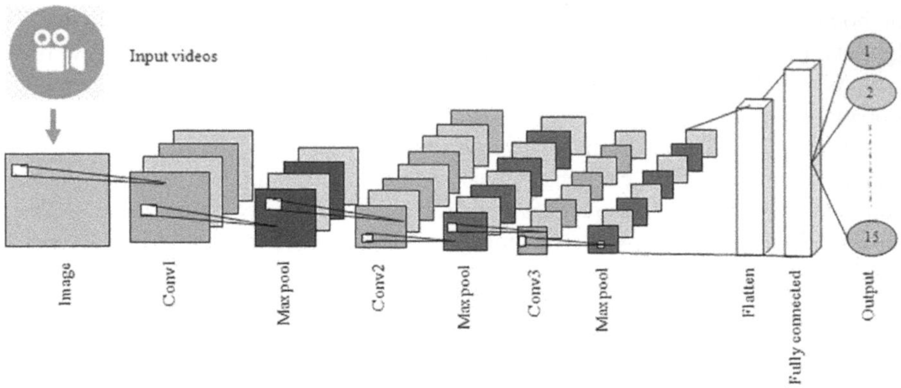

FIGURE 14.2 An FDCNN model.

This can be calculated as follows:

$$C\left(x_{u,v}\right)=\sum_{i=-\frac{n}{2}}^{\frac{n}{2}}\sum_{i=\frac{m}{2}}^{\frac{m}{2}} f_k(i,j)x_{u-i,v-j} \tag{14.1}$$

where f_k is a filter, $n \times m$ the kernel size, and the input image is x.

14.3.3.2 Subsampling Layers

Subsampling or pooling layers shrink the map size into a smaller size. Figure 14.3 shows the example of a max-pooling layer. The following four steps implement the pooling function:

- Pick a window size (usually 2 or 3)
- Pick a stride (usually 2)
- Move your window across your filtered images.
- The maximum value is taken from each window.

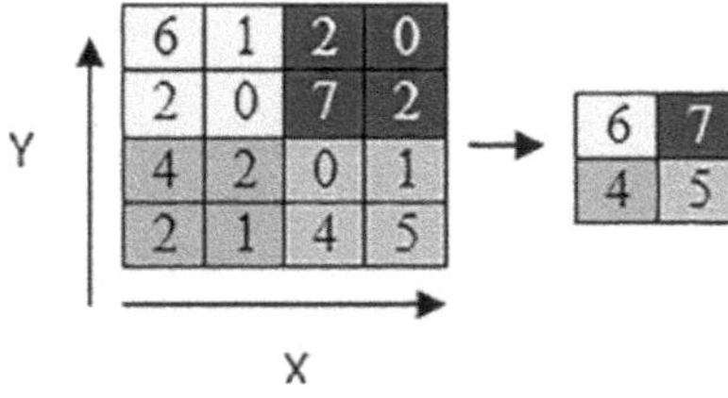

FIGURE 14.3 Example of max-pooling.

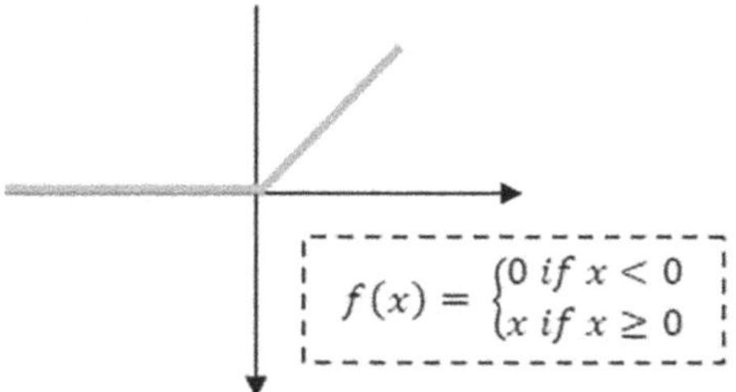

FIGURE 14.4 Rectified linear unit.

It can be calculated using Equation (14.2):

$$M(x_i) = \text{Max}\left\{ x_{i+k,\, i+l} \middle| |k| \leq \frac{m}{2}, |l| \leq \frac{n}{2} \; k, k \in N \right\} \tag{14.2}$$

14.3.3.3 Rectified Linear Unit

A rectified linear unit is an activation function, while the input is below zero, and the output is zero. Figure 14.4 shows an example of a rectified linear unit. It is calculated as follows:

$$R(x) = \max(0, x) \tag{14.3}$$

14.3.3.4 Fully Connected Layer

Fully connected layer is the same as a neural network in which all neurons in this layer are connected with each neuron in the previous layer. It can be calculated as

$$F(x) = \sigma(W * x) \tag{14.4}$$

14.3.3.5 Softmax Layer

Backpropagation can be done in this layer. The networks backpropagate the error and increase the performance. If N is a size of the input vector, $S(x)$: R → [0, 1]N. It is calculated by

$$S(x)_j = \frac{x^{xi}}{\sum_{i=0}^{N} e^{xi}} \tag{14.5}$$

where $1 \leq j \leq N$

14.3.3.6 Output Layer

The size of the output layer is equal to the number of classes. It represents the class of the input image.

$$C(x) = \{\; i | \exists i \; \forall j \neq i : x_j \leq x_i \} \tag{14.6}$$

14.3.4 Feedforward Deep Convolutional Neural Network

The input videos are converted into frames and saved in a separate folder as training and validation sets. Now, the raw images are the input for the first layer. An FDCNN consists of multiple convolutional layers, each of which performs the function that is discussed earlier. Figure 14.5 shows an FDCNN model. The input image is of size 150 × 150 × 3; where 3 represents the color channel. In this network, the size of the filter is 3 × 3 for all layers, and the filter is called as the weight. Multiplying the original pixel value with weight value is called sliding or convolving. These multiplications are summed to produce a single number called receptive field. Each receptive field produces a number. Finally, get the feature map with a size of (150 × 150 × 3). In the first layer, 32 filters are applied and have 32 stacked feature maps in this stage. Then, the subsampling (or max pooling) layer reduces the spatial (feature) size of the representation with a size of (75 × 75 × 32). In the second layer, 64 filters are applied and have 64 stacked feature maps. Then, the max pooling layer reduces the feature dimension to (37 × 37 × 64). In the third convolutional layer, 128 numbers of filters are applied and have 128 stacked feature maps. Then, the output of the maxpooling layer reduces the feature dimensions to (18 × 18 × 128). All max pooling layers are located with a size of 2 × 2. Finally, fully connected layers with 512 hidden units are placed, and the output classes have 15 neurons as per classes and show predicted emotions.

14.3.5 VGG16 Model

Figure 14.6 describes the architecture of the Visual Geometric Group (VGG)16 model. The VGG16 model consists of five blocks with 16 layers. The first two blocks have two convolution layers and one maxpooling layer. The remaining blocks have three convolution layers and one maxpooling layer. Finally, fully connected layer has one flatten layer and three dense layers. The sequences of an RGB (red, green, blue) image with size 256 × 256 are the inputs for this model. After each block, the size of the input image is different, which is defined in the architecture diagram.

14.4 PERFORMANCE EVALUATION METRICS

14.4.1 Dataset

An emotion dataset (University of York) containing five different emotions (happy, angry, sad, untrustworthy, and fear) is performed by 29 actors (Figure 14.7).

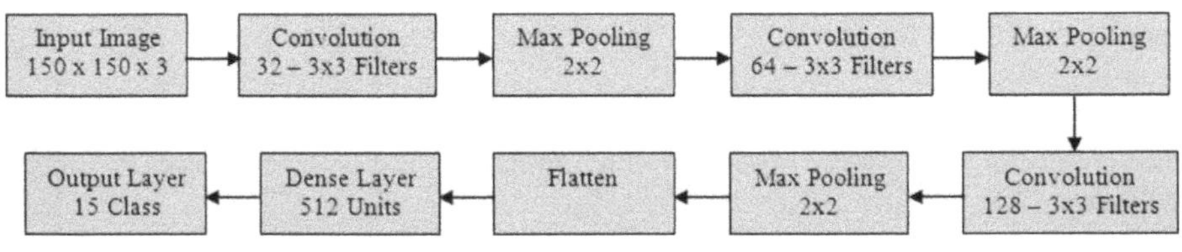

FIGURE 14.5 Architecture of an FDCNN model.

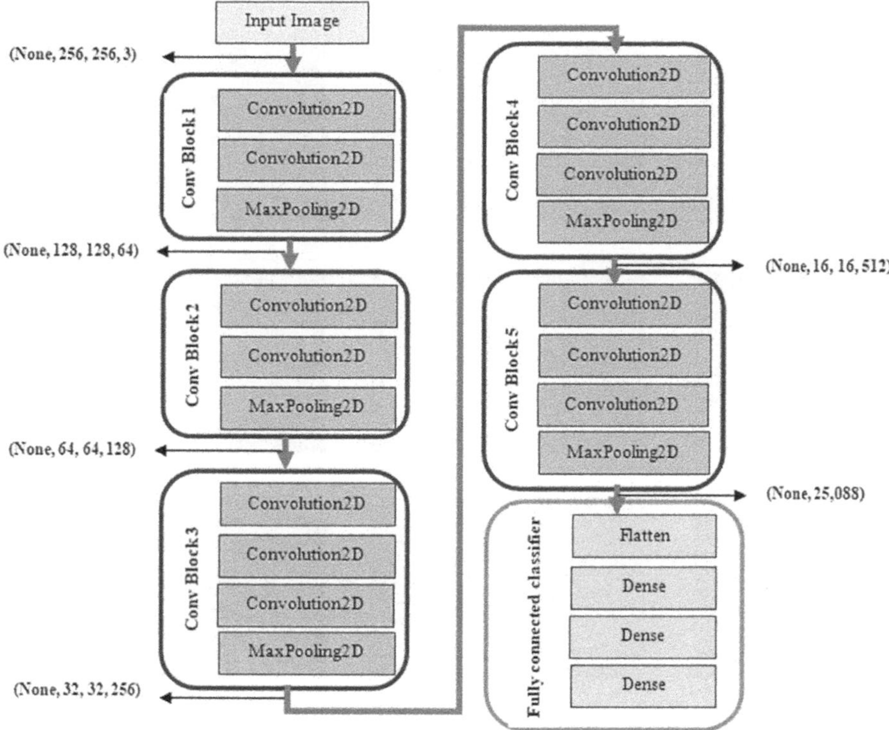

FIGURE 14.6 Architecture of VGG16.

The sequences were taken over a static (black) background with a frame size of 1,920 × 1,080 pixels at a rate of 25 fps [61].

All the experiments were implemented using Windows 10 operating system with Intel core i5 3.3 GHz processor with Anaconda Python and Jupyter notebook. The dataset is trained with a batch size 64 and 20 epochs.

14.4.2 Performance Evaluation Metrics

The confusion matrix is a table with actual classifications as columns and predicted ones as rows (Figure 14.8).

- True Positives (TP)—when the data point of actual class and predicted was (True).
- True Negatives (TN)—when the data point of actual class and predicted was (False).
- False Positives (FP)—when the data point of actual class (False) and the predicted is (True).
- False Negatives (FN)—when the data point of actual class (True) and the predicted is (False).

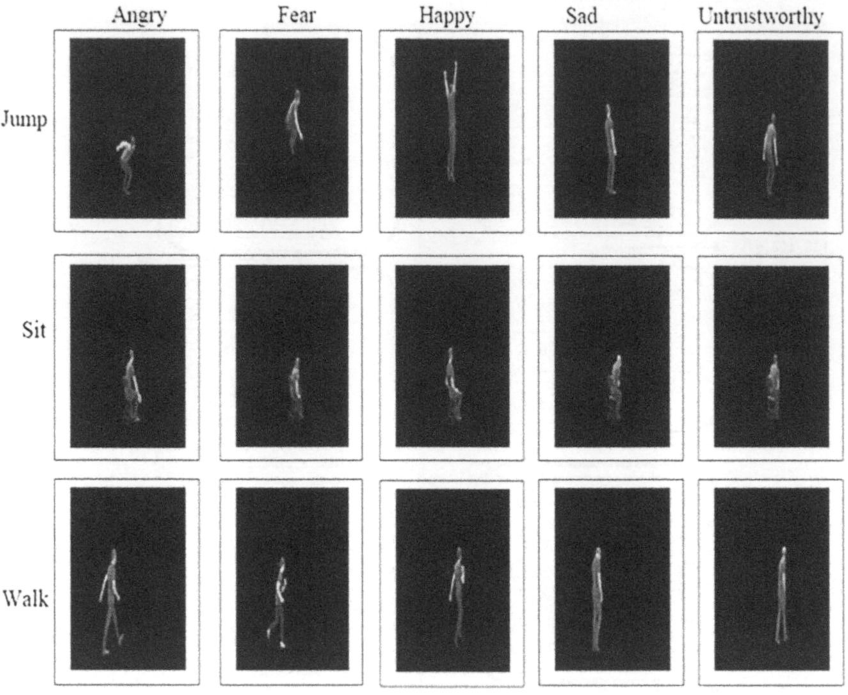

FIGURE 14.7 An example frame of five emotions and three actions.

Predicted Class \ Actual Class		
	True Positive	False Positive
	True Negative	False Negative

FIGURE 14.8 Confusion matrix.

The performance evaluation of the proposed work can be calculated using Accuracy, Recall, F-Score, Specificity, and Precision. Accuracy in classification problems is the number of correct prediction made by the model over all kinds of prediction made, which can be calculated using Equation (14.7).

$$\text{Accuracy} = \frac{\text{TP} + \text{TN}}{\text{TN} + \text{FP} + \text{TP} + \text{FN}} \tag{14.7}$$

Recall specifies how extraordinary an emotion is recognized accurately.

$$\text{Recall} = \frac{\text{TP}}{\text{TP} + \text{FN}} \tag{14.8}$$

The symphonies mean of Precision and Recall is called as F-score.

$$\text{F-score} = 2\frac{\text{Precision} \times \text{Recall}}{\text{Precision} + \text{Recall}} \tag{14.9}$$

Specificity shows the evaluation of how great a strategy is in recognizing a negative emotion accurately.

$$\text{Specificity} = \frac{\text{TN}}{\text{TN} + \text{FP}} \tag{14.10}$$

At last, Precision shows the proportion of classification, which can be calculated by Equation (14.11).

$$\text{Precision} = \frac{\text{TP}}{\text{TP} + \text{FP}} \tag{14.11}$$

where TP and TN are the quantity of true positive and true negative prediction of the class, and FP and FN are the quantity of false positive and false negative expectations.

14.5 RESULTS AND DISCUSSIONS

Table 14.1 describes the FDCCN model that shows the individual recognition accuracy of all classes. The correlation among happy jump, untrustworthy jump, and angry jump are closer to one another. The performance of this network is good in

TABLE 14.1
Individual Accuracy (%) for 15 Class Emotions of an FDCNN Model

	AJ	AS	AW	FJ	FS	FW	HJ	HS	HW	SJ	SS	SW	UJ	US	UW
AJ	95	0	0	1	0	0	2	0	0	0	0	0	2	0	0
AS	0	96	0	0	1	0	0	2	0	0	1	0	0	0	0
AW	0	0	96	0	0	1	0	0	1	0	0	0	0	0	2
FJ	2	0	0	94	0	0	1	0	0	2	0	0	1	0	0
FS	0	1	0	0	95	0	0	1	0	0	2	0	0	1	0
FW	0	0	3	0	0	95	0	0	2	0	0	0	0	0	0
HJ	0	0	0	1	0	0	96	0	0	2	0	0	1	0	0
HS	0	1	0	0	1	0	0	97	0	0	0	0	0	1	0
HW	0	0	0	0	0	1	0	0	94	0	0	3	0	0	2
SJ	4	0	0	1	0	0	1	0	0	94	0	0	0	0	0
SS	0	3	0	0	0	0	0	1	0	0	95	0	0	1	0
SW	0	0	4	0	0	2	0	0	0	0	0	92	0	0	2
UJ	0	0	0	2	0	0	3	0	0	4	0	0	91	0	0
US	0	3	0	0	2	0	0	1	0	0	2	0	0	92	0
UW	0	0	2	0	0	5	0	0	1	0	0	1	0	0	91

Note: AJ = AngryJump, AS = AngrySit, AW = AngryWalk, FJ = FearJump, FS = FearSit, FW = FearWalk, HJ = HappyJump, HS = HappySit, HW = HappyWalk, SJ = SadJump, SS = SadSit, SW = SadWalk, UJ = UntrustworthyJump, US = UntrustworthySit, UW = UntrustworthyWalk.

TABLE 14.2
Individual Accuracy (%) for 15 Class Emotions of VGG16 Model

	AJ	AS	AW	FJ	FS	FW	HJ	HS	HW	SJ	SS	SW	UJ	US	UW
AJ	96	0	0	1	0	0	1	0	0	0	0	0	2	0	0
AS	0	96	0	1	0	0	0	1	0	0	1	0	0	1	0
AW	0	0	95	0	0	1	0	0	1	0	0	0	1	0	2
FJ	2	0	0	96	0	0	1	0	0	2	0	0	1	0	1
FS	0	1	0	0	96	0	0	1	0	0	2	0	1	1	0
FW	1	0	2	0	0	95	0	0	2	0	0	0	0	0	0
HJ	0	0	0	1	0	0	96	0	0	2	0	0	1	0	0
HS	0	1	1	0	1	0	0	96	0	0	0	0	0	1	0
HW	0	0	0	0	0	1	0	0	94	0	0	3	0	0	2
SJ	4	0	0	1	0	0	1	0	0	96	0	0	0	2	0
SS	0	3	0	0	0	2	0	1	0	0	97	0	0	1	0
SW	1	0	4	0	0	2	0	0	0	2	0	95	0	0	2
UJ	2	0	0	2	0	0	1	0	0	1	0	0	94	0	0
US	0	0	0	0	2	0	0	1	0	0	2	0	0	95	0
UW	0	0	2	0	0	0	0	0	1	0	0	1	0	0	96

Note: AJ = AngryJump, AS = AngrySit, AW = AngryWalk, FJ = FearJump, FS = FearSit, FW = FearWalk, HJ = HappyJump, HS = HappySit, HW = HappyWalk, SJ = SadJump, SS = SadSit, SW = SadWalk, UJ = UntrustworthyJump, US = UntrustworthySit, UW = UntrustworthyWalk.

recognizing emotions like happy walk, sad walk, sad sit, and happy sit than other emotions. Table 14.2 describes the correlation among all 15 classes and shows the individual recognition accuracy of each class using VGG16 model. Tables 14.1 and 14.2 are the individual recognition accuracy of all classes, and the values in those tables do not exceed 100%. Table 14.3 shows the performance measure of precision, recall, and f-measure values of FDCNN and VGG16 model. The value 1 means best, and the value 0 means worst. In this experiment, the values obtained in Table 14.3 are mostly close to 1.

14.6 CONCLUSION

In this chapter, an FDCNN and VGG16 model for recognizing human emotions from body movements on sequence of frames were proposed. This model is representing deep features to extract saliency information at multiple scales. The proposed method is evaluated on a challenging benchmark emotion dataset (University of York). The emotions such as Angry Jump, Angry Sit, Angry Walk, Fear Jump, Fear Sit, Fear Walk, Happy Jump, Happy Sit, Happy Walk, Sad Jump, Sad Sit, Sad Walk, Untrustworthy Jump, Untrustworthy Sit, and Untrustworthy Walk are used in this work. Among the two models, VGG16 performed better recognition of emotions. The performance of this model is better than the baseline models.

Future work aims at developing research applications to recognize the emotions of children with autism spectrum disorder (ASD). The autism children express their

TABLE 14.3
Performance Measure of an Emotion Dataset with FDCNN and VGG16

	Precession		Recall		F-Measure	
Model	**VGG16**	**FDCNN**	**VGG16**	**FDCNN**	**VGG16**	**FDCNN**
AJ	0.937	0.891	0.828	0.821	0.933	0.881
AS	0.881	0.854	0.897	0.885	0.909	0.877
AW	0.842	0.825	0.921	0.913	0.911	0.901
FJ	0.881	0.862	0.897	0.886	0.919	0.891
FS	0.871	0.869	0.932	0.924	0.949	0.876
FW	0.895	0.883	0.891	0.883	0.907	0.879
HJ	0.884	0.873	0.935	0.927	0.915	0.913
HS	0.862	0.854	0.899	0.879	0.919	0.902
HW	0.843	0.839	0.928	0.911	0.931	0.911
SJ	0.931	0.930	0.858	0.849	0.927	0.886
SS	0.852	0.844	0.887	0.879	0.915	0.897
SW	0.892	0.882	0.931	0.922	0.951	0.879
UJ	0.933	0.924	0.818	0.821	0.973	0.880
US	0.891	0.887	0.890	0.881	0.899	0.878
UW	0.852	0.831	0.910	0.928	0.915	0.909

Note: AJ = AngryJump, AS = AngrySit, AW = AngryWalk, FJ = FearJump, FS = FearSit, FW = FearWalk, HJ = HappyJump, HS = HappySit, HW = HappyWalk, SJ = SadJump, SS = SadSit, SW = SadWalk, UJ = UntrustworthyJump, Us = UntrustworthySit, UW = UntrustworthyWalk.

emotion through facial and body movements. ASDs are neurodevelopmental disorders in which multiple genetic and environmental factors play roles. Symptoms of deficit in social communication and restrictive, repetitive behavioral patterns emerge early in a child's development. ASD is a lifelong neurodevelopmental disorder involving core deficit in interpersonal communication and social interactions, as well as restricted, repetitive mannerisms and interests (American Psychiatric Association, 2013). Children with ASD struggle with significant relationships and behavioral challenges and, in most cases, have serious implications for social inclusion in adulthood.

REFERENCES

1. R. Plutchik, *The Emotions*, University Press of America, 1991. Lanham, Maryland
2. F. E. Pollick, H. M. Paterson, A. Bruderlin, A. J. Sanford, "Perceiving Affect from Arm Movement," *Cognition*, Vol. 82, Issue 2, pp. 51–61, 2001.
3. P. N. Lopes, P. Salovey, R. Straus, "Emotional intelligence, personality, and the perceived quality of social relationships," *Personality and Individual Differences*, Vol. 35, Issue 3, pp. 641–658, 2003.
4. M. A. Goodrich, A. C. Schultz, "Human-Robert interaction: A survey," *Foundation and Trends in Human Computer Interaction*, Vol. 1, Issue 3, pp. 203–275, 2007.
5. H. R. Markus, S. Kitayama, "Culture and the self: Implementations for cognition, emotion, and motivation," *Psychological Review*, Vol. 98, pp. 224–253, 1991.

6. Q. Memon, "On assisted living of paralyzed persons through real-time eye features tracking and classification using Support Vector Machines," *Medical Technologies Journal*, Vol. 3, Issue 1, pp. 316–333, 2019.
7. J. Arunnehru, M. Kalaiselvi Geetha, "Automatic human emotion recognition in surveillance video," *Intelligent Techniques in Signal Processing for Multimedia Security*, Editors: Dey, Nilanjan, Santhi, V, Springer-Verlag, pp. 321–342, 2017. Berlin
8. D. Holden, J. Saito, T. Komura, "A deep learning framework for character motion synthesis and editing," *SIGGRAPH '16 Technical Paper*, July 24–28, Anaheim, CA, ISBN: 978-1-4503-4279-7/16/07, 2016.
9. H. Gunes, C. Shan, S. Chen, Y. L. Tian, "Bodily expression for automatic affect recognition," *Emotion Recognition: A Pattern Analysis Approach*, John Wiley and Sons, Editors: Amit Konar, Aruna Chakraborty, pp. 343–377, 2015. Hoboken, New Jersey
10. L. Zhang, S. Wang, B. Liu, "Deep learning for sentiment analysis: A survey," 2018. https://arxiv.org/pdf/1801.07883.
11. H. Brock, "Deep learning - Accelerating next generation performance analysis systems," in *12th Conference of the International Sports Engineering Association*, Brisbane, Queensland, Australia, pp. 26–29, 2018.
12. Z. Ali, Q. Memon, "Time delay tracking for multiuser synchronization in CDMA networks," *Journal of Networks*, Vol. 8, Issue 9, pp. 1929–1935, 2013.
13. Y. LeCun, Y. Bengio, G. Hinton, "Deep learning," *Nature*, Vol. 521, pp. 436–444, 2015.
14. N. Elfaramawy, P. Barros, G. I. Parisi, S. Wermter, "Emotion recognition from body expressions with a neural network architecture," *Session 6: Algorithms and Learning*, Bielefeld, Germany, 2017.
15. P. Khorrami, T. Le Paine, K. Brady, C. Dagli, T. S. Huang, "How deep neural networks can improve emotion recognition on video data," 2017. https://arxiv.org/pdf/1602.07377.pdf.
16. H. Ranganathan, S. Chakraborty, S. Panchanathan, "Multimodal emotion recognition using deep learning architectures," 2017. http://emofbvp.org/.
17. P. Barros, D. Jirak, C. Weber, S. Wermter, "Multimodal emotional state recognition using sequence-dependent deep hierarchical features," *Neural Networks*, Vol. 72, pp. 140–151, 2015.
18. Z. Al-Kassim, Q. Memon, "Designing a low-cost eyeball tracking keyboard for paralyzed people," *Computers & Electrical Engineering*, Vol. 58, pp. 20–29, 2017.
19. E. Correa, A. Jonker, M. Ozo, R. Stolk, "Emotion recognition using deep convolutional neural networks," 2016.
20. S. E. Kahou, V. Michalski, K. Konda, R. Memisevic, C. Pal, "Recurrent neural networks for emotion recognition in video," *ICMI 2015*, USA, pp. 9–13, 2015. doi:10.1145/2818346.2830596.
21. F. Noroozi, C. A. Corneanu, D. Kamínska, T. Sapínski, S. Escalera, G. Anbarjafari, "Survey on emotional body gesture recognition," *Journal of IEEE Transactions on Affective Computing*, 2015. PrePrints, DOI Bookmark: 10.1109/TAFFC.2018.2874986
22. M. Hu, H. Wang, X. Wang, J. Yang, R. Wang, "Video facial emotion recognition based on local enhanced motion history image and CNN-CTSLSTM networks," *Journal of Visual Communication and Image Representation*, Vol. 59, pp. 176–185, 2018. doi:10.1016/j.jvcir.2018.12.039.
23. X. Sun, P. Wu, S. C. H. Hoi, "Face detection using deep learning: An improved faster RCNN approach," *Neurocomputing*, Vol. 299, pp. 42–50, 2018.
24. O. Gupta, D. Raviv, R. Rasker, "Illumination invariants in deep video expression recognition", *Pattern Recognition*, 2017. doi:10.1016/j.patcog.2017.10.017.
25. G. Batchuluun, R. Ali, W. Kim, K. R. Park, "Body movement based human identification using convolutional network," *Expert Systems with Applications*, 2018. doi:10.1016/j.eswa.2018.02.016.

26. F. Letsch, D. Jirak, S. Wermter, "Localizing salient body motion in multi-person scenes using convolutional neural network," *Neurocomputing*, pp. 449–464, 2019.
27. J. Gu, Z. Wan, J. Kuen, "Recent advances in convolutional neural network," *Pattern Recognition*, Vol. 77, pp. 354–377, 2018.
28. D. Tran, L. Bourdev, R. Fergus, L. Torresani, M. Paluri, "Learning spatiotemporal features with 3D convolutional networks," *IEEE International Conference on Computer Vision (ICCV)*, pp. 4489–4497, 2015. Santiago, Chile
29. T. Guha, R. K. Ward, "Learning sparse representations for human action recognition," *IEEE Transactions on Pattern Analysis and Machine Intelligence*, Vol. 34, Issue 8, pp. 1576–1588, 2012.
30. H. Wang, C. Yuan, W. Hu, C. Sun, "Supervised class-specific dictionary learning for sparse modeling in action recognition," *Pattern Recognition*, Vol. 45, Issue 11, pp. 3902–3911, 2012.
31. F. Zhu, L. Shao, "Weakly-supervised cross-domain dictionary learning for visual recognition," *International Journal of Computer Vision*, Vol. 109, Issue 1–2, pp. 42–59, 2014.
32. F. Zhu, L. Shao, "Correspondence-free dictionary learning for cross-view action recognition," in *ICPR*, pp. 4525–4530, 2015. Piscataway, NJ
33. Y. LeCun, B. Boser, J. S. Denker, D. Henderson, R. E. Howard, W. Hubbard, L. D. Jackel, "Backpropagation applied to handwritten zip code recognition," *Neural computation*, Vol. 1, Issue 4, pp. 541–551, 1989.
34. A. Krizhevsky, I. Sutskever, G. E. Hinton, "Imagenet classification with deep convolutional neural networks," in *Advances in Neural Information Processing Systems*, pp. 1097–1105, 2014., 25. Doi: 10.1145/3065386.
35. Y. Du, W. Wang, L. Wang, "Hierarchical recurrent neural network for skeleton based action recognition," in *Proceedings of the IEEE Conference on Computer Vision and Pattern Recognition*, pp. 1110–1118, 2015. Boston, MA
36. F. Zhu, L. Shao, J. Xie, Y. Fang, "From handcrafted to learned representations for human action recognition: A survey", *Image and Vision Computing*, 2016. Volume 55 Issue P2, pp: 42-52
37. A. Cortes, V. Vapnik, "Support-vector networks," *Machine learning*, Vol. 20, Issue 3, pp. 273–297, 1995.
38. G. E. Hinton, R. R. Salakhutdinov, "Reducing the dimensionality of data with neural networks," *Science*, Vol. 313, Issue 5786, pp. 504–507, 2006.
39. P. Smolensky, "Information processing in dynamical systems: Foundations of harmony theory," *Report, DTIC Document*, 1986.
40. L. Sun, K. Jia, T. H. Chan, Y. Fang, G. Wang, S. Yan, "DL-SFA: Deeply-learned slow feature analysis for action recognition," in *Proceedings of the IEEE Conference on Computer Vision and Pattern Recognition*, pp. 2625–2632. 2015, 10.1109/CVPR.2014.336. Columbus, United States
41. D. Tran, L. Bourdev, R. Fergus, L. Torresani, M. Paluri, "Learning spatiotemporal features with 3D convolutional networks," *IEEE International Conference on Computer Vision (ICCV)*, pp. 4489–4497, 2015. Santiago, Chile
42. T. Guha, R. K. Ward, "Learning sparse representations for human action recognition," *IEEE Transactions on Pattern Analysis and Machine Intelligence*, Vol. 34, Issue 8, pp. 1576–1588, 2012.
43. H. Wang, C. Yuan, W. Hu, C. Sun, "Supervised class-specific dictionary learning for sparse modeling in action recognition," *Pattern Recognition*, Vol. 45, Issue 11, pp. 3902–3911, 2012.
44. F. Zhu, L. Shao, "Weakly-supervised cross-domain dictionary learning for visual recognition," *International Journal of Computer Vision*, Vol. 109, Issue 1–2, pp. 42–59, 2014.

45. F. Zhu, L. Shao, "Correspondence-free dictionary learning for cross-view action recognition," in *ICPR*, pp. 4525–4530. 2014, Piscataway, NJ
46. J. Wang, J. Yang, K. Yu, F. Lv, T. Huang, Y. Gong, "Locality-constrained linear coding for image classification," in *Computer Vision and Pattern Recognition (CVPR), IEEE Conference on*, pp. 3360–3367, 2010. San Francisco, CA, USA
47. A. Cohen, I. Daubechies, J.-C. Feauveau, "Biorthogonal bases of compactly supported wavelets," *Communication in Pure Applied Mathematics*, Vol. 45, Issue 5, pp. 485–560, 1992.
48. L. Liu, L. Shao, X. Li, K. Lu, "Learning spatio-temporal representations for action recognition: A genetic programming approach," *IEEE Transactions on Cybernetics*, Vol. 46, Issue 1, pp. 158–170, 2016.
49. Y. Zhang, Y. Zhang, E. Swears, N. Larios, Z. Wang, Q. Ji, "Modeling temporal interactions with interval temporal bayesian networks for complex activity recognition," *IEEE Transactions on Pattern Analysis and Machine Intelligence*, Vol. 35, Issue 10, pp. 2468–2483, 2013.
50. F. M. Khan, S. C. Lee, R. Nevatia, "Conditional Bayesian networks for action detection," in Advanced Video and Signal Based Surveillance (AVSS), *10th IEEE International Conference*, pp. 256–262, 2013. Krakov, Poland
51. C. Y. Park, K. B. Laskey, P. C. Costa, S. Matsumoto, "A process for human aided multi-entity Bayesian networks learning in predictive situation awareness," in Information Fusion (FUSION), *19th International Conference on*, pp. 2116–2124, 2016.
52. L. Deng, D. Yu, "Deep learning," *Signal Processing*, Vol. 7, pp. 3–4, 2014.
53. A. Ivakhnenko, "Polynomial theory of complex systems," *IEEE Transactions on Systems, Man, and Cybernetics*, Vol. SMC-1, Issue 4, pp. 364–378, 1971.
54. G. E. Hinton, S. Osindero, Y. W. Teh, "A fast learning algorithm for deep belief nets," *Neural Computation*, Vol. 18, Issue 7, pp. 1527–1554, 2006.
55. Q. V. Le, W. Y. Zou, S. Y. Yeung, A. Y. Ng, "Learning hierarchical invariant spatio-temporal features for action recognition with independent subspace analysis," in *Computer Vision and Pattern Recognition (CVPR)*, pp. 3361–3368, 2011. Colorado, USA
56. P. Foggia, A. Saggese, N. Strisciuglio, M. Vento, "Exploiting the deep learning paradigm for recognizing human actions," in *Advanced Video and Signal Based Surveillance (AVSS)*, pp. 93–98, 2014. Seoul, Korea
57. M. Hasan, A. K. Roy-Chowdhury, "Continuous learning of human activity models using deep nets," in *European Conference on Computer Vision*, pp. 705–720, Springer. Zurich, Switzerland, 2014
58. L. Ballan, M. Bertini, A. Del Bimbo, L. Seidenari, G. Serra, "Effective codebooks for human action representation and classification in unconstrained videos," *IEEE Transactions on Multimedia*, Vol. 14, Issue 4, pp. 1234–1245, 2012.
59. A. Krizhevsky, I. Sutskever, G. E. Hinton, "Imagenet classification with deep convolutional neural networks," in *Advances in Neural Information Processing Systems*, pp. 1097–1105. 2012, San Francisco, CA
60. Y. Du, W. Wang, L. Wang, "Hierarchical recurrent neural network for skeleton based action recognition," in *Proceedings of the IEEE Conference on Computer Vision and Pattern Recognition*, pp. 1110–1118. Boston, MA, USA, 2015
61. Y. LeCun, B. Boser, J. S. Denker, D. Henderson, R. E. Howard, W. Hubbard, L. D. Jackel, "Backpropagation applied to handwritten zip code recognition," *Neural Computation*, Vol. 1, Issue 4, pp. 541–551.

Index

K

L

M

N

O

P

Q

R

S

T

U

V

W

Milton Keynes UK
Ingram Content Group UK Ltd.
UKHW031533071024
449327UK00005B/86